D1631756

Sorting and Sort Systems

HAROLD LORIN
IBM Systems Research Institute

▲▼ ADDISON-WESLEY PUBLISHING COMPANY
Reading, Massachusetts · Menlo Park, California
London · Amsterdam · Don Mills, Ontario · Sydney

ISBN 0-201-14453-0
ABCDEFGHIJ-HA-798765

To the Systems Research Institute,
my colleagues and students

THE SYSTEMS PROGRAMMING SERIES

*The Program Development Process Part I—The Individual Programmer	Joel D. Aron
The Program Development Process Part II—The Programming Team	Joel D. Aron
*The Design and Structure of Programming Languages	John E. Nicholls
Mathematical Background of Programming	Frank Beckman
Structured Programming	Harlan D. Mills Richard C. Linger

*An Introduction to Database Systems	C. J. Date
Compiler Engineering	Patricia Goldberg
Interactive Computer Graphics	Andries Van Dam
*Sorting and Sort Systems	Harold Lorin
Compiler Design Theory	Philip M. Lewis Daniel J. Rosenkrantz Richard E. Stearns

Recursive Programming Techniques	William Burge
Compilers and Programming Languages	J. T. Schwartz John Cocke

*Published

Foreword

The field of systems programming primarily grew out of the efforts of many programmers and managers whose creative energy went into producing practical, utilitarian systems programs needed by the rapidly growing computer industry. Programming was practiced as an art where each programmer invented his own solutions to problems with little guidance beyond that provided by his immediate associates. In 1968, the late Ascher Opler, then at IBM, recognized that it was necessary to bring programming knowledge together in a form that would be accessible to all systems programmers. Surveying the state of the art, he decided that enough useful material existed to justify a significant codification effort. On his recommendation, IBM decided to sponsor The Systems Programming Series as a long term project to collect, organize, and publish those principles and techniques that would have lasting value throughout the industry.

The Series consists of an open-ended collection of text-reference books. The contents of each book represent the individual author's view of the subject area and do not necessarily reflect the views of the IBM Corporation. Each is organized for course use but is detailed enough for reference. Further, the Series is organized in three levels: broad introductory material in the foundation volumes, more specialized material in the software volumes, and very specialized theory in the computer science volumes. As such, the Series meets the needs of the novice, the experienced programmer, and the computer scientist.

Taken together, the Series is a record of the state of the art in systems programming that can form the technological base for the systems programming discipline.

The Editorial Board

Preface

This book is written for programmers. Its style and content are based upon an image of a professional who desires a complete but pragmatic knowledge of sorting and sort systems, and does not wish to learn a specific programming language, advanced statistics, or a hypothetical machine in order to obtain this knowledge. The intent of the book is to prepare a programmer to create sort programs. For the programmer who wishes a more intensive or formal appreciation of particular methods, this text will provide the basis for reaaing the more formidable literature in the field. In any event, to keep his knowledge up to date a programmer must continually peruse that literature.

Anouilh, in the introduction to his play *Becket,* says that he is not a serious man. He is a playwright and not a historian. He proves this by writing an excellent play based upon the historical absurdity that the Plantagenet Count of Anjou (Henry II of England), one of the most educated men in twelfth-century Europe, was a barbarian compared to his Saxon friend Thomas à Becket. Pure historians are critical of this play in a way not unlike the way computer scientists may criticize this book. Yet the play is more useful for gaining insight into the affairs of men than are the histories of the period. I hope this informal book will give similar insight into the processes of sorting.

Because of the particular idiosyncrasies of the text, I would like to discuss at the beginning how I hope the book will be used and why certain things were done. I have grossly simplified formulas. I have avoided artificial brevity wherever possible and taken the approach that more pages read quickly are better than few words read slowly.

I believe it is useful to read the book cover to cover, and by doing so, to master an overview of the area of sorting. I think the reader will find the prose natural and discursive. Each method is introduced and supported with an example. Then the properties and variations of the method are informally discussed. Wherever possible, coding is provided in the Appendix. There is a difference in flavor between the discussion of internal methods (Part I) and external methods (Part II). In the first part, I provide a great amount of detail because it is highly probable that the reader will program an internal sort. On the other hand, the optimum merges are usually produced in "laboratories" by very specialized people. Part II will enable the reader to talk to these people and (if made necessary by some blow of fate) to develop a respectable tape or disk merge. The reason for this approach is that the design details involved in creating an optimum merge are very specific to a machine, configuration, and operating system. No general discussion of fundamental techniques can make an expert.

The reader with a pressing problem—the necessity of programming an effective sort—can use the book in another way. If his desire is to sort 100 or so elements on an internal list, he might read the first two chapters and select a sort. He might just use "sifting" as described. If he wishes more powerful sorts, he can avoid the first chapters and read about QUICKSORT, merges, and tournament sorts, and then select one of these good sorts to please his taste and accommodate his situation.

I have included many simple sorts in the book because they form a conceptual foundation for more complex methods, because they are useful in and of themselves for small lists, and because they are useful as the internal algorithms for external sorts. For a number of reasons, the discussion about factors which influence the performance of a sort is given at the end of Chapter One. First, I believe it is more meaningful to describe some simple sorts before discussing performance. Second, I believe the nonalgorithmic influence on performance should be given as quickly as possible. This makes the end of Chapter One the obvious place to put this discussion. The reader asks, "How do I select the most appropriate sort?" after reading of three methods. The considerations raised by the end of Chapter One fortify his appreciation of the methods which follow.

The issue of formulas is important. I do not wish to give an illusion of precision and I do not want to require the reader to know probability theory and statistics. A formula such as

$$C = 2(N + 1) \ \log \left(\frac{n + 1}{M + 2} \right), \qquad N > M,$$

may be a theoretically valid expression for the number of comparisons in QUICKSORT for a file in random order. It is not as easy to use as the approximation $2N \log N$. Furthermore, since few files are truly random, the precise result may actually be misleading. Another common formula has the form

$$C = \frac{N(N-1)}{2.}$$

This can be approximated by $N^2/2$ with only 1 percent error for $N = 100$ and 0.1 percent for $N = 1000$. Since the primary purpose of these formulas is to permit different sort methods to be evaluated against the parameters of a sorting requirement, it is sufficient that the formulas discriminate adequately among candidate methods. For this reason, I have selected an easy-to-use approximation wherever possible for simple sorts, taking care to consider the behavior of the sort in actual use. For the more effective sorts, I provide formulas current in the literature. Numerous references are provided to assist the reader who wishes to investigate source material more thoroughly. A valuable general reference for the serious student of computer science is *The Art of Computer Programming,* Volume 3: Sorting and Searching, by Prof. D. E. Knuth, published by Addison-Wesley Publishing Company, 1973 [21].

When the reader has completed this book he will be ready, if that is his wish, to become an apprentice systems programmer in a Sort Laboratory, and/or to do some yeoman sort development for his own problems.

New York   H. L.
February 1975

Contents

CHAPTER 9
INTERNAL MERGING

CHAPTER 10
DISTRIBUTIVE SORTS

CHAPTER 11
COMPARISON OF INTERNAL SORTS

PART 2
EXTERNAL SORTING

CHAPTER 12
THE SORT PHASE OF AN EXTERNAL SORT

CHAPTER 13
TAPE MERGING

CHAPTER 14
POLYPHASE TAPE MERGING

CHAPTER 15
CASCADE AND COMPROMISE TAPE MERGES

CHAPTER 16
OSCILLATING AND CRISSCROSS MERGES

CHAPTER 17
TAPE MERGE OVERVIEW

Part 1
Internal Sorting

1
Basic Sort Concepts and Basic Sorts

1.1 INTRODUCTION

Sorting is the process of arranging items "in order." The arrangement of items is undertaken so that calculations which require data in a particular sequence can operate efficiently, so that output reports can be meaningfully presented, and so that successor processes may have useful input. In normal human usage of data the order of data is very important since order suggests critical relationships. It is also true that contemporary hardware operates most efficiently on ordered data. The concept of order is reasonably intuitive. There are many "natural" orders, such as alphabetic order for a list of names, or ascending value for a list of numbers.

When we think of order, our basic understanding is of true physical order, such as in the arrangement of a deck of punch⌐d cards or records on a magnetic tape. But the output of a sort does not necessarily involve actual rearrangement. Order in a file may be represented in other ways, particularly by the use of an index. Consider a collection of manuals, each of which has a form number. We may create an index card file of this collection, in which each card contains the name of a manual, its location on our shelves, the date of publication, and some classification of topics. From time to time we may wish to review our collection. For example, we may wish to compare our manual numbers with a comprehensive list of published manuals in numerical order. The card file may be normally in order by topic code or date of publication. To compare the cards with the list, we reorder the index cards, not the publications.

It is not always obvious whether it is better to sort indices or records. In fact, it is not always obvious whether it is necessary to sort at all. Consider a tape computer with a very small memory. We want to produce a summary report by product within a branch office. We have a tape containing reports of individual sales for all products and all branch offices. There will be some tendency for sales records to cluster within branch offices, and all the sales records for a particular branch office may be together and even in sequence. If we know that this is true, there is no reason to execute a sort program. A simple sequence check may be sufficient to ensure that each record is properly processed. Items that are out of sequence may be reordered in small groups or simply set aside for a later run. On the other hand, if the data is randomly ordered or we do not know what its order is likely to be, a sort is the fastest and cheapest way to guarantee its sequence.

The nature of the computer also affects the decision to sort. Assume that the sales tape is in random order. If we decide not to sort, then, if our computer memory can store only the total for one product for one branch office, we will have to search the tape for its entire length looking for records which contribute to this count. If 100 different products were sold in 50 sales offices, 5000 full inspections of the tape would have to be made. Surely the process of sorting before producing the report is more efficient. If there is room in memory for 100 totals, then each traversal of the tape will produce a complete office summary, and only 50 passes will be made. But this is still less efficient than processing a sequential tape.

However, if there is room in memory for 5000 totals, a sort is unnecessary. Each record as it comes in can be added to an appropriate running total stored in memory, and there is no advantage to sorting the tape. The sort can be eliminated with very little, perhaps not any, increase in time to the report-producing process.

The decision to sort, though often obvious, as in this tape example, is sometimes rather subtle. It is especially likely to be so in disk environment where there is direct access to data locations. The cost of not sorting the data is possible excessive arm movement between accesses to records during processing, additional rotational delays, and consequent increased elapsed times for the process that uses the data. When it is obvious that a sort will reduce total time, it is still not obvious that a record sort is better than an index sort. This decision requires analysis and serious investigation of the device and the usage pattern of the data.

The considerations involved in the decision to sort or not to sort include analysis of the various outputs required, the nature of the devices and their channel arrangements, the size of primary store, the frequency of access, etc. The decision to provide a major sort step in an information system is an important one—one that should be made only after serious study of the hardware and data parameters justifies the fundamentally nonproductive sort. In essence, the rule is: Sort only when there is no way to avoid it.

1.1.1   Records, Fields, and Keys

The unit of data that information-processing systems characteristically process is called a *record*. A record is a cluster of information about an event or structure. Each element of information in a record, such as man number or unit price or gross value, is called a *field* of the record. The collection of fields identifies and describes what the record represents. Records are organized into *files* or *data sets*. Sorting is the process of rearranging records or indices to records so that their positions relative to one another in the file follow an order determined by a known key.*

In its simplest form and on the assumption that the sort will lead to physical order, sorting two records consists of comparing the key fields of the two records and determining which has the lower value. The sort then rearranges the records to place the one with the low key before the record with the high key.

Any of several rules can be used to decide which key is lower. The obvious rules are *alphabetic order, numeric order,* and *alphanumeric order.* (In the last rule, keys may have intermixed letters and digits; the conventions of a system determine the order of numerics and alphabetics.)

The key of a record may actually be one field or it may be some combination of fields distributed through the record. If more than one field is defined to be part of the key and the fields are noncontiguous, the key is often referred to as a *split-key.*

A sequence within a sequence may be desired. For example, a file may be ordered by employee number within location number. The field representing location number would be called the *major key,* and the field representing man number would be called the *minor key.* It is possible to define levels of keys beyond this, of course, and all levels other than the first are called minor keys. Major and minor keys can be treated as a single split-key in a record.

In an information-processing system, more than one arrangement of a file may be required for the convenience of the system. For the purposes of one sort, one field may be a key, but another sort might use a different field as a key.

A record may consist of nothing but a key field. This is not uncommon in scientific processing. The one-field record will be assumed in the illustrations of sorting methods in this text in order to simplify the initial presentation of the sorts by eliminating extraneous problems in data movement. Since the methods apply equally well to records containing keys and nonkey fields, the assumption is not restrictive.

---

* A *key* is a field which contains the value used in the ordering rules to sequence the file.

1.2   SORTING PROCEDURES

There are many different sorting algorithms and many approaches to programming them. The problem of recognizing that one sorting procedure is distinct from another is a variant of the problem of determining when two programs are equivalent. In the selection of the material for this text, sorting procedures which have been treated as distinct in the past continue to be treated as such here [3, 11, 25].

A sort is ultimately a program, and a sorting procedure will be subject to a diversity of encodings. The possibility exists that differences in performance between two distinct sort algorithms may at times be less than the difference between "good" and "bad" encodings of the same algorithm. This may also be true of the differences in performance between higher-level language and hand-coded versions of the same algorithm.

The procedures described in the text are illustrative only. They should be taken as the start, not the conclusion, of sort development. In addition to necessary accommodations, for details of equipment and data, as needed, the methods themselves should be considered for refinement, for combination, or for extension.

Traditionally, sort methods are classified into *internal* methods and *external* methods [23]. An internal method is one that can be applied with acceptable performance only to those lists of data that are completely contained in the primary storage (memory) of the processor. An external method is one that is reasonably applied to files of data that are too large to fit in primary storage and must therefore rely on external bulk storage devices (tape, disk, drum) during the sorting process. The word *list* is frequently used for a collection of records contained in primary storage. The external techniques may be highly efficient even when applied to lists represented wholly in primary storage.

In the process of external sorting, parts of a file are read into storage, ordered internally, and then rewritten on the external devices. This process occurs a number of times. Internal methods are used to rearrange the data developed from pass to pass. Thus, when one speaks of sorting on tapes, one implies not only the process of reading and writing tapes but also an internal sort which orders and combines elements from these tapes as they are read.

1.3   ATTRIBUTES OF INTERNAL SORTS

Sorts may be characterized by the complexity of the method used to determine a sequence of comparisons, by the mixture of methods used, by the basic way order is imposed, and by the amount of space required.

1.3.1 Linear or Nonlinear

The complexity of the method used to determine a sequence of comparisons depends on how much structural information is used to organize the records. A linear algorithm imposes no structure on the list it has to order. The list is treated as a linear succession of elements, and the sequence of comparisons and transfers performed on it reflects this fact. Comparands are selected in sequence by advancing element by element, up or down the list. By contrast, nonlinear methods impose a structure on the lists they sort. They achieve efficiency by treating their lists as binary trees or by otherwise partitioning and selecting comparand sequences in a more complex way than that of the linear sorts. All "efficient" sorts, sorts which approach the theoretical minimum number of comparisons, are nonlinear.

1.3.2 Simple or Combined

A simple sort uses the same technique throughout the sorting process. There is no variation in method in response to conditions in the data or at some defined point in the procedure. By contrast, an algorithm may be combined. The developer of a sort may determine that the best method for his data involves a combination of techniques, with switches from one to another depending, for example, on the number of items in a sublist. Simple algorithms are not necessarily unsophisticated; many of the tree-oriented techniques are "simple" in the sense in which that word is used here.

1.3.3 Comparative or Distributive

A *comparative* method is one in which the way data is ordered depends on comparison of the relative magnitude of keys on the list. Most of the sorting methods we will discuss are comparative methods. The alternative, a *distributive* method, inspects each key either character by character or as an entity. The interim placement of a key depends not on relative magnitude but on a given characteristic when the key is matched against some standard. Distributive sorts can be useful when much is known about the data to be sorted; for example, the relative frequency of keys within certain ranges, the length of keys, and the number of records. When less is known, the comparative sorts are preferable, since they have less sensitivity to the *distribution* of the particular set of values being sorted. They are sensitive to the particular *order* of the values they receive.

1.3.4 Minimal Storage or Nonminimal Storage

A minimal-storage technique requires space for only the program code and the list to be ordered. (Some algorithms which require some trivial additional storage are also considered to be minimal-storage techniques.) By contrast a nonminimal-storage technique requires space for a second copy of the initial list. The advantage of minimal-storage sorting is that it allows the sorting of larger lists in primary storage, though at the expense of sort time. Certain algorithms are minimal- and nonminimal-storage versions of what is essentially the same process.

1.4 BASIC ALGORITHMS

The text will open by discussing seven very fundamental internal sorting algorithms. The first three are presented in this chapter as a basis for an introduction to comparative sorts. The remaining four are presented in Chapter 2. The seven methods are:

Linear Selection	Standard Exchange
Linear Selection	Basic Exchange
with Exchange	Sifting
Linear Selection	Linear Insertion
with Counting	

These are very elementary algorithms, whose use is limited to situations where, for one reason or another, the speed or space of a sort is not important, or where the number of records to be sorted is very small.* They are linear, simple, comparative sorts. Our interest in them is partially tutorial and partially pragmatic because they can be useful, simple tools.

1.5 LINEAR SELECTION

Straight Linear Selection is a nonminimal-storage technique.

1.5.1 Linear Selection

A *pass* of Linear Selection involves selecting the element with the lowest key on the list to be sorted and placing it on a growing output list. Additional passes are made until the output list is complete. At completion, the output list represents the ordered original list. At the start of each pass, the first element on the list is assumed to be the element with the lowest key. The key of this element and its address are transferred to working storage. From

---

* They can be preferred methods for small lists.

there it is compared with the key of the second element. If it is lower, it is then compared with the third. It is compared with all linear successors until a lower element on the list is discovered. Then the key of the newly found lower element with its address replaces the initial one in working storage as the comparand for further testing.

The key of the smallest element yet encountered is always in working storage to be used for comparison against remaining keys on the list. When the end of the list is reached, the smallest element on the list is known to be the one whose address and key are in the working storage. This element is then transferred from the original list to the current position in the output area. A dummy value (such as a field of all 9s) known to be higher than any key on the list is substituted for the key of each element in the original list as it is transferred, in order to preclude the possibility that the element could again be selected.

Representing the address of the selected element in working storage makes it possible not only to locate the element for transfer, but also to determine where to place the dummy key. When the key field is the only field of the element, the transfer to the output list may be made directly from working storage since the entire record is represented in working storage.

Each pass adds the current lowest member of the list being sorted to the output list. Thus, pass 1 delivers the lowest, pass 2 the second lowest, etc., and the ordered list is formed in the output area. The sort ends when all members of the input list have been delivered to the output list.

Initial list		Working storage
1.	3	LOW = 3
		SOURCE = 1
2.	11	
		Sequence of comparisons
3.	6	
		Key 3 : key 11
4.	4	
		Key 3 : key 6
5.	9	
		Key 3 : key 4
6.	5	
		Key 3 : key 9
7.	7	.
		.
8.	8	.
		Key 3 : key 10
9.	10	
		Key 3 : key 2* (Lower value found)
10.	2*	
11.	1	

Fig. 1.1 Linear Selection, pass 1.

1.5.2 Example

Consider the list of elements (represented by their keys) in Fig. 1.1. At the start of the first pass, the key of the first member of the list is put in working storage location LOW, and its address is put in SOURCE. The initial comparison of the pass is of key 3 in working storage with key 11 in position 2. The comparison finds that key 3 is lower, and it remains the assumed lowest key. A simple linear pointer to the list (initially set at position 2) is advanced to prepare for the next comparison.

The second comparison is between key 3 and key 6 in position 3. This comparison finds key 3 again low, and consequently the next comparison is between key 3 and key 4. The sequence of comparisons involving key 3 ends when one is made against position 10 containing key 2, the first value in the list that is lower than 3. The key, 2, and location, 10, of the new lower element (see Fig. 1.2) become the new contents of LOW and SOURCE.

Initial list	Working storage
3	LOW = 2
	SOURCE = 10
11	
6	Sequence of comparisons
4	Key 2 : key 1* (Lower value found)
9	
5	
7	
8	
10	
2	
1*	

Fig. 1.2 Linear Selection, end of pass 1.

Comparing now continues between key 2 and its linear successors. The first such comparison finds a value lower than key 2. The new value and location are placed in LOW and SOURCE. At this point the pass ends because there are no further elements on the list. The key in LOW is transferred to the output area. If this list had records with fields other than the key field, then the transfer would be from the list, addressed through SOURCE, to the output area. With transfer accomplished, a dummy value *(z)* is placed on the list, and a new pass is undertaken. Figure 1.3 shows the status of the input list and the output area at the end of five passes.

Pass	1	2	3	4	5
	3	3	z	z	z
	11	11	11	11	11
	6	6	6	6	6
	4	4	4	z	z
List	9	9	9	9	9
	5	5	5	5	z
	7	7	7	7	7
	8	8	8	8	8
	10	10	10	10	10
	2	z	z	z	z
	z	z	z	z	z
Output	1	1	1	1	1
		2	2	2	2
			3	3	3
				4	4
					5

Fig. 1.3 Status of list after each of first five passes (z = dummy value).

1.5.3   Discussion

Since each pass contributes one selected element to the output list, there must be as many passes as there are elements on the list. If N is the number of elements on the list, there will be N passes.

Since each pass starts with the first element in the list, each pass requires $N - 1$ comparisons, regardless of the initial order of the list. The number of comparisons is not sensitive to the order of initial data. Therefore the best, the worst, and the expected (average) number of comparisons are the same for this method. The total number of comparisons is always $N(N - 1)$. The reader will find N^2 a practical approximation.

The number of transfers from the input list to the output list is N. However, this value does not represent the work the method must perform, because it does not include the work of moving keys and addresses to LOW and SOURCE. If the size of the record is small, the time taken for this transfer will be proportionately great because the time required to move an address and key will be close to the time required to move a record. The number of key/address transfers depends on initial order in the list. In a list completely ordered, there would be no replacement of values in LOW during a pass; in a list inversely ordered, there would be $N - 1$ replacements.

The data space requirement for Linear Selection is usually given as twice the space required to hold the original list. Linear Selection has the property of developing a growing ordered list as it progresses from pass to pass. If there is no need to store the entire ordered list in core, it is possible to reduce the size of the working space by beginning to write output *while the list is being formed.*

1.6   LINEAR SELECTION WITH EXCHANGE: USE OF EXCHANGES

A minimal-storage version of Linear Selection is introduced here as a first example of exchange during the process of sorting. An *exchange* is the interchange of the positions of two records on a list, depending on the result of a test of their relative magnitude. When a record on a list is found to have a lower key than that of a record occurring earlier on the list, they exchange positions. The variation in the performance of exchange methods is due to the sophistication with which the sequence of comparisons and exchanges is determined. Often the number of exchanges is reduced by postponing an exchange until the end of a pass. This technique is used in the method we will next describe.

1.6.1   Linear Selection with Exchange

At the beginning of the first pass, the first element of the list is assumed to have lowest key of the list. The key with its address is moved into working storage and from there compared with all linear successors until a lower key is found. The lower key with its address becomes the contents of the working storage area.

Comparing continues with the new contents of working storage. Whenever a key lower than the key in working storage is found in the list, a replacement takes effect. In this way the lowest key so far discovered is always in working storage. The pass ends when the end of the list is reached. Up to this point, the process is identical to straight Linear Selection. The next step distinguishes the Exchange sort.

At the end of the pass the record whose key is in the working area exchanges position with the record at the top of the list. The lowest value on the list now occupies the first position. The second pass is identical except that, with the first element established as lowest, the second key on the list is assumed to be next lowest. The first position is excluded from the process. At the end of the second pass the second lowest keyed record of the list is ranked in the second position of the list. The third pass starts comparing the key in the third position, etc. The procedure terminates when the $(N - 1)$th record is put in position.

Initial list		Working storage
1.	3	LOW = 3
		SOURCE = 1
2.	11	
		Sequence of comparisons
3.	6	
		Key 3 : key 11
4.	4	
		Key 3 : key 6
5.	9	.
		.
6.	5	.
		Key 3 : key 10
7.	7	
		Key 3 : key 2* (Lower value found)
8.	8	
9.	10	
10.	2*	
11.	1	

Fig. 1.4 Linear Selection with Exchange, pass 1.

Initial list		Working storage
1.	3	LOW = 2
		SOURCE = 10
2.	11	
		Key 2 : key 1* (Lower value found)
3	6	
4.	4	
5.	9	
6.	5	
7.	7	
8.	8	
9.	10	
10.	2	
11.	1*	

Fig. 1.5 Linear Selection with Exchange during pass 1.

1.6.2   Example

Figure 1.4 shows the initial list. The sequence of comparisons for pass 1 involves the contents of position 1, key 3, with all linear successors until a lower key is found. Then LOW and SOURCE are changed. No change on the physical list is involved; there is a change only of LOW and SOURCE, as shown on Fig. 1.5. With key 2 in LOW, comparing continues

Initial list		Working storage
1.	1	LOW = 1
		SOURCE = 11
2.	11	
3.	6	
4.	4	
5.	9	
6.	5	
7.	7	
8.	8	
9.	10	
10.	2	
11.	3	

Fig. 1.6 Linear Selection with Exchange, pass 1; exchange of keys 1 and 3.

	After comparison of			
Initial list	Key 11 : key 6	Key 6 : key 4	Key 4 : key 2	Final list
1	1	1	1	1
11 LOW = 11	11 LOW = 6	11 LOW = 4	11 LOW = 2	2
6 SOURCE = 2	6 SOURCE = 3	6 SOURCE = 4	6 SOURCE = 10	6
4	4	4	4	4
9	9	9	9	9
5	5	5	5	5
7	7	7	7	7
8	8	8	8	8
10	10	10	10	10
2	2	2	2	11
3	3	3	3	3

Fig. 1.7 Linear Selection with Exchange, pass 2; exchange of keys 2 and 11.

with its linear successor in position 11. This comparison results in an exchange, since key 1 is lower (and also last on the list). The list is transformed to that of Fig. 1.6. The first pass now ends. The reader will notice that Figs. 1.1 and 1.4 are identical, stressing the similarity between the exchange and nonexchange versions of Linear Selection.

Subsequent passes are identical, except that the initial value placed in working storage is selected from a different position for each pass—position 2 for pass 2, position 3 for pass 3, etc. Figure 1.7 shows the transformations during the second pass. Successive passes continue in this way, the list to be compared growing shorter each time, until $(N - 1)$ positions have been ranked. The final position then also properly contains the highest key; so the procedure ends.

1.6.3   Discussion

The striking feature of the method is the shrinkage of the list from pass to pass. Each successive pass involves one fewer comparison, so that there are $(N - 1), (N - 2), (N - 3), \ldots , 2, 1$ comparisons on succeeding passes. The total number of comparisons is the average number of comparisons per pass times the number of passes. This is $(N - 1)(N)/2$, approximated by $N^2/2$, and is not affected by any condition of the data. The number of exchanges is the number of passes. Again, no count has been made of the number of stores in LOW and SOURCE.

An interesting variation of the method involves suppressing LOW and SOURCE. Each time a lower key is found, an exchange takes place with the contents of the position at the top of the current unsorted list. Instead of one exchange per pass, there is an exchange for each discovery of a lower key. For lists with records that are large relative to key size, the effort of making the additional full exchange would be a heavy burden. The method as earlier described reduces all but $(N - 1)$ exchanges to key and address movements. For lists where the key is the record, the two implementations are roughly equivalent.

There is a secondary effect of exchanging directly on the list. The list itself is subject to intermediate reorderings. The effect may be to increase the number of exchanges because of the tendency for small numbers in the early positions of the list to "sink" to the bottom. There may be permutations of data for which this sinking has some advantage but, unless the programmer has thorough knowledge of the data, the method of delaying exchanges should be preferred.

1.7   LINEAR SELECTION WITH COUNTING

The method of sorting by counting is described in the literature as a procedure for arranging an internal list of numbers. Actually, it is not a method of sorting but a technique which can be used with various methods in order to reduce the number of exchanges or to eliminate exchanging entirely. It is a form of indexing in which a relative position count of each element is developed during the process of comparing. The next section describes the technique in connection with Linear Selection.

1.7.1   Counting as a Technique

The space used for Linear Selection with Counting will include an output area (as for Linear Selection) to hold the final ordered list. The size of the output area is subject to the same considerations put forth with Linear Selection. In addition, space must be provided for a counter location for each element on the list. The manipulation of these counter values forms a set of indices for the positions of elements on the ordered list.

On each pass, a key is compared with its linear successors. Each time a key is found to be higher than another, a count is made in its associated counter. Each time it is found to be equal or lower, a count is made in the counter associated with the higher comparand. At any time, therefore, the counter for an element contains a count of the keys known to be *lower*.

On the first pass, the first key on the list is compared with all other keys. The count of keys that are lower is developed in its associated counter. All keys higher will have a count of 1 in their associated counters. On the second pass, the first key is ignored. The second key is compared with all successor keys and counts are recorded. The process continues for $(N - 1)$ passes. At this point the relative position of all elements is known. Placing the keys in the output list according to their count produces an ordered list.

1.7.2   Example

Consider the list in Fig. 1.8. During the first pass, key 3 will be compared with all successor keys. All keys other than 2 and 1 are higher than key 3 and consequently have a count of 1 in their counters. Key 3 has a count of 2 since it is greater than two keys of the list. On the second pass, key 11 (position 2) is compared and found to be higher than the nine remaining keys on the list. To the value of 1, left in the counter at the end of the first pass, nine counts are added, one by one, during the second pass, for a total count of 10 in the counter associated with key 11 (see Fig. 1.9). On

	Key	Count
1.	3	2
2.	11	1
3.	6	1
4.	4	1
5.	9	1
6.	5	1
7.	7	1
8.	8	1
9.	10	1
10.	2	0
11.	1	0

Fig. 1.8 Counting with Linear Selection, end of pass 1.

	Key	Count
1.	3	2
2.	11	10
3.	6	1
4.	4	1
5.	9	1
6.	5	1
7.	7	1
8.	8	1
9.	10	1
10.	2	0
11.	1	0

Fig. 1.9 Counting with Linear Selection, end of pass 2.

	Key	Count
1.	3	2
2.	11	10
3.	6	5
4.	4	1
5.	9	2
6.	5	1
7.	7	2
8.	8	2
9.	10	2
10.	2	0
11.	1	0

Fig. 1.10 Counting with Linear Selection, end of pass 3.

	Key	Count
1.	3	2
2.	11	10
3.	6	5
4.	4	3
5.	9	8
6.	5	4
7.	7	6
8.	8	7
9.	10	9
10.	2	1
11.	1	0

Fig. 1.11 Counting with Linear Selection, final pass.

the third pass (Fig. 1.10) key 6 is the comparand. Since it is higher than four of the remaining elements on the list, its count is augmented to 5. The counts of keys 7, 8, 9, and 10 are augmented to reflect that they are greater than key 6. When the last comparison pass is completed, all counters are set, and transfers can then be made to the output list. Final counts are shown in Fig. 1.11.

The output list is exactly the same size as the initial list; therefore the address of any element position in the output list can be calculated. The distribution to the output list takes the value of the counter of a key and adds it to the origin address of the output area, to develop the address of the key on the output list. The process involves a single pass down the count list.

1.7.3   Discussion

The number of comparisons is $(N)(N - 1)/2$ (approximately $N^2/2$). There are $(N - 1)$ passes with an average of $N/2$ comparisons in each pass. The number of counts is exactly the same as the number of comparisons, since for each comparison there is a count. The number of transfers is N.

The method of counting can be varied. For example, if every counter is initialized to the origin of the output list and each count adds the length of a record, the value of the counter will directly represent the destination address of the record.

1.8   CONSIDERATIONS IN SORT IMPLEMENTATION

The purpose of this section is to introduce the nature of those considerations involved in choosing a sort method. These topics are introduced at this time so that a reader may approach the subsequent discussions of methods with a general awareness of parameters which affect performance.

Three sorting methods, all variations of Linear Selection, have been described. What considerations are relevant if one method is to be chosen? What factors other than the sort algorithm influence the time required to sort data?

One must consider how important the performance of the sort is. There are situations in which no real benefit will accrue to the programmer for the effort invested in developing an excellent sort. The most primitive sorting schemes will be adequate, because sort time may overlap something else or be too small to be important.

There will be situations in which the programmer is reasonably concerned about sort performance but does not consider it absolutely critical; he is here interested only in developing a sort that will perform within a certain limit.

If one is really anxious to develop a very good sort, the process of developing a sort is complex and consists of many stages. The first requirement for excellent sort selection is a thorough understanding of the data to be sorted. When the data is understood, we have enough information to determine whether to use a comparative or a distributive technique, what space may be required, and whether it is useful to investigate sorting only keys.

The next step is to find a general method that seems efficient in terms of its expected comparisons and data movement. The method should be modeled in a higher-level language. If it is not clear which method is best, more than one method should be modeled and run against test data to determine whether tendencies exist in the data which will impact a sort in ways not yet known. One may try combinations of methods. Although much is known about sorting algorithms, much is not known, and it is possible that some subtle combination of data characteristics may represent a bad case for a usually good method.

Once a method has been selected, it must be coded in machine language for best performance.* One must consider such details as the use of the machine registers, core-to-core transfers versus transfers through registers, the relative speed of transfer for elements of various size, the relative speed of comparisons of different types on different modes of data, the mapping of the data on the addressing structure of the machine, and the amount of core space available.

The following sections suggest development considerations beyond the basic properties of the sorting method.

1.8.1   Data Considerations

The characteristics of data which influence the selection of a sort are the following:

1. Size of list to be sorted

2. Length of keys and existence of split-keys

3. Mode of keys (binary, decimal, floating, etc.)

4. Distribution of keys of list

   a) Range of values

   b) Clustering of values

   c) Duplication

   d) Permutations of keys

5. Size of records and variability of record length

6. Eventual use of data

---

* If machine level coding is not desirable for other reasons, the tradeoff between sort performance and cost of sort development must be resolved in the light of these other factors.

1.8.1.1   Size of list

The number of elements determines:

1. Whether there will be a need for external sorting;
2. Whether minimum storage must be used;
3. Whether the overhead associated with elaborate methods can be justified.

Certain sorting methods are sensitive to N as a particular value (a perfect power, a perfect square, etc.). The amount of work required to sort N's that are not powers or squares is often disproportionately close to the work required to sort the next higher perfect square or power.

1.8.1.2   Length of key and existence of split-keys

The length and location of key field(s) will determine how much time is required to perform comparisons. The impact of various key sizes and placements in a record is very dependent on central processing unit (CPU) characteristics. The important considerations are the following:

1. Does the key fit into a directly manipulatable structure of the machine?
2. Is it the only field of a word, or must it be extracted from a word and compared under "mask"? Is there a time penalty for extracting or masking?
3. Is the key spread across multiple words? Does access involve multiword manipulation, field definition, shifting, etc.? Are there time penalties associated with putting keys in usable form?
4. Is compare time a direct function of key size? On a character machine, is the compare time a function of number of characters in the key? Can a key be divided so that a comparison of high-order characters will suffice to order most keys?

If key preparation is required, it is possible to perform all key manipulation by first appending a constructed *control key* to the record. It is possible to dynamically construct a usable key on each reference for each comparison. The choice of either technique depends on available space and timing characteristics of the CPU.

1.8.1.3   The mode of the keys

The time to compare depends on the code representation of the keys and the availability and timing of compare commands for various modes. If no "decimal compare" instruction exists, it may be necessary to convert to pure binary. The time involved in key transformation, one-time or repetitive, is added to sorting time.

1.8.1.4   Distribution and permutation

Most discussion of sort performance assumes "random" data. This means that given a range of possible key values, the probability of any value in the range is as great as that of any other. Permutation, a particular ordering of a set of numbers, is random if any initial ordering of the numbers is as probable as any other. There are $N!$ possible permutations of a list of N distinct numbers.

The implementer of a sort program will often profit from testing the assumption that his data is random. He may discover that a fixed procedure of collecting the data has introduced strong bias or partial ordering. He may be able to develop a method to take advantage of this bias or to protect a method against an undesirable bias in the data.

One important feature of distribution is duplication. When it is necessary, as in some applications, to recognize duplicate keys, a test for *equal* must be included in the compare function. Each comparison must determine whether a key is higher or lower than or equal to another. When we speak of N "comparisons," we mean N composite comparisons. On some machines testing for equality takes an additional instruction that will actually double compare time; others will reuse a comparand in registers, and time is therefore not quite doubled. Some architectures using "condition code" logic require only a trivial increment in time to execute a high/low/equal comparison over the time needed for a simple high/low. When it is not necessary to recognize duplications, it is important to prevent the "direction" of comparisons from forcing exchanges of equals.

1.8.1.5   Size and variability of records

The size of records determines (with the number of records) whether the list will fit in primary storage at all and thus whether minimal-storage techniques are needed to fit the list. Record size also determines the profitability of using detached or nondetached key sorts. *A detached key sort* sorts a separate list of *control tags.* Each tag contains the key and its address on the initial list. A *nondetached key sort* forms a list of addresses, which it uses to reference keys on the initial list. Both techniques avoid the movement of initial records during the sort. If the sort method involves a large number of exchanges or transfers and if the record size is large relative to key size, the time to build the auxiliary lists may be absorbed by the saving in record movement time.

Storage space may be saved for methods requiring additional work areas by placing only tags or addresses rather than records in those areas. If key manipulation (see Section 1.8.1.2) is required, it will be part of the process of forming detached keys. In some situations it will be possible to avoid any record movement. The processes which follow the sort can index the initial list through the ordered table of tags or addresses.

Several problems develop with the direct manipulation of records of variable size.

1. It is very inefficient to refer to records that are not neighbors. This limitation is very important to nonlinear methods.
2. Exchanges cannot be made between records of different size.

Detached or nondetached key sorting can be profitably applied to variable data. The variability of record size is a factor only on the last pass, when the records are finally ordered.

When keys are widely distributed through a record and are large relative to average record size, and when record size varies considerably as a result of field sizes (as opposed to number of fields), it may be profitable to consider the restructuring of records into a compressed, fixed-size form with an associated map allowing final reconversion [1].

1.8.1.6   Eventual use of data

The sorted list may be used by another routine which expects the list to be in storage. It may be output to a device for processing by a later program. It may be output to a device for presentation to a human user. The nature of the next process to use the result of the sort determines whether it is necessary to physically order the list or simply leave an index. Eventual use affects the amount of storage space required for some methods and limits the amount of record rearrangement that may be undertaken by the sort for its own convenience.

1.8.2   Program Relationships

Sorting is surrounded by other program processes. Sometimes these processes are subsumed into the sort program; sometimes they remain external to it. For example, advancing down a list is usually treated as part of the sort function. In the insertion sorts, however, the function of providing a "next" item for the sort may be external to the sort routine itself.

The variations in sort program construction are innumerable, and the principles (such as they are) of good program design must apply. Whether it is profitable to have an exchange subroutine external to a compare subroutine, to have a data-advance subroutine external to a compare subroutine, to put record/key preparation in the compare function, to isolate it on a preliminary pass—all are considerations which affect the implementation of a sort and its eventual performance.

A sort exists as a program or a subroutine, and as such has a relation with other programs; it lives in a specific program environment which will determine the "flavor" of the sorting program. Often the program environment, by determining source of data, destination of data, amount of core space available, relative importance of certain functions, etc., will imply what sorting methods may be used, or place limits on their use.

1.8.3 Machine Characteristics

The desirability of one sort over another and the specific implementation of a sort are fundamentally determined by the machine that will run the sort. This is why, as noted above, coding in machine language is preferred to coding in higher-level language. The aspects of machine architecture that affect sort performance and influence implementation are almost innumerable. A partial list follows.

1. Number of registers
2. Compare logic and compare timings
3. Efficiency of index register decrements and limit tests
4. Presence of special search instructions
5. Masking and extraction capabilities
6. Size of data fields most efficiently moved
7. Storage sizes

Usually a sort algorithm that is good on one machine will be good on another. When the differences between sorts are marginal, some feature of the machine may well contribute to the choice between them. For example, the presence of a fast linear search instruction on the UNIVAC 1100 Series might tend to make linear methods more interesting for those machines than for machines without such instruction.

2
Exchange Sorts and Linear Insertion

2.1 EXCHANGE SORTING

Exchange sorting is the generic term used to describe a family of minimal-storage sorts which interchange elements of a list when an earlier member is found to be higher than a later member. The scan of the file may work from top to bottom or from bottom to top, or it may alternate from pass to pass.

There are a number of defined variations which are distinguished from one another by the specific sequences in which positions of the list are compared. Common to all variations of elementary exchange methods is the comparison of an element with its nearest neighbor and the potential movement of an element with a higher key one position toward the bottom of the list and an element with a lower key one position toward the top. Three simple forms of exchange sorting are *Pair Exchange, Standard Exchange,* and *Sifting.*

2.1.1 Pair Exchange: The Method

The Pair Exchange method (also called "Odd–Even Transport") consists of a variable number of "odd" and "even" passes until the list is in order. During the odd-numbered passes, every element in an odd-numbered position is compared with its neighbor in the next even-numbered position, and the larger of the two takes the even position. The scan continues down the list until the last odd-numbered element $(N - 1)$ on the list is compared to the last even-numbered element (N). If the list has an odd number of elements, then the last element is not involved in a comparison. A count of exchanges is maintained throughout each pass.

Even-numbered passes compare even positions with succeeding odd positions, and exchanges are made as necessary to ensure that the larger of any pair occupies the odd position. In this way, high-value keys migrate to the bottom of the list. There are as many passes as are required to move the number farthest from its final position into that position. There is then a final pass, needed to recognize order. It leaves an exchange count of 0 during the pass. The method requires at least two passes, one odd and one even.

2.1.1.1   Example

We will again use the list of Chapter 1 (see Fig. 2.1). On the first pass the compare sequence will be as shown in the column headings to the right of the initial list. The starred numbers are the numbers participating in the comparison; the arrow indicates that an exchange has been made as a result of the comparison.

	Initial list	Comparison of positions				
		1 : 2	3 : 4	5 : 6	7 : 8	9 : 10
1.	3	3*	3	3	3	3
2.	11	11*	11	11	11	11
3.	6	6	4*↘	4	4	4
4.	4	4	6*↗	6	6	6
5.	9	9	9	5*↘	5	5
6.	5	5	5	9*↗	9	9
7.	7	7	7	7	7*	7
8.	8	8	8	8	8*	8
9.	10	10	10	10	10	2*↘
10.	2	2	2	2	2	10*↗
11.	1	1	1	1	1	1
EXCOUNT		0	1	2	2	3

Note: EXCOUNT = cumulative count of exchanges. Asterisks indicate elements in comparison; arrows indicate elements in exchanges.

Fig. 2.1 Pair Exchange, pass 1.

Figure 2.2 shows the second pass. The comparisons and their results are indicated as in Fig. 2.1. During the second pass, the movement of key 1, the most displaced element of the list, begins. Figure 2.3 shows the positions at the end of the remaining passes. Note that, from pass to pass, key 1 moves

	Initial list	Comparison of positions				
		2 : 3	4 : 5	6 : 7	8 : 9	10 : 11
1.	3	3	3	3	3	3
2.	11	4*	4	4	4	4
3.	4	11*	11	11	11	11
4.	6	6	5*	5	5	5
5.	5	5	6*	6	6	6
6.	9	9	9	7*	7	7
7.	7	7	7	9*	9	9
8.	8	8	8	8	2*	2
9.	2	2	2	2	8*	8
10.	10	10	10	10	10	1*
11.	1	1	1	1	1	10*
EXCOUNT		1	2	3	4	5

Note: EXCOUNT = cumulative count of exchanges. Asterisks indicate elements in comparison; arrows indicate elements in exchanges.

Fig. 2.2 Pair Exchange method, pass 2.

	List at end of pass 2	3	4	5	6	7	8	9	10	11	12
1.	3	3	3	3	3	3	3	2	2	1	1
2.	4	4	4	4	4	4	2	3	1	2	2
3.	11	5	5	5	5	2	4	1	3	3	3
4.	5	11	6	6	2	5	1	4	4	4	4
5.	6	6	11	2	6	1	5	5	5	5	5
6.	7	7	2	11	1	6	6	6	6	6	6
7.	9	2	7	1	11	7	7	7	7	7	7
8.	2	9	1	7	7	11	8	8	8	8	8
9.	8	1	9	8	8	8	11	9	9	9	9
10.	1	8	8	9	9	9	9	11	10	10	10
11.	10	10	10	10	10	10	10	10	11	11	11
EXCOUNT		3	3	3	2	3	3	3	2	1	0

Fig. 2.3 Pair Exchange method, remaining passes.

	Initial list	First pass								
		After comparison of								
		2 : 3	3 : 4	4 : 5	5 : 6	6 : 7	7 : 8	8 : 9	9 : 10	10 : 11
1.	3	3	3	3	3	3	3	3	3	3
2.	11	6	6	6	6	6	6	6	6	6
3.	6	11	4	4	4	4	4	4	4	4
4.	4	4	11	9	9	9	9	9	9	9
5.	9	9	9	11	5	5	5	5	5	5
6.	5	5	5	5	11	7	7	7	7	7
7.	7	7	7	7	7	11	8	8	8	8
8.	8	8	8	8	8	8	11	10	10	10
9.	10	10	10	10	10	10	10	11	2	2
10.	2	2	2	2	2	2	2	2	11	1
11.	1	1	1	1	1	1	1	1	1	11
	EXCOUNT =	1	2	3	4	5	6	7	8	9

Exchanges/Comparisons = 9/10
Note: Arrows indicate exchanged elements.

Fig. 2.4 Standard Exchange method.

up one position, and that the sort ends when key 1 is ranked. The significance of key 1 lies not in the fact that it is the lowest key on the list but in
that, in this example, it starts off farthest from its final position. The sort
ends when the most displaced key has been ranked and no further exchanges
occur (EXCOUNT = 0).

2.1.1.2   Discussion: Number of comparisons

The number of comparisons for each pass is roughly $N/2$. The number of
comparisons for a complete sort depends on the number of passes, which
in turn depends on the displacement of the key farthest from its rank position.

Usually the number of passes is one more than the absolute value of
the difference between the rank of the most displaced element and its final
rank. This number has a nominal maximum value of $N + 1$, since the
greatest possible displacement of a key is $N - 1$ and it is possible, as in the
example, not to move that element on the first pass. The maximum number
of passes, however, is usually given as N. The minimum number of passes
which occurs when the list is in order is two. Therefore an approximation
of the maximum number of comparisons is $N^2/2$ and the minimum number
of comparisons is N.

What may one expect as an average number of passes and, consequently, as an average number of comparisons for this method? Since the
number of passes is a function of the maximum displacement, it is necessary
to determine what that displacement may be.

To investigate the expected maximum displacement, one must look at
all possible permutations of a list and determine the number of permutations
which have various displacement values. To do this with even as few as
11 numbers is impractical. An approximation of the expected number of
comparisons can be achieved using the formula $(N^2/2) - (N/2)\sqrt{3N/2}$.
Here $N^2/2$ is the maximum number of comparisons based upon $N/2$ comparisons for N passes. The subtrahend reduces the number of comparisons
by expecting that $\sqrt{3N/2}$ passes will not occur, that the farthest displaced
number will be ranked in $(N - \sqrt{3N/2})$ passes, and therefore $(N/2)\sqrt{3N/2}$
comparisons will not occur. Notice that for $N = 100$ the maximum number
of comparisons is 5000 and the expected number is 4350. Since $N^2/2$ grows
much more rapidly than $(N/2)\sqrt{3N/2}$, the expected number of comparisons comes closer and closer to the maximum as N grows, and the method
should be avoided for lists of any serious size. For a list of 100 the method
exceeds the worst comparison expectations of a "good" method by a factor

of 3 : 1; for a list of 1000, by a factor of 24 : 1.* For a list of 25, the method is roughly in the same "ballpark" as known "good" methods. For a list of 50 the method has already nearly twice the comparisons.*

2.1.1.3   Discussion: Number of exchanges

The number of exchanges will vary from zero for an initially ordered list to slightly less than the number of comparisons for a list in reverse order. Since the final pass involves no exchanges, the maximum number of exchanges is $N(N-1)/2$. The expected number of exchanges is $N(N-1)/4$. $N^2/2$ and $N^2/4$ are reasonable approximations, of course.

2.1.1.4   Comment

This method is probably the least effective of the exchange sorts. Although the expected number of comparisons is marginally less than that of Standard Exchange (to be discussed next), it requires additional overhead in the odd/even pass control and consequently involves more procedure space and more elapsed time. It is, however, discussed in Chapter 20 as an effective sort for parallel processors.

2.1.2   Standard Exchange: The Method

The Standard Exchange method (also called "Bubble Sort") undertakes to place one element of a list in its final rank position during each pass over the list. Thus the first pass places the highest key record in the last position, the second pass places the next highest in the penultimate position, etc. The method may also be reversed, so that it ranks from low to high.

On the first pass, the first member of the list is compared with its immediate successor, and if the successor is smaller, the two are exchanged. The larger element, now in the second position, is compared with the element in the third position. An exchange is made, if necessary, to place the larger of these in the third position. Position 3 is then compared with position 4, position 4 with 5, etc. When position $N-1$ is compared with position N, the pass ends.

---

* The approximation for expected number of comparisons is a very much simplified version of an expression used by K. Iverson in *A Programming Language.* (New York: Wiley, 1962.) The statements comparing comparisons between this method and a "good" method are based upon an expectation of $2N*\log_2 N$ comparisons for those methods.

If $k - 1$ is the earlier element on the list and k the later, then in every $(k - 1):k$ comparison, the larger comparand ends up in position k. The element in motion toward the bottom is the element currently assumed to be highest on the list. When a comparison finds the kth element higher, no exchange takes place. Every comparison consequently involves the assumed highest $(k - 1)$ with its immediate successor (k).

The second pass is identical to the first, except that the ranked position is excluded from the sequence. Each later pass excludes an additional ranked position, shortening the list.

A count of exchanges is kept for each pass. A pass that gives rise to no exchanges terminates the sort.

2.1.2.1   Example

Consider the usual list of numbers (Fig. 2.4). The first pass, shown in detail, gives the status of the list after each comparison. The positions compared are given in the column headings. After the comparison between position 2 and position 3 (that is, $2:3$), key 6 and key 11 are exchanged. After $3:4$, key 11 and key 4 are exchanged. The progress of the highest key down the list can be easily followed in this example, since key 11 is the largest key. At the end of the pass $(10:11)$, key 11 is in ranked position. There have been nine exchanges during the pass. As earlier, the count of exchanges kept by the algorithm for all passes is represented as EXCOUNT.

At the end of the second pass, the two highest elements are ranked, at the end of the third pass the highest three, etc. The example requires the maximum number of passes for completion.

2.1.2.2   Discussion: Comparisons

The number of comparisons for the method depends on the number of passes required to order the data. For each pass there will be $K - 1$ comparisons, where K represents the number of unranked members of the list at the beginning of each pass. For the first pass, of course, K is equal to N. The number of comparisons is given by the arithmetic progression

$$(N - 1) + (N - 2) + (N - 3) + ... ,$$

where the sum is a function of the number of terms and the number of terms represents the number of passes. Since there will be a minimum of one pass, the minimum number of comparisons is $N - 1$, or N, roughly. At worst, there will be N passes, and so the maximum number of comparisons will be $(N)(N - 1)/2$ (approximately $N^2/2$).

The expected number of comparisons will be close to that for pair exchange (see Section 2.1.2.4). A full discussion for readers interested in

the analysis of the algorithm exists in Knuth, *Sorting and Searching* [21].
An approximation of the expected number of comparisons, $(N^2/2) - (3N/4)$,
is given in Fig. 2.7.

2.1.2.3   Discussion: Exchanges

The number of exchanges varies with the order of the list. There are no
exchanges for an ordered list. The maximum number of exchanges occurs
when the list is in reverse order; there is then an exchange for every comparison, or $(N)(N - 1)/2$ exchanges. The expected number of exchanges is
$(N)(N - 1)/4$. The approximations $N^2/2$ and $N^2/4$ are given in Table 2.7.

2.1.2.4   Comparison of Pair Exchange and Standard Exchange

One will usually expect fewer passes but more comparisons with Standard
than with Pair Exchange, because the number of passes in Standard Exchange depends on maximum positive displacement; in Pair Exchange it
depends on the greatest displacement in either direction. The number of
permutations having a maximum positive displacement of greater than a
given number is fewer than the number of permutations having a displacement greater than that number in either direction. The average number
of comparisons per pass increases to offset the reduced number of passes.
However, because the probability of fewer passes is higher with Standard
Exchange than with Pair Exchange, Standard is more likely to have a good
case. This characteristic, combined with its simpler procedural aspects (the
reader may compare PL/1 algorithms in the appendix), makes Standard
Exchange generally preferable to Pair Exchange although the maximum
number of comparisons is identical and the *expected* number is marginally
higher.

2.1.2.5   Comparison of Standard Exchange and Linear Selection with Exchange

The Standard Exchange method is similar to Linear Selection with Exchange. A fundamental difference between the methods, however, is that
in Linear Selection with Exchange the number of passes is fixed, whereas
in Standard Exchange the number of passes is variable and depends on
the initial order of the data. If there are no exchanges during the pass, the
list is known to be in order, and the procedure may end.

In comparing Linear Selection with Exchange to Standard Exchange,
we must determine whether the savings involved in avoiding an exchange
of records for each inversion can be expected to compensate for the potential
need for more passes to sort the list.

2.1.3   Sifting: The Method

Sifting (also called "linear insertion with exchange" or "shuttlesort") is by far the best of these methods. It differs from the other exchange methods in that it does not maintain fixed compare sequences. In addition, as a result of the pattern of compare sequences, the division into distinct "passes" also disappears.

Sifting operates exactly as Standard Exchange does until an exchange is made. The comparand with the lower key is then forced up the list as far as it can go. It is compared "backwards" with its linear predecessors to the top of the list. As long as it is lower in key value than a predecessor it is exchanged, and another comparison is undertaken. When the element moving upward encounters an element of lower key, the process stops, and the top-to-bottom comparing resumes from the position at which the initial exchange was made.

Let us call the top-to-bottom comparisons "primary" and the bottom-to-top comparisons "secondary." Any primary comparison may give rise to a number of secondary comparisons. If a primary comparison involves positions 6 and 7, then the chain of secondary comparisons can have as many as five comparisons. That maximum occurs if the initial key in position 7 is lower than all keys of the list up to that position. The actual length of a sequence of secondary comparisons is a function of the value of the climbing element relative to that of each member of the preceding ordered list.

The sort ends when a primary comparison attempts to involve an $(N + 1)$th element.

2.1.3.1   Example

The usual list of numbers is given in Fig. 2.5. The method will initially undertake a primary comparison of positions 1 and 2. Since no exchange is made, the next step will be a primary comparison of positions 2 and 3, which causes an exchange of keys 6 and 11. Now, a secondary comparison sequence is undertaken to force key 6 as far up the list as it can go. The next comparison, therefore, is of positions 2 and 1. Since key 3 is lower than key 6, there is no exchange, and the secondary sequence ends. It would have ended in any case, since an attempt to continue the sequence would have discovered the top of the list.

The next step is a resumption of primary comparison. Since the last primary was $2 : 3$, the next is $3 : 4$. Position 3 contains key 11, which had been placed there as a result of the last primary exchange. The $3 : 4$ primary finds an inversion, and an exchange is made and followed by a secondary sequence of $2 : 3$, $1 : 2$, in order to position key 4. Note that the list is ordered up to the point of the last primary comparison.

Figure 2.5 shows the remaining primary and secondary comparisons up

	Initial	P 1 : 2	P 2 : 3	S 1 : 2	P 3 : 4	S 2 : 3	S 1 : 2
1.	3	3*	3*	3*	3	3	3*
2.	11	11*	6*	6*	6	4*	4*
3.	6	6	11*	11	4*	6*	6
4.	4	4	4	4	11*	11	11
5.	9	9	9	9	9	9	9
6.	5	5	5	5	5	5	5
7.	7	7	7	7	7	7	7
8.	8	8	8	8	8	8	8
9.	10	10	10	10	10	10	10
10.	2	2	2	2	2	2	2
11.	1	1	1	1	1	1	1

	P 4 : 5	S 3 : 4	P 5 : 6	S 4 : 5	S 3 : 4	S 2 : 3	P 6 : 7	S 5 : 6	S 4 : 5
1.	3	3	3	3	3	3	3	3	3
2.	4	4	4	4	4	4*	4	4	4
3.	6	6*	6	6	5*	5*	5	5	5
4.	9*	9*	9	5*	6*	6	6	6	6*
5.	11*	11	5*	9*	9	9	9	9	7*
6.	5	5	11*	11	11	11	7*	7*	9
7.	7	7	7	7	7	7	11*	11	11
8.	8	8	8	8	8	8	8	8	8
9.	10	10	10	10	10	10	10	10	10
10.	2	2	2	2	2	2	2	2	2
11.	1	1	1	1	1	1	1	1	1

	P 7 : 8	S 6 : 7	S 5 : 6	P 8 : 9	S 7 : 8	P 9 : 10	S 8 : 9	S 7 : 8	S 6 : 7
1.	3	3	3	3	3	3	3	3	3
2.	4	4	4	4	4	4	4	4	4
3.	5	5	5	5	5	5	5	5	5
4.	6	6	6	6	6	6	6	6	6
5.	7	7	7*	7	7	7	7	7	7
6.	9	8*	8*	8	8	8	8	8	2*
7.	8*	9*	9	9	9*	9	9	2*	8*
8.	11*	11	11	10*	10*	10	2*	9*	9
9.	10	10	10	11*	11	2*	10*	10	10
10.	2	2	2	2	2	11*	11	11	11
11.	1	1	1	1	1	1	1	1	1

Note: Asterisks designate keys involved in the given comparison. Arrows indicate exchanges made as result of that comparison. (P = primary comparison; S = secondary).

Fig. 2.5 Sifting method.

to the point where key 2 has achieved position 6. This secondary chain will involve five more comparisons, ending with the placement of key 2 in position 1 as the result of the 1 : 2 comparison. The final primary, 10 : 11, will then be undertaken and a secondary sequence of nine comparisons and exchanges will move key 1 into position 1, thus ordering the list.

2.1.3.2　Discussion: Number of comparisons

The minimum number of comparisons, $N - 1$, occurs when the initial list is fully ordered. The maximum number of comparisons occurs when the list is in reverse order. Then each of the $N - 1$ primary comparisons generates a secondary sequence that extends the entire distance from the point of the primary comparison to the top of the list. The average size of a secondary sequence will be $(N - 1)/2$. Therefore the total number of comparisons is $(N - 1)^2/2$. We may consider N and $N^2/2$ usable approximations for the minimum and the maximum number of comparisons, respectively. To determine the expected number of comparisons, consider that the average distance covered by a search for any position on a list will be one-half the length of the list. The "sifting" process is analogous to searching an ordered list of numbers for the correct entry position of a new number. There is no better assumption than that the new element should be at the center of the list. Therefore the average length of a secondary sequence is approximately $(N - 1)/4$ since each sequence will be half its maximum length. When there are $N - 1$ primary comparisons, there will be approximately $(N - 1)^2/4$ secondary comparisons. To this must be added $(N - 1)$ primary comparisons; and the total number of expected comparisons will then be $N - 1 + ((N - 1)^2/4)$; $N + N^2/4$ will not mislead.

2.1.3.3   Number of exchanges

The minimum and maximum number of exchanges are, respectively, zero when the list is in order and equal to the number of comparisons when the list is in reverse order. The expected number of exchanges follows the expected number of comparisons very closely. At most, there can be $2N - 2$ comparisons unaccompanied by exchange. These are the primaries and the last secondaries. If every primary gives rise to an exchange, the maximum number of exchanges results. Since we expect approximately $N^2/4$ secondary comparisons, we can expect roughly $N^2/4$ exchanges.

2.2   INSERTION SORTING

Insertion sorting is a generic name for a group of sorts based on successive insertion of "new" elements into a growing ordered list. There are three essential variations to the methods: Linear Insertion, Centered Insertion, and Binary Insertion. These sorts differ in the method used to find the proper place to insert an element. The simplest is Linear Insertion. As its name implies, this views the already existing list as a simple linear list, to be inspected from the top or the bottom, element by element, until the proper position for the new element is found.

This method is commonly used when a list of all known elements must be kept in order at all times and when additions are introduced dynamically by a process external to the sort. The sort operates each time it receives a new element, placing that element on the list in order and releasing control. A real-time on-line system has the characteristics which make insertion a useful technique.

The relationship between the process generating the numbers to be sorted and the sort is such that the sort is entered repeatedly. Such repeated entry may involve some linkage overhead, such as parameter passing and formal-procedure entrance and exit. This overhead must be recognized and included in any analysis of the time involved in using an insertion method in this way. It is possible to organize an insertion sort as one unified process (see Section 2.1.3).

2.2.1   Linear Insertion: The Method (See also Section 2.1.3)

A space equal in size to the anticipated size of the final list of all elements is allocated to the sort. A counter representing the size of the list is established with an initial value of zero. This counter is used to control the length of search for the proper position of an element entering the list. The sort is entered for each element, and one "call" of the sort subroutine places one item on the list and increments the list counter by 1.

The first element is placed in the top position of the output area. The next element attempting to join the list is compared with the first element. If the key of the new element is greater, it is put in position below the initial member. If the key of the new element is smaller, the initial element is transferred to position 2, and the new element is placed in position 1.

All succeeding new elements are compared to each member of the list in order from the first position until a higher element is found. That higher element and all succeeding members of the list are then moved one position toward the bottom of the list. This operation frees the location into which the new element is inserted.

2.2.2   Example

Linear Insertion is illustrated in Fig. 2.6. On the first call to the sort, the length of the output list is zero. The first element to be placed on the output list is key 3. The column headed "First call" shows key 3, indicated by an arrow, placed on the list at the origin of the output area, location TOSORT. The length of the list is augmented to 1. (As a convention, all members placed on the list are indicated by an arrow, and all members transferred are indicated by an asterisk.)

On the second entry to the sort, length = 1. The new key, 11, is compared with key 3, found to be higher, and placed in the second position. Length is set equal to 2 for the third entry. Key 6 is found to be higher than key 3, is then compared with key 11, and is found to be lower. Since key 6 belongs between key 3 and key 11, the position now occupied by key 11 must be made available to key 6. Key 11 is moved down one space, and key 6 is properly inserted on the list.

First call	Second call	Third call
Length = 0, new = 3	Length = 1, new = 11	Length = 2, new = 6
TOSORT = 3 ←	TOSORT = 3	TOSORT = 3
0	11 ←	6 ←
0	0	11*
0	0	0
0	0	0
0	0	0
0	0	0
0	0	0
0	0	0
0	0	0

Fourth call	Fifth call	Sixth call	Seventh call
Length = 3, new = 4	Length = 4, new = 9	Length = 5, new = 5	Length = 7, new = 7
TOSORT = 3	TOSORT = 3	TOSORT = 3	TOSORT = 3
4 ←	4	4	4
6*	6	5 ←	5
11*	9 ←	6*	6
0	11*	9*	7 ←
0	0	11*	9*
0	0	0	11*
0	0	0	0
0	0	0	0
0	0	0	0

Eighth call	Ninth call	Tenth call	Eleventh call
Length = 7, new = 8	Length = 8, new = 10	Length = 9, new = 2	Length = 10, new = 1
TOSORT = 3	TOSORT = 3	TOSORT = 2 ←	TOSORT = 1 ←
4	4	3*	2*
5	5	4*	3*
6	6	5*	4*
7	7	6*	5*
8 ←	8	7*	6*
9*	9	8*	7*
11*	10 ←	9*	8*
0	11*	10*	8*
0	0	11*	10*
0	0	0	11*

Note: Asterisks denote moved keys. Arrows denote element placed.

Fig. 2.6 Linear insertion.

All succeeding entries to the sort are handled in similar fashion. Key 4 causes the transfer downward of keys 11 and 6, so that 4 may be inserted in position between keys 3 and 6; key 9 causes the transfer for key 11, etc.

2.2.3 Discussion: Number of Comparisons

The number of comparisons on any pass of the sort depends on the number of items on the list. The output area has length N to hold all expected items. Since the items arrive one at a time, however, the area is not full until the sort is over. In general, the expected number of items on the list is $N/2$

since the list varies in size from 0 to N. In $N-1$ of the calls to the sort, there will be comparisons of an element with other elements, since the first call merely places an element on the list. In the worst case, each new element will go to the bottom of the list, and the number of comparisons in each pass is $N/2$, the average length of the list for all passes. The worst case for comparisons, therefore, is that in which the elements of the list to be ordered arrive in order. Here for the first time we have observed a sensitivity to order in which the better the order of the data, the more comparisons will be undertaken. There will be $(N)(N-1)/2$ comparisons ($\simeq N^2/2$) when the elements arrive in perfect order.

If the elements arrive in reverse order, only $N-1$ comparisons will be required. Each new element will, in one comparison, find the element first on the list to be higher.

In general, half the list in existence at any given time will be searched before a higher key is found. The expected number of comparisons is therefore $(N)(N-1)/4$. We may use $N^2/4$ as an approximation.

2.2.4　Discussion: Number of Data Movements

Unlike the exchanging methods discussed in previous sections, insertion sorting methods involve record transfers from one position to another, not position interchanging. Each data movement is therefore roughly half the work for insertion that it is for exchange. The precise difference between the exchange and the transfer functions is, of course, machine-specific. On a machine with a fast, variable-field-size, memory-to-memory move instruction, the transfer of a set of records down a list can be very efficient. All records that are to be moved can be moved with one instruction, and the relative efficiency of making room for the insertion is very high when compared with exchange.

The minimum number of transfers, 0, occurs when the list is presented in order. The maximum number of transfers occurs when the list is presented in reverse order; then the first element is moved $N-1$ times, the second element $N-2$ times, the third element $N-3$, etc. The average number of positions moved by an element is $N/2$. For $N-1$ elements (the last element is not moved), there will be $(N)(N-1)/2$ transfers. The expected number is half this, or $(N)(N-1)/4$　(roughly $N^2/4$).

2.3　SUMMARY OF BASIC COMPARATIVE METHODS

The basic comparative methods have now been described. Figure 2.7 summarizes their performance characteristics, providing the reader with an initial guide to choosing among them.

Sort (Space requirement)	Comparisons (Min. expected, max.)	Data movements (Transfers or exchanges)	Other functions
Linear Selection (N to $2N$)	N^2, N^2, N^2	N transfers	$N^2/4$ key movements
Linear Selection with Exchange (N)	$N^2/2$, $N^2/2$, $N^2/2$	N exchanges	$N^2/4$ key movements
Linear Counting (N to $3N$)	$N^2/2$, $N^2/2$, $N^2/2$	N transfers	$N^2/2$ counts
Pair Exchange (N)	N, $N^2/2 - (N/2)\sqrt{3N/2}$, $N^2/2$	0, $N^2/4$, $N^2/2$ exchanges	
Standard Exchange for $N < 100$ (N)	N, $N^2/2 - 3N/4$, $N^2/2$	0, $N^2/4$, $N^2/2$ exchanges	
Sifting (N)	N, $N + N^2/4$, $N^2/2$	0, $N^2/4$, $N^2/2$ exchanges	
Insertion (N to $2N$)	N, $N^2/4$, $N^2/2$	0, $N^2/4$, $N^2/2$ transfers	

Fig. 2.7 Summary of characteristics of seven basic comparative methods of sorting. Approximate results for sorting a list of N items.

It is not the intent of Fig. 2.7 to provide a guide to precise calculation of comparisons and transfers. The sorting methods described in Chapters 1 and 2 should not be used when it is important to select a sort with an eye to minimizing or predicting performance rigorously. Therefore, approximations have been used in the summary table. Thus, Fig. 2.7 provides a rough guide to the general performance of the methods discussed. Other sources provide more detail [3, 7, 12, 17, 21].

The very rough and tentative character of Fig. 2.7 cannot be overemphasized. Careful evaluation of the precise nature of each function on a specific machine is only a first step toward intelligent procedure. The very wide gap between "expected," maximum, and minimum must be filled by the programmer's appreciation of his data and its initial order or disorder.

3
Shell's
Sorting
Method

3.1 INTRODUCTION

On July 28, 1959, D. L. Shell [27] described a minimal-storage sorting method which has become generally known as Shellsort, although it is sometimes called Merge-Exchange [18]. Much attention has been given to it in the literature, and there is still active discussion of its properties. In this chapter we will describe the approach, discuss the effect of variations in the computation of a critical parameter, and describe a coding variation. Readers who want more detailed information are referred to the bibliography [10, 14, 27].

3.2 THE METHOD

Shell's sort is an extension of sifting. The last pass of a Shellsort is identical to the sifting method described in Chapter 2. Preceding passes use the same technique, but instead of comparing and exchanging immediate neighbors, they compare elements a given distance apart. For example, position 1 is compared to position 5, position 5 to 9, 9 to 13, etc. When an inversion is found, a chain of secondary comparisons involves those elements which were part of the sequence of primary comparisons. For example, if an inversion between positions 9 and 13 is found, the potential secondary sequence after the exchange comprises the comparisons 5 : 9 and 1 : 5.

For each pass, the distance between elements to be compared is determined by reducing a computed initial distance. It is reduced to 1 for the last pass. The purpose of the pattern in the earlier passes is to reduce the number of secondary comparisons and exchanges necessary in the later passes. The method has the effect of allowing elements to take long jumps toward final rank in the early passes, rather than moving one position at a time. In the last pass, numbers will tend to be close to position and consequently will require little movement to final rank.

The method varies in the computation of the initial distance between elements and the specific reduction of that distance from pass to pass.

3.3   EXAMPLE

In this discussion we will adopt the partitioning technique first outlined by Shell. A list of length N is conceptually divided into $N/2$ partitions $((N - 1)/2$ if N is odd), each of which contains two elements. If the number of elements on the list is odd, then one of the partitions will contain three elements. The elements of each partition are $N/2$ locations away from each other. With eleven items on the list, there will be five partitions, with a distance of five positions between the elements of a partition. Consequently the first partition contains the element in position 1, that in position 6, and that in position 11; the second partition contains the elements in positions 2 and 7, ... , and the last partition contains those in 5 and 10.

The initial pass of the method orders the elements of each partition. The sequence of comparisons moves from partition to partition as the list is scanned. For example, the primary sequence is 1:6, 2:7, 3:8; that is, the first element of partition 1 is compared with the second element of partition 1, and then the first element of partition 2 is compared with the second element of partition 2, etc. Figure 3.1 shows the sequence of comparisons and exchanges and the order of the list after certain comparisons and at the end of the first pass.

Up to 6:11, every comparison involves a new partition, and each element has been compared only once. The 6:11 comparison involves a test of two elements of the first partition, one of which (6) has been used previously. Therefore the key contained in position 6 is known to be the larger of the two previous elements. If the 6:11 comparison finds the contents of position 11 high, then partition 1, 6, 11 is in order. However, if the key in position 11 is low, then an exchange between positions 6 and 11 must take place. Column 4 of Fig. 3.1 shows the condition of the list after this exchange (of keys 1 and 5) occurs.

At this point no information is available about the relative magnitude of the key just moved into location 6 and the key of the first element of

	Primary			Secondary
1	2 After	3 After	4 After	5 After
Initial list	2 : 7	5 : 10	6 : 11	1 : 6
1. 3	3	3	3	1
2. 11	7	7	7	7
3. 6	6	6	6	6
4. 4	4	4	4	4
5. 9	9	2	2	2
6. 5	5	5	1	3
7. 7	11	11	11	11
8. 8	8	8	8	8
9. 10	10	10	10	10
10. 2	2	9	9	9
11. 1	1	1	5	5

Sequence of comparisons and exchanges
(by position)

Comparisons	Exchanges
1 : 6	—
2 : 7	2, 7
3 : 8	—
4 : 9	—
5 : 10	5, 10
6 : 11	6, 11
1 : 6 (secondary)	1, 6

Note: Arrows show exchanges.

Fig. 3.1 Pass 1 of SHELLSORT with $N/2$ partitions.

the partition, in location 1. In order to determine that the partition is fully in order, it is necessary to make another 1 : 6 comparison. A comparison of this type, resulting from an exchange of values lower in the partition, is called "secondary," as in sifting.

The secondary comparison between positions 1 and 6 causes an exchange between the elements, and as a result, partition 1 is ordered. All partitions are now ordered, and column 5 of Fig. 3.1 shows the list at the end of the first pass over the data.

At the end of the pass, partitions are redefined by changing the distance between elements in a partition. Shell suggests that the distance for the second pass be half the initial distance (with the result truncated to an integer). Using this formula, we define the partition distance for the second pass as 2.

The second pass orders the new partitions in the manner of the first pass. Figure 3.2 shows the sequence of primary and secondary comparisons

Comparisons		Exchanges
Primary	Secondary	
1 : 3		
2 : 4		2, 4
3 : 5		3, 5
	1 : 3	
4 : 6		4, 6
	2 : 4	2, 4
5 : 7		
6 : 8		
7 : 9		7, 9
	5 : 7	
8 : 10		
9 : 11		9, 11
	7 : 9	7, 9
	5 : 7	5, 7
	3 : 5	

	Initial list	Final list
1.	1	1
2.	7	3
3.	6	2
4.	4	4
5.	2	5
6.	3	7
7.	11	6
8.	8	8
9.	10	10
10.	9	9
11.	5	11

Fig. 3.2 Pass 2 of SHELLSORT (distance = 2).

for the second pass and the status of the list at the end of the second pass. Note that the method does not guarantee that exchanges always move an element closer to position. In fact, key 10 moves up to position 7 before it returns to position 9.

At the end of the second pass partitions are redefined again. The distance is again reduced to one-half the previous distance. For this example the distance is now 1. The final pass of Shellsort always has a distance of 1 and is a simple sift. Figure 3.3 shows the sequence of primary and secondary comparisons on the way to the ordered list.

Three passes have ordered the list. The number of passes is the number of divisions by 2 required to reduce N to 1. Thus, with Shell's algorithm, each new distance is the "floor" (integer quotient truncated) of one-half the previous distance. When N is a power of two, Shell's algorithm requires $\log_2 N$ passes.

Comparisons		Exchanges
Primary	Secondary	
1 : 2		
2 : 3		2, 3
	1 : 2	
3 : 4		
4 : 5		
5 : 6		
6 : 7		6, 7
	5 : 6	
7 : 8		
8 : 9		
9 : 10		9, 10
	8 : 9	
10 : 11		

Initial list		Final ordered list
1.	1	1
2.	3	2
3.	2	3
4.	4	4
5.	5	5
6.	7	6
7.	6	7
8.	8	8
9.	10	9
10.	9	10
11.	11	11

Fig. 3.3 Final pass of SHELLSORT (distance = 1).

3.4 THE CALCULATION OF DISTANCE

The distribution of the work of the method—the amount of effective preparation which the early passes accomplish—is closely connected with the sequence of distances between the elements of partitions from pass to pass.

Several alternative methods for calculating distances have been suggested. Hibbard [14] has developed an algorithm which ensures that the distance between elements will always be an odd number. The function provided by Hibbard (in ALGOL) is

$$2^{\text{entier } \log_2 N} - 1.*$$

For eleven numbers, entier $\log_2 N$ is 3, and 2^3 is the largest integer power of 2 that is less than 11. The value of 2^3 is 8, and the first Hibbard distance

---

* "Entier" is an ALGOL function equivalent to "floor."

is $2^3 - 1 = 7$. Subsequent Hibbard distances are calculated in the same way as Shell's, that is, as the floor of one-half the previous distance.

A comparison of the Hibbard and Shell distance sequences on a list of 16 items demonstrates the impact of distance selection. The Hibbard version requires three passes. The Shell version requires a lengthy fourth pass. Figure 3.4 compares the results of the two versions. The initial distance of 7 for 16 numbers comes from the requirement that the exponent give a number less than N. Thus (entier $\log_2 N$) is 3 for $N = 16$ as it was for $N = 11$. There is a basic difference in the distribution of work between the last pass and the earlier passes. Note that the Hibbard sequence makes 18 of its 37 exchanges (about half) during the last pass, whereas the Shell sequence makes 29 of its exchanges (about two-thirds) during the last pass. The number of exchanges made during the last pass is a inverse measure of the amount of effective ordering done by previous passes and, consequently, of the effectiveness of the method.

	Hibbard		Shell	
	Comparisons	Exchanges	Comparisons	Exchanges
Pass 1	11	8	8	5
Pass 2	22	11	14	4
Pass 3*	32	18	19	6
Pass 4	—	—	43	29
	65	37	84	44

*Final Hibbard Pass (D = 1). Hibbard, D = 7, 3, 1. Shell, D = 8, 4, 2, 1.

Fig. 3.4 Comparative results with Hibbard and Shell distances.

As the final pass of Shell's version is entered, there are two partitions, that of the even positions, 2, 4, 6, 8, ... , and that of the odd positions, 1, 3, 5, 7, ... Each partition is individually sorted, but there is no meaningful order relative to each other. This condition will always occur when N is a power of 2, because the distance values will always be even. There will never be comparisons between odd-numbered elements and even-numbered elements. As a consequence, the final pass must merge two independent strings.

In the example, with $N = 16$, Hibbard outperforms Shell. Hibbard requires one fewer pass, 19 fewer comparisons, and 7 fewer exchanges.

Other sequence-generation techniques have been described. In their comment on Shell's method, Frank and Lazarus [10] discuss the need for keeping distance odd. They suggest reducing distance faster in order to increase the number of secondary comparisons and decrease the number of primaries.

The optimum sequence is not known. The trick is to find the number that balances the number of passes, the distribution of work in each pass, the expected length of secondaries, etc. There are sequences other than those mentioned here. In a study of the method, Knuth [21] has investigated several values (including Fibonacci numbers) for DISTANCE (I) always divisible by DISTANCE ($I + 1$), and those discussed here. He concludes that there is some "witchcraft" involved in determining the best sequence. The expected number of comparisons has not been determined for the general case. The minimum found for the expected number of comparisons is $N \log_2 N$.

3.5   A DELAYED-EXCHANGE VERSION

The description of the method given in Section 3.2 includes the assumption that an exchange follows each comparison which finds a higher element "above" the rising element of a partition. A method for reducing the number of exchanges is provided in an algorithm published by Hibbard [14]. The algorithm differs from that previously described in that it uses temporary storage to hold a comparand during a sequence of comparisons.

Each primary comparison, key_i: key_i + dist., involves the transfer of key_i + dist. to temporary storage. The comparison of position 1 with position 4, for example, actually involves a transfer from position 4 to temporary storage and a comparison of the key in temporary storage with the key in position 1. If key_i is higher, it is moved to position key_i + dist. If key_i + dist. is higher, no move is necessary. The transfer to temporary storage effectively frees the position of the later element of the partition.

When key_i is higher than key_i + dist., it is transferred from its position to the "free" position, and the method initiates a secondary comparison sequence. The contents of temporary storage (key_i + dist.) is the record climbing up the partition. Each record key compared with temporary storage and found to be higher is transferred down the list into the current free location. Each transfer frees the position of the transferred item. When an element of the list with key lower than the key of temporary storage is found, the contents of temporary storage are inserted into the most recently freed list position, which is its proper place.

The use of temporary storage becomes efficient when secondary comparison sequences are of length 2 or greater. Since this may be expected to be true of random data in lists of any appreciable size, the delayed-exchange version may, in general, be superior to other Shellsorts.

4

Structure
in
Sorting

4.1 THE IDEA OF STRUCTURE

All of the more efficient comparative internal sorts impose some partitioning or some nonlinear structure on the list to be sorted, in order to reduce the number of comparisons. The particular structure which is imposed in one large class of sorts is that of the *tree*. This section will introduce some fundamental notions of trees and tree structures, to support discussions of the sorting methods introduced in following sections.

Any algorithm which operates on nonlinear structures depends on a means of developing addresses for the various items which are by some criterion "next." Often imbedded pointers are used. Sometimes address-formation algorithms, involving division or multiplication of addresses, are used to move from one element to another.

4.2 BINARY TREES

Figure 4.1 shows a structure called a *binary tree*. The numbers outside the circles represent relative position numbers of the elements represented by the circles. The numbers inside the circles are the values to be sorted. Elements of a tree are called *nodes*. On a list of numbers to be sorted, each element is a node of the tree. The lines between nodes are *links*. They represent some means of moving from one node to another. A node which has a link to another node will be a parent, or predecessor (PRED), a left successor (LSUC), or a right successor (RSUC). In Fig. 4.1, node 1 is the parent of nodes 2 and 3. Node 2 is a left successor of node 1; node 3 is a right successor. Nodes 2 and 3, in turn, are parents of nodes 4, 5, 6, and 7.

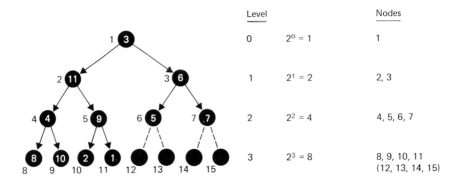

Fig. 4.1 Structure and tabular description of binary tree.

A binary tree must have one and only one root node, which has *no* predecessors. Any node other than a root node may have only one predecessor. A node may have *only two successors.* The last requirement distinguishes a binary tree from other trees [13].

Trees are sometimes shown in a list format so that the parents and successors are explicitly shown in an entry. The list of Fig. 4.1 represents the tree in the upper portion of the figure. All links are explicit in the entry for each node. Some further characteristics follow.

1. The LSUC of any node is always that node whose position number is twice its own position number. The LSUC of node 1 is node 2. The LSUC of node 2 is node 4. The LSUC of node 4 is node 8.

2. The RSUC of any node is always that node whose position is twice its own position *plus* 1. The RSUC of node 1 is node 3. The RSUC of node 2 is node 5. The RSUC of node 3 is node 7, etc.

3. The position of the nth LSUC for any node i is $i*2^n$. The root node is node 1. The position number of the first LSUC of the root node is $2^1 \times 1 = 2$. The position number of the second LSUC of the root node (the second level LSUC of the root node) is $2^2 \times 1 = 4$. The exponent is the length of the path (in links) from one node to another via LSUCs. The multiplier i adjusts for initial position. What is the distance from node 1 to node 8? It is the number of links in the path from node 1 to node 8. There are three such links; 2^3 gives the position node 8. What is the position number of the third LSUC of node 3? $2^3 \times 3 = 24$.

4. The exponent, n, also gives the number of nodes which exist at the level whose leftmost node is n, if the tree is complete at all levels. Given that a node 2^n exists, there are 2^n nodes at the level of 2^n if the tree is complete at that level. If the node 2^n exists, there are at least 2^n nodes on the tree. The sum of the nodes for all levels closer to the root is

$$\sum_{i=0}^{n-1} 2^n = 2^n - 1;$$

therefore, if there is one node at level 2^n, there are $(2^n - 1) + 1 = 2^n$ nodes altogether. If there is no node 2^{n+1}, then there is a maximum of $(2^{n+1} - 1)$ nodes on the tree. Our tree has $n = 3$ levels, with a maximum of 15 nodes and 11 actual nodes.

5. The position number of an immediate RSUC has been given as twice the position number of its predecessor plus one. The node number of the rightmost node that is n levels away from the root is $2^{n+1} - 1$. The node numbers of any nodes at level n range from 2^n to $2^{n+1} - 1$.

With these rules it is not necessary to explicitly represent the position numbers of nodes on the tree in order to access them. It is possible to move from any node to any other node by applying the positionality and addressing rules just given. For example, if one were at node 4 and wished to move to its rightmost successor, node 9, one could compute its position (node number) by multiplying four by two and adding one. To move from a node to its *parent* involves integer *division* by two.

Note that any node within the tree can be the root of another tree, a subtree. A subtree is named by the node number of its selected root node. This example has subtrees 2, 3, 4, 5, 6, 7, 8, 9, 10, 11. Given a tree of size n, there are $(n - 1)$ subtrees. For any node to be a member of a subtree it must have a path to the root node of the subtree or be itself the root. Thus 4, 8, and 9 are members of subtree 4, but 11 is not a member of subtree 4.

4.3   VALUE OF A BINARY TREE

A path of a tree has a length which is the number of links in the path. In a binary tree, the length of the path of all LSUCs from the root to the endpoint node at the lowest level is the height of the tree expressed as the number of levels.

The concept of height is useful in developing the total value of the tree. It is defined in various ways. The definition used here comes from Windley [29] with a trivial modification. The value of a tree is developed by assigning to the root node a value of 0. Every other node is assigned the value of the level on which it exists. The immediate successors of the root node have value 1, nodes at the next level have value 2, etc. The value of the tree is the sum of the values of all nodes. This value is given as

$$\sum_{L=0}^{k} N(L) \cdot V(L),$$

where k is the number of the levels of the tree, N is the number of nodes at the level, and V is the value of the level.

Note that the value of any subtree can be computed in an identical manner by assigning a level value of 0 to the root node of the subtree. The value of the tree in Fig. 4.2 is 22.

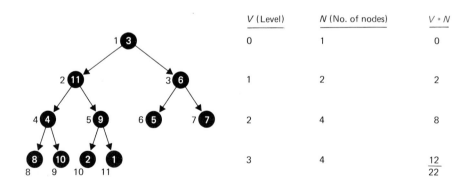

V (Level)	N (No. of nodes)	V * N
0	1	0
1	2	2
2	4	8
3	4	12
		22

Fig. 4.2 Value of a tree.

4.4   FIRST USE OF A TREE

A first example of the use of the tree-structure concept in the ordering of data will show how to build a binary tree representing a list so that as it grows it represents order (Fig. 4.3). Explicit pointers will be used and space for entries of LSUC, RSUC, PRED are assumed.

Position	K	LSUC	RSUC	PRED
1.	3	Ⓝ	2	Ⓧ
2.	11	3	Ⓝ	1
3.	6	Ⓝ	Ⓝ	2

Fig. 4.3 Growing a tree.

Initially select the first item on the list (key 3) as the root of the tree. To place key 11 on the tree, adopt the convention that a number lower than the value of a node will be its LSUC and a number higher than a node will be its RSUC. Key 11 therefore becomes the RSUC of the root node whose key is 3. The RSUC field of the entry for key 3 will hold the position number of key 11.

To place the third item of the list (key is equal to 6), compare it with the root node. Because it is larger, it must be an RSUC of that node. The RSUC position is not vacant, however. The address of the item currently occupying that position is in the RSUC entry of the key 3 node. Key 6 is compared with key 11. It is found to be lower, and consequently it becomes the LSUC of the node representing key 11. To effect this, the LSUC field of node 2 is set to 3 and the PRED field of the key 6 item is set to 2. The table formed in this way contains the explicit links between the nodes of the tree that contain keys. No other nodes are shown.

How would the order 3, 6, 11 be deduced from the structure just formed? There are a number of algorithms for going from node to node while tracing order. For example, start with the root item on the list. In Fig. 4.3 there is no lower number on the list. This is reflected by the absence of an LSUC. Key 3 at the root node must be the lowest item on the list so far.

Now examine the RSUC of the node to determine whether a number higher than 3 exists. It does, and we move on to that RSUC node at position 2. Examine the LSUC field of that node to see if a number lower than it exists on the list. There is an LSUC at position 2 (key 11). This LSUC is position 3 (key 6).

Determine that there is no number between 3 and 6, by discovering no LSUC at the key 6 node. One now knows that key 6 is the lowest on the list after key 3. Now determine whether a number exists between 6 and 11, by examining the RSUC of key 6. There is no higher number.

Since there is only a PRED address at key 6, return to key 11 (the parent of key 6). Examine the RSUC of key 11 to determine whether any higher number exists. None does. Return to the predecessor of key 11, discover that it is the root node, and complete the tour of the tree. A critical detail in the tour, of course, is remembering which LSUCs and RSUCs have already been used. The algorithm here used to deduce order is summarized as follows:

1. Inspect the LSUC field of a node. If it is empty or marked "used" (already followed), this node is the next smallest on the list. If it is valid, inspect the node to which it points. Follow LSUCs until an empty or used LSUC is found.
2. When an empty or used LSUC is found, inspect the RSUC field. If it is empty or used, follow the PRED field to the parent node. Repeat, beginning with Step 1.
3. When a valid RSUC exists, repeat, beginning with Step 1 at the RSUC node. Inspection of an RSUC occurs only when a next lowest is found.
4. The terminating condition is the inspection of the node with an empty RSUC and the root node as parent.

Figure 4.4 shows the completed tree for the list of the 11 numbers used throughout the text. Its odd shape is due to the fact that our positioning rules required a five-level tree to order only 11 items.

The building of this tree can be accomplished without explicit links. Begin by choosing the first element of the list as the root node. Place it in location 1 of any empty area used to represent the tree. Each subsequent element will seek out its appropriate node by examining the root node and other candidate locations.

Call a location C. If an element exists at C, perform a comparison. If the element at C is higher than a new element, then the next comparison must be made at the LSUC of C. The LSUC of C is, by definition, twice C. If the element at C is lower, then the next comparison must be made at $2C + 1$, the RSUC. Consequently, a series of comparisons can be generated by calculating $2C$ or $2C + 1$. In this case, no explicit pointers are required.

Whether this procedure is more or less efficient than the use of explicit links depends on coding details, such as the cost of inspecting a location for availability as opposed to determining a valid LSUC, RSUC field. In

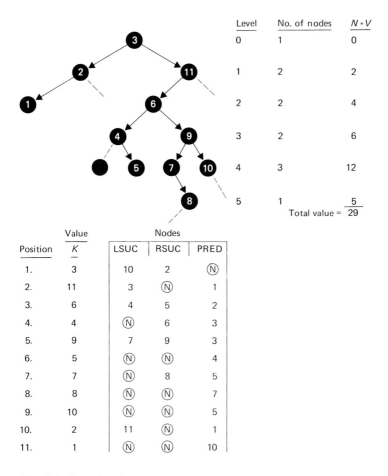

Level	No. of nodes	N * V
0	1	0
1	2	2
2	2	4
3	2	6
4	3	12
5	1	5

Total value = 29

	Value	Nodes		
Position	K	LSUC	RSUC	PRED
1.	3	10	2	Ⓝ
2.	11	3	Ⓝ	1
3.	6	4	5	2
4.	4	Ⓝ	6	3
5.	9	7	9	3
6.	5	Ⓝ	Ⓝ	4
7.	7	Ⓝ	8	5
8.	8	Ⓝ	Ⓝ	7
9.	10	Ⓝ	Ⓝ	5
10.	2	11	Ⓝ	1
11.	1	Ⓝ	Ⓝ	10

Fig. 4.4 Completed tree.

general, however, the area needed to represent the tree is unpredictable and will characteristically be larger than the area required to build the explicit tree. The specific amount of space required is dependent on the shape of the tree. A tree with five levels (not including root node) will require roughly 2^5 element locations. Since the shape of the tree depends on permutation and is not usually known, it is impossible to accurately predict how much space is needed. Figure 4.5 is a representation of Fig. 4.4 mapped on the tree implied by the five levels of Fig. 4.4. This tree is formed without using link addresses. Each node number in this tree corresponds to a memory address. The highest occupied address contains key 8, which joined the tree when it was compared, first, with key 3 in location 1. Being higher, it next

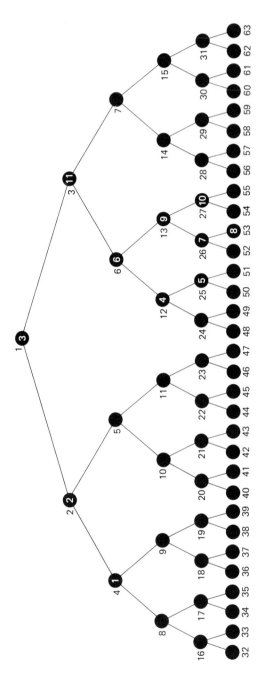

Fig. 4.5 Five-level tree.

looked at location 3 containing key 11, then at location 6 containing key 6. Next, location 13 containing key 9 was inspected. From there, in accordance with the algorithm, location 26 with key 7 was inspected and key 8 became a candidate for location 53. Finding no element at that location, key 8 finally settled into position.

The use of explicit links allows a considerable and predictable compression of the tree space (at the cost of link fields). However, there are implicit tree-oriented sorting techniques, which we will discuss in the next sections, that do not require additional space or that require so little that they are considered minimal-storage techniques.

4.5 NUMBER OF COMPARISONS AND THE SHAPE OF A TREE

Figure 4.6 shows the number of comparisons needed to place the eleven items in Figs. 4.4 and 4.5. It also shows the route each item took in being placed. The total number of comparisons is 29. The *computed value* of this tree is 29 (from Fig. 4.4). The number of comparisons to form a tree is always the total value of the tree.

Key	Number of comparisons	Comparand(s)
3	0	—
11	1	3
6	2	3, 11
4	3	3, 11, 6
9	3	3, 11, 6
5	4	3, 11, 6, 4
7	4	3, 11, 6, 9
8	5	3, 11, 6, 9, 7
10	4	3, 11, 6, 9
2	1	3
1	2	3, 2
	29	

Fig. 4.6 Building the tree.

Tree formation is dependent on the permutation of the data which the sort receives. The variation in the shape of the tree reflects the initial order of data; see Fig. 4.7 for the tree and the number of comparisons for an ordered list. The tree forms a straight line, and the number of comparisons is 55. This number, which is the maximum, holds for a list in reverse order. The reason the maximum number of comparisons is achieved for both reverse and natural order is that the formed tree has either no left branch or no right branch, and it is consequently impossible for any new member of the tree to bypass a comparison with any earlier element.

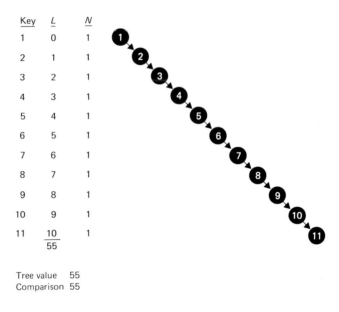

Key	L	N
1	0	1
2	1	1
3	2	1
4	3	1
5	4	1
6	5	1
7	6	1
8	7	1
9	8	1
10	9	1
11	$\dfrac{10}{55}$	1

Tree value   55
Comparison 55

Fig. 4.7 Ordered-data tree.

The long line indicates that every element is an RSUC or an LSUC, and the tree will have maximum height. Every item coming into the tree will compare with one item on all levels existing on the tree at the time of its entry. Note this from Figs. 4.5 and 4.6. A maximum number of comparisons is required to form a tree when the tree being formed has a maximum height. The maximum tree value is derived from the observation that there will be $N - (N - 1)$ comparisons for the first item after the root, $N - (N - 2)$ for the second. The average number of comparisons (along the straight line of the graph) is $N/2$. This will occur $(N - 1)$ times. The maximum number of comparisons is therefore $(N^2 - N)/2$.

The minimum number of comparisons is achieved when the tree has minimum value. In a binary tree minimum value is achieved when it has minimum height. It will have minimum height when there is a maximum of proper nodes, that is, nodes with both descendants in place. Figure 4.8 shows that, when the binary tree is perfectly formed (all nodes are blank or proper), it has a minimum number of levels and a minimum value.

The minimum number of levels (excluding the root) for a tree is the integer logarithm (to the base 2) of N. Thus, 10,000 elements will have 13 levels. The minimum number of comparisons can be determined by computing the value of the optimum tree for the given value of N. The reader may approximate the minimum number of comparisons using $N(\log_2 N) - N$.

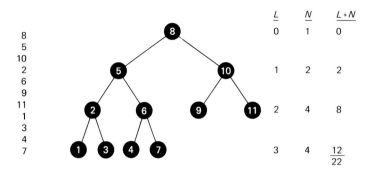

L	N	$L*N$
0	1	0
1	2	2
2	4	8
3	4	12
		22

Fig. 4.8 Minimum-value tree.

The expected number of comparisons involved in building a tree is a function of the shape of the tree and the probability of a tree's having a particular shape. The expected number of comparisons is the product of the probability of any permutation times the comparisons required for that permutation, across all permutations.

Figure 4.9 shows how the shape of the tree is affected by order for all permutations of four numbers. Note, from the table at the bottom of the figure, that the number of favorable permutations is high relative to the total number of permutations. For the 24 (4!) permutations, 12 require a minimum number of comparisons. Given "random" data, where any permutation of the data is equally probable, the expected number of comparisons is lower for a method which has a high probability of encountering a good case than for a method which has a low probability of encountering a good case. The randomness of data, of course, will affect the behavior of a method.

Since we can expect good cases, the expected number of comparisons should be closer to the minimum $N(\log_2 N) - N$ than to the maximum $(N^2 - N)/2$.

The value $N \log_2 N$ is widely used in the literature as an approximation of the minimum expected number of comparisons for a comparative sort. What this means is that, given any sorting algorithm, computing its expected number of comparisons for all permutations, and ignoring special handling for duplicate keys, one can expect no less than $N \log_2 N$ comparisons. The approximation errs on the high side. The reader must distinguish between the "minimum number of comparisons" and the "minimum expected number."

A method characterized as good regarding the average number of comparisons should have on the order of $N \log_2 N$ comparisons. The expected number of comparisons for tree-based algorithms is often expressed in the

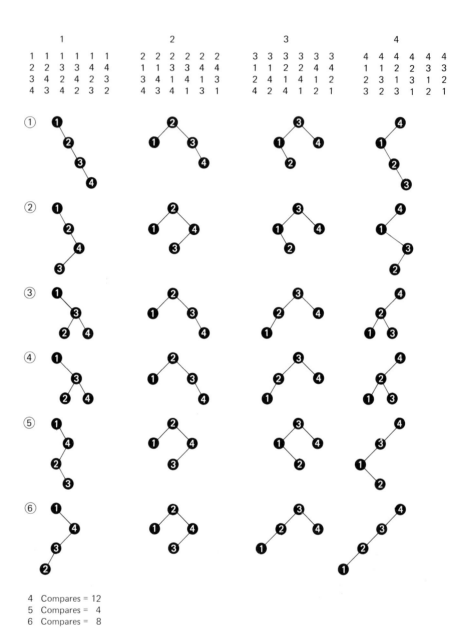

1						2						3						4					
1	1	1	1	1	1	2	2	2	2	2	2	3	3	3	3	3	3	4	4	4	4	4	4
2	2	3	3	4	4	1	1	3	3	4	4	1	1	2	2	4	4	1	1	2	2	3	3
3	4	2	4	2	3	3	4	1	4	1	3	2	4	1	4	1	2	2	3	1	3	1	2
4	3	4	2	3	2	4	3	4	1	3	1	4	2	4	1	2	1	3	2	3	1	2	1

4   Compares = 12
5   Compares =   4
6   Compares =   8

Fig. 4.9 Permutations of tree shape.

form $\alpha N \log_2 N$, where the value of α varies due to method or to refinements of methods. The range of α is from 1.1 to 2.0. Lower values of α are often achieved at the cost of some preliminary overhead.

4.6   PARTITIONING

Since each tree contains a number of subtrees, one can view the arrangement of trees as the arrangement of a family of subtrees. The consequence is that the arranging operation runs as one dependent on K to one dependent on N, where K represents the size of a subtree.

It is not necessary to evoke the concept of a tree in order to use the concept of partitioning. Given any sorting method whose number of comparisons is dependent on N, the partitioning of N into sublists is an advantage.

Consider any linear selection or exchange technique. A partitioning of a list of N items into K partitions of size S would reduce the total number of comparisons from the order of N^2 to $S^2 + S^2 + S^2$ plus whatever operations are required to achieve interpartition order. Given 100 items and sorting as 10 sublists, the total number of comparisons would be on the order of 10×10^2 or 1000 comparisons, as opposed to roughly 10,000. Additional comparisons are required to order the 10 partitions into one partition. The reader is referred to the discussion of high-order selection in Chapter 8.

4.7   OTHER TREES

It is possible to build trees in which nodes have more than two successors. The restriction that a node may have only one predecessor must still hold, but it can have any number of successors. Figure 4.10 shows a perfectly legitimate tree. We will consider trees of this type in the section associated with merging and similar techniques.

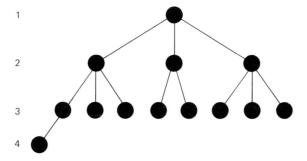

Fig. 4.10 Nodes with multiple successors.

5
Tournament
Sorts

5.1 GENERAL

Treesorts in general become interesting when we want fast sorting in minimal storage or when we wish to build very long strings in preparation for an external merge. Two algorithms are presented here. The Quicksort family of sorts is discussed in Chapter 7. All subsequent comparative algorithms are tree-oriented to some degree. A common analogy used in describing the nature of the treesort method is the classical tennis (or invitational college basketball) tournament. Consider Fig. 5.1, which shows five players who will compete in the initial round. In order to pair each player with another for a game, it is necessary initially to have an even number of players. With an odd number, one player must "sit out" the first round. This player is, in effect, "ceded" a victory in the first round of play. Figure 5.1 shows that player 1 is ceded the first round of play. Players 3 and 4 will play a game, and player 2 will compete with player 5. The winners will advance to the next round.

In general, if there are N players and N is an even number, there will be $N/2$ games in the first round. Figure 5.1 shows that players 2 and 3 won their respective games. These two and player 1, who was ceded the first round, are eligible for the second round of play. There is again an odd number of players, and it is again necessary to cede. This round is ceded to player 2, and a game is scheduled between players 1 and 3. Player 1 wins the game and advances to the final round, a game with player 2. The winner of that game, player 1, is the champion. In terms of sorting, the lowest member of the list has been selected.

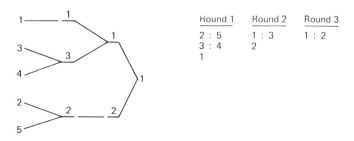

Round 1	Round 2	Round 3
2 : 5	1 : 3	1 : 2
3 : 4	2	
1		

Fig. 5.1 Tournament graph.

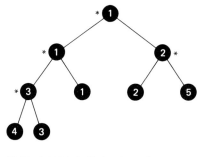

Note: Asterisks indicate advanced
elements.

Fig. 5.2 Tournament graph of Fig. 5.1 as a binary tree. Nodes with asterisks
hold winners.

We recognize the graph of the tournament of Fig. 5.1 from the sporting
pages of the newspapers. Redrawn as Fig. 5.2, it becomes a binary tree.
The root node represents the lowest member of the list, the "winner of the
tournament."

Let us now consider some general statements about the tournament.

1. The result is independent of the method used to determine ceding. If
we had ceded to different elements, number 1 would still have won the
tournament (see Figs. 5.1 and 5.2).

2. The number of rounds of the tournament depends on the number of
players. In general there will be (ceiling $\lceil \log_2 N \rceil$ rounds. For five
entrants there are three rounds: $\lceil \log_2 5 \rceil = 3$.*

---

* The symbol $\lceil \ \rceil$ represents "ceiling." Ceiling (cf. "floor") means "take the log
and round *up* to the next integer."

3. Each round involves half as many "games" (comparisons) as the previous round. Ceding may be avoided only when the initial number of elements is a perfect power of 2.

4. Given any N, the championship tournament will have $N - 1$ games (comparisons). If ceding is discounted, the winner will participate in $\log_2 N$ contests, one at each level or round.

The reader will be reminded of Linear Selection, which took $(N - 1)$ comparisons in each pass to select a "winner." In fact, the Tournament sort is a "selection" sort; it has the property of distinct passes, each one choosing the lowest element of a remaining list. However, Linear Selection required $(N - 1)$ comparisons in each pass. Its exchange variant required $(K - 1)$, where K was reduced by 1 in each pass from $(N - 1)$ to 1. The subsequent passes of the Tournament sort require considerably fewer comparisons to select a pass winner. Each of these "ranks" one contestant, so that at the end of all passes we have an ordered list. Each "consolation" pass uses the information about relative rank generated by a previous pass, so that it is not necessary to undertake a fresh tournament for second place, third place, etc.

5.2 TOURNAMENT SORT: USING WORKING STORAGE

This Tournament sorting method is essentially Algorithm 143, published in *Communications of the ACM,* December 1962. That algorithm, written by A. F. Kaupe, Jr., is based on Algorithm 113 by R. W. Floyd (*CACM,* August 1961). It is one of the possible approaches to Tournament sorting, which, as the name implies, is similar to the pairing of opponents in sporting events.

This method is not a perfect minimal-storage version of the treesort, although it approximates minimal storage when long records with short keys are to be sorted. It is inherently a detached-table sort. Space is required to represent $(2N - 1)$ keys and $(2N - 1)$ addresses. It must be contiguous space, arranged so that the elements (or tags) to be sorted occupy the last N logical positions of the working area. The first $(N - 1)$ logical positions must be free. The layout in storage space for five elements is represented in Fig. 5.3. (There will be further discussion of space in Section 5.2.2.) The addresses in the figure are used to find the winner when it is to be transferred from the list.

The first pass develops a tree structure by comparing keys and selecting a "winner" (the lower comparand) to be advanced in the tournament. The winner is moved into its parent position, whose address is the integer portion of half the winner's current address. For example, the lower of the keys in positions 8 and 9 will be moved into position 4. The empty positions

	Keys		Addresses
1.	—		—
2.	—		—
3.	—		—
4.	—		—
5.	1		5
6.	2	Initial	6
7.	5	list	7
8.	4		8
9.	3		9

N list size
$2N - 1$ spaces for list and work area
$2N - 1$ spaces for addresses

Addresses may be organized in various ways.
Here they are assumed to be embedded in
the record.

Fig. 5.3 Storage space layout for five elements.

in the enlarged work space ensure that addresses representing parent slots will be directly available.

The first pass starts with a comparison of the last positions on the list. The lower key is moved into the parent position. The original address of the lower key is moved into a corresponding position in address space. Comparing continues up the list $(N : N - 1, \ N - 2 : N - 3$, etc.) until an element is transferred into location 1, representing the root node of the tree. When a comparison involves an element previously transferred up the list (for example, $5 : 4$), a new level of the tree is developed. In Fig. 5.4, the numbers inside the nodes are keys, and the numbers outside the nodes are the relative list addresses of the nodes. Comparisons have formed the first level of the tree. A tournament "round" is over, because all subsequent comparisons will be at the next level of the tree. At this stage, there are three subtrees in Fig. 5.4, but we do not yet know their relative position in the file structure.

A new round begins whenever a comparison involves an element from a position which was initially a parent node. Because node 4 received key

	Work area	Addresses	Comparison sequence (positions)	Movements (lower to parent position)
First level of tree ② 3 ③ 4 1.	—	—	9 : 8	9 → 4
2.	—	—	7 : 6	6 → 3
Original ① ② ⑤ ④ ③ 3.	2	6		
4.	3	9		
Position 5 6 7 8 9 5.	1	5		
6.	2	6		
7.	5	7		
8.	4	8		
9.	3	9		

Fig. 5.4 Tree and space after first round.

3 as a result of the 9 : 8 comparison, the 5 : 4 comparison in the second round involves a node higher on the tree. As a result of the 5 : 4 comparison, position 2 receives a key. The winner of the 5 : 4 comparison is effectively advanced to the next round, and thus a new level of the tree is formed. The 3 : 2 comparison is the final round, which advances the winner, the lowest key in the list, to the root node; Fig. 5.5 shows the final tree. The list in Fig. 5.5 represents the contents of the work space after the first pass. The record representation of the root node is placed in an output area, and the position from which the winner came is filled with a key *(Z)* higher than the range of keys which may occur in the data. Other nodes are left as they stand, even though some keys appear in several places.

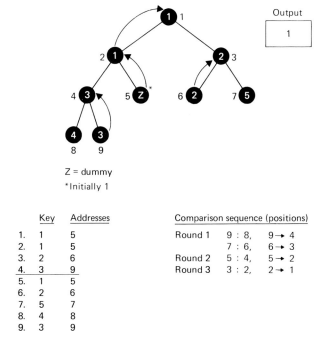

Z = dummy

*Initially 1

	Key	Addresses		Comparison sequence (positions)			
1.	1	5		Round 1	9 : 8,	9 → 4
2.	1	5			7 : 6,	6 → 3
3.	2	6		Round 2	5 : 4,	5 → 2
4.	3	9		Round 3	3 : 2,	2 → 1
5.	1	5				
6.	2	6				
7.	5	7				
8.	4	8				
9.	3	9				

Fig. 5.5 Tournament sort, final tree, first pass.

The second pass begins with a comparison of the positions originally holding the previous winner and its sibling. The previous winner, position 5, now contains a high dummy; and the sibling, position 4, must be lower. It is advanced up the tree into the parent position, 2. From the parent position it is compared with a sibling node at that level, and the winner

Initial list, pass 2		
1.	1	5
2.	1	5
3.	2	6
4.	3	9
5.	Z	5
6.	2	6
7.	5	7
8.	4	8
9.	3	9

Final list, pass 2		
1.	2	6
2.	3	9
3.	2	6
4.	3	9
5.	Z	5
6.	Z	6
7.	5	7
8.	4	8
9.	3	9

Output	Area
1	
2	

Comparison sequences by positions

Round 1   5:4,   4→2
Round 2   3:2,   3→1

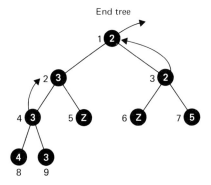

End tree

Fig. 5.6 Tournament sort, pass 2.

	TOSORT		WORKITM	WORKADD
1.	3	1.	—	—
2.	11	2.	—	—
3.	6	3.	—	—
4.	4	4.	—	—
5.	9	5.	—	—
6.	5	6.	—	—
7.	7	7.	—	—
8.	8	8.	—	—
9.	10	9.	—	—
10.	2	10.	—	—
11.	1	11.	3	11
		12.	11	12
		13.	6	13
		14.	4	14
		15.	9	15
		16.	5	16
		17.	7	17
		18.	8	18
		19.	10	19
		20.	2	20
		21.	1	21

Fig. 5.7 Tournament sort, initialization.

is advanced. Figure 5.6 shows the list at the beginning and at the end of pass 2, the tree at the end, and the compare sequences. At the end of pass 2, the contents of the root node are added to the output list. Additional passes will extend the output list in the correct sequence of keys.

Care must be taken in computing the address for the initial comparison of a pass other than the first. Given the position of the winner, one must determine whether its sibling is at that position plus 1 or at that position minus 1. When N is odd, the sibling is always at position minus 1 when the relative position of the winner is odd and at position plus 1 when it is even. When N is even, this is reversed. The comparison between a dummy value and a sibling can be replaced by some other technique for starting the sibling's move up the tree.

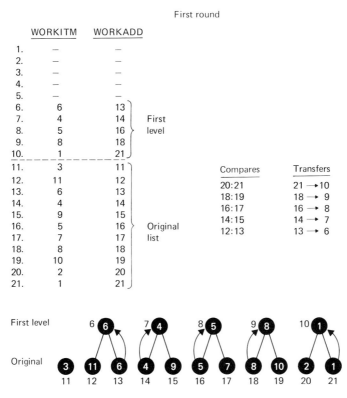

First round

	WORKITM	WORKADD			Compares	Transfers
1.	—	—				
2.	—	—				
3.	—	—				
4.	—	—				
5.	—	—				
6.	6	13	First			
7.	4	14	level			
8.	5	16				
9.	8	18				
10.	1	21				
11.	3	11			20:21	21 → 10
12.	11	12			18:19	18 → 9
13.	6	13			16:17	16 → 8
14.	4	14			14:15	14 → 7
15.	9	15			12:13	13 → 6
16.	5	16	Original			
17.	7	17	list			
18.	8	18				
19.	10	19				
20.	2	20				
21.	1	21				

Note: Arrows indicate elements advanced.

Fig. 5.8 Tournament sort, end of first pass, round one.

5.2.1 Example

A more extensive example may be justified because of the conceptual importance of the method. Consider Fig. 5.7. There are three labeled columns, TOSORT, WORKITM, and WORKADD. TOSORT contains the usual list. WORKITM and WORKADD represent two areas as they look after an initializing process which acquires space and transfers the keys of TOSORT to locations in WORKITM, and the TOSORT addresses to WORKADD. (WORKADD may be a field of WORKITM.)

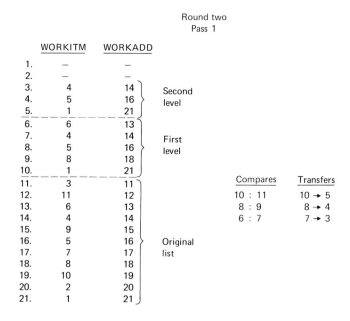

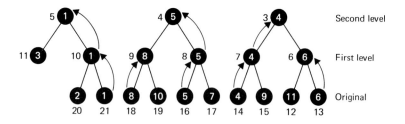

Note: Arrows indicate elements advanced.

Fig. 5.9 Tournament sort, second round.

Figure 5.8 shows the condition of the work area at the end of the first round of pass 1 and the partial tree that is formed. Entries in positions 6 through 10 on the list represent first-round winners.

The next comparisons will be 11 : 10. Position 10 is the parent node of positions 20 and 21 and contains key 1, the contents of position 21. The comparison 11 : 10 is therefore a comparison involving a first-round winner, and it is the first of the second-round comparisons. Position 11 has effectively been ceded the first round. This automatically raises the ceded node to the

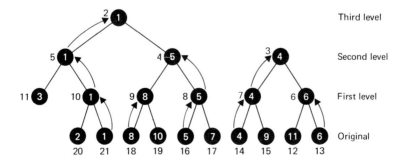

Note: Arrows indicate elements advanced.

Fig. 5.10 Tournament sort, third round.

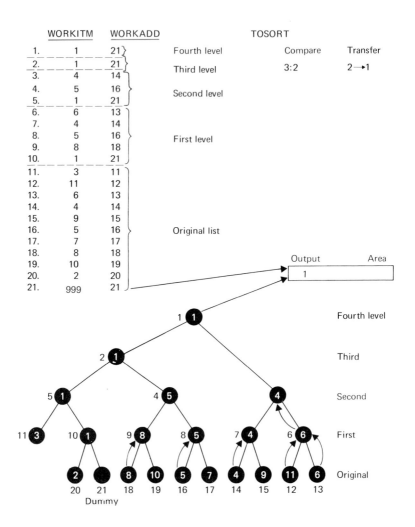

Fig. 5.11 Tournament sort, final round.

next level of the tree, where it will be in the first comparison of the next round. The second round will continue until position 5 is involved in a comparison. At that time the list will look like Fig. 5.9, and the tree will have grown as indicated. The positions of numbered nodes have been reversed in this diagram in order to facilitate the representation of the 10 : 11 comparison.

The third round, shown in Fig. 5.10, cedes to position 3. The final round is given in Fig. 5.11. The list has a dummy in the initial position of the winner, which is indicated on the tree. With the complete tree formed, the winner is ready to be transferred to an output area. The completed list represents the tree structure.

Although, for the purpose of demonstration, we have in this discussion made distinctions between the rounds, the algorithm does not do so. As can be seen from the summary of Fig. 5.11, the comparison sequence moves naturally up the list in a straightforward manner until position 1 of the list is used as a parent address. This ends the pass. No intermediate round-to-round controls are required.

At the beginning of the second pass, the address of the winner is known. The first comparison of pass 2 will involve this winner and its sibling. The comparison between positions 20 and 21 will force the contents of position 20 up the tree. Key 2 of position 20 rises into position 10, the vertex of the 10, 20, 21 triplet. Figure 5.12 shows the list with key 2 so advanced.

	WORKITM	WORKADD		
1.	1	21		
2.	1	21		
3.	4	14		
4.	5	16		
5.	1	21		
6.	6	13		
7.	4	14		
8.	5	16		
9.	8	18		
→10.	2	20*		
11.	3	11		
12.	11	12	Comparison	Transfer
13.	6	13	21 : 20	20 → 10
14.	4	14		
15.	9	15		
16.	5	16		
17.	7	17		
18.	8	18		
19.	10	19		
20.	2	20		
21.	999	21		

Note: Bracket, arrow, asterisk represent comparison and transfer in Figs. 5.12, 5.13, 5.14, and 5.15.

Fig. 5.12 Tournament sort, placing key 2.

The sibling node of position 10 is position 11. The next comparison is that of positions 10 and 11. Figure. 5.13 shows the list after that comparison. Key 2 is in position 5; it is following the route key 1 followed in the first pass. Figure 5.14 shows the list at the end of pass 2. The winner, key 2, has been placed in output. A dummy has been inserted, and the third pass is about to be undertaken. It took four comparisons to select a winner in the second pass.

	WORKITM	WORKADD		
1.	1	21		
2.	1	21		
3.	4	14		
4.	5	16		
5.	2	20*	Comparison	Transfer
6.	6	13	11 : 10	10 → 5
7.	4	14		
8.	5	16		
9.	8	18		
10.	2	20		
11.	3	11		
12.	11	12		
13.	6	13		
14.	7	14		
15.	9	15		
16.	5	16		
17.	7	17		
18.	8	18		
19.	10	19		
20.	2	20		
21.	999	21		

Fig. 5.13 Tournament sort, placing key 2.

	WORKITM	WORKADD		
1.	2	20	Comparisons	Transfers
2.	2	20*	5 : 4	5 → 2
3.	4	14	3 : 2	2 → 1
4.	5	16		
5.	2	20		
6.	6	13		
7.	4	14		
8.	5	16		Pass Summary
9.	8	18		
10.	2	20	Comparisons	Transfers
11.	3	11	21 : 20	20 → 10
12.	11	12	11 : 10	10 → 5
13.	6	13	5 : 4	5 → 2
14.	4	14	3 : 2	2 → 1
15.	9	15		
16.	5	16		
17.	7	17		
18.	8	18		
19.	10	19		
20.	999	20		
21.	999	21		

Fig. 5.14 End of pass 2.

The number of comparison being equal, on all passes other than the first, to the number of levels on the tree, only one test per level is necessary to find a winner. Thus, $\lceil \log_2 N \rceil$ comparisons are required per pass after pass 1.

Pass 3 is shown in Fig. 5.15. The lists shown are, from left to right, the initial list for the pass, the transformed lists after each comparison and transfer, and the final list of the pass.

Third pass

Comparisons (positions)

	Initial list		21 : 20		11 : 10		5 : 4		3 : 2		Final list	
	Key	Address	Key	Address	Key	Address	Key	Address	Key	Address	Key	Address
1.	2	20	2	20	2	20	2	20	3	11*	3	11
2.	2	20	2	20	2	20	3	11*	3	11	3	11
3.	4	14	4	14	4	14	4	14	4	14	4	14
4.	5	16	5	16	5	16	5	16	5	16	5	16
5.	2	20	2	20	3	11*	3	11	3	11	3	11
6.	6	13	6	13	6	13	6	13	6	13	6	13
7.	4	14	4	14	4	14	4	14	4	14	4	14
8.	5	16	5	16	5	16	5	16	5	16	5	16
9.	8	18	8	18	8	18	8	18	8	18	8	18
10.	2	20	999	20*	999	20	999	20	999	20	999	20
11.	3	11	3	11	3	11	3	11	3	11	3	11
12.	11	12	11	12	11	12	11	12	11	12	11	12
13.	6	13	6	13	6	13	6	13	6	13	6	13
14.	4	14	4	14	4	14	4	14	4	14	4	14
15.	9	15	9	15	9	15	9	15	9	15	9	15
16.	5	16	5	16	5	16	5	16	5	16	5	16
17.	7	17	7	17	7	17	7	i7	7	17	7	17
18.	8	18	8	18	8	18	8	18	8	18	8	18
19.	10	19	10	19	10	19	10	19	10	19	10	19
20.	999	20	999	20	999	20	999	20	999	20	999	20
21.	999	21	999	21	999	21	999	21	999	21	999	21

Output ◄─────── Area

1	
2	
3	

Fig. 5.15 Tournament sort, pass 3.

The positions involved in each comparison are given in the column headings. The associated tree shows dummy values at entry into the pass as shaded areas and the condition of the tree at the end of the pass but before the dummying of the winner. Note that the first comparison, 20 : 21, is of two dummies. By convention, the LSUC from this comparison is the winner and is advanced to the parent. This rule results in the placement of a dummy in position 10. The round continues without confusion.

The sort ends when all items have been transferred to an output list. There are N passes; the first pass builds the tree, and $(N - 1)$ subsequent passes fill it with dummies.

5.2.2 Discussion: Space

The space requirement for this version of the Tournament sort is determined by the program environment in which the method is used and by the size of the key relative to the size of the record.

The sort is inherently a key sort. If the keys are the records, then the space required is a function of the manner in which addresses are to be represented. In a word machine a list of 11 one-word elements could take 42 words of space: 21 words for key working space and 21 for address working space. It is possible to concatenate short keys and addresses in one word as well. The initializing phase of the sort would transfer the keys and form the tags before the first pass began. Total space requirement relative to list size when keys = records is therefore high. For multi-field records of significant size, the relative size of space required to represent $(2N - 1)$ tags (key and addresses) becomes less significant.

It is possible to maintain lists of addresses only. Access to keys would be through the address table. The address on the table would be advanced from "node" to "node" on the basis of comparison results. The method would be changed only to the extent that a separate representation of keys is not required but techniques of indirect addressing are. Considerable attention has been paid to the representation and control of address lists in connection with this method.

The size of the ultimate output area depends, for multi-field records, on the subsequent use of the data. It is also affected by the machine instruction set (which may permit records to be gathered from various locations in memory with a single command), by the presence of "gather-write" on the machine, and by the buffering requirements for the output devices [5].

Further variations on the theme of the "tournament" are described in Section 5.3 and in Chapter 11, where the principle of "replacement" is added to the method.

5.2.3   Number of Comparisons and Transfers

The number of comparisons for the Tournament sort is $(N-1)\ +$ $(N-1)\lceil\log_2 N\rceil$. On the first pass $(N-1)$ comparisons are required, and the subsequent $(N-1)$ passes each require $\lceil\log_2 N\rceil$ comparisons. When the value of N is a power of 2, this number of comparisons is exact. Otherwise it is a good approximation, and will be less than the number required for the next higher power because of the occurrence of ceding. The ceiling of the logarithm should be used in these calculations.

The number of transfers of records and the time involved in transferring are functions of the specific organization of the tournament. For the method we have described, there will always be one transfer of a key/address pair for each comparison and an end-of-pass transfer of either a record or an address, or both, at the end of each pass. The total data movement time, excluding the initialization phase, is therefore C (where C is the number of comparisons) times key/address transfer time, plus N final pass transfers:

$$\text{Comparisons} \times \text{Tag transfer} + (N \times \text{Record transfer}).$$

5.2.4   Comments

The programmer calculating time with this method must not fail to include such factors as initialization time to build the tag table and address-calculation time to locate parent nodes. These are significant costs. Consider that the development of parent addresses from comparand addresses can cost $\log_2 N$ divisions (an ominous word) per pass and can overwhelm the method, causing it to operate significantly more slowly than methods which have more comparisons.

There are numerous coding approaches to address development. On binary machines a shift instruction can be substituted for a division by 2. The treesort algorithm in Appendix A uses the approach of maintaining the parent position as a pointer and comparing twice its value with twice its value plus 1. The programmer is well advised to pay some attention to the addressing techniques since an address derivation is required for each comparison and transfer. Use of a technique to avoid the division or multiplication implied in finding node positions is surely worth considering for most machines. One such technique is the establishment of separate parent and comparand pointers with initial values of $2N-1$ and $N-1$, respectively. The problem of reducing address derivation costs is part of the discussion of Section 5.4.

5.3   A MINIMUM-STORAGE TOURNAMENT

An approach to the use of tree structure that requires no additional space but depends entirely on exchanging has been presented by R. W. Floyd as Algorithm 245 in the *Communications of the ACM*. Direct reliance on a binary tree structure is maintained, but the relationship between a parent and its descendants is changed. This method is of interest to programmers who are willing to relax time constraints in order to reduce core requirements to a minimum.

5.3.1   Method

The first pass develops a tree on which every vertex (except the root node) contains a key which is higher than the keys in either descendant. For any triplet of nodes $K, 2K, (2K + 1)$, the highest key value is placed in node K. See Fig. 5.16. The triplets of this tree are listed under the heading "Triplets." Working from the bottom of the tree (the end of the list), one makes comparisons between sibling endpoints and parents. The siblings are compared, and the higher is compared to the parent. If the parent is lower, positions are exchanged.

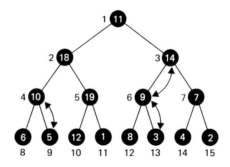

Triplet
positions

7,14,15
6,12,13
5,10,11
4,8,9
3,6,7
2,4,5
1,2,3   (excluded in first pass)

Note: Arrows show vertex ordering exchanges.

Fig. 5.16 Triplets of a tree.

At one point in the journey up the list, comparisons begin for siblings that are themselves parents (vertices). If a parent is found to be low, it may be necessary to sink the parent more than one level in order to keep all vertices larger than all descendants. Note the sequence of comparisons concerning the triplet (3,6,7) in Fig. 5.16. The initial contents of node 3 is key 3. This key is displaced by key 14 in position 6. Additional comparisons (13 : 12, 12 : 6) are required to ensure the ordering of the triplet (6,12,13). The "sinking" element descends the tree for as many levels as are necessary to find its proper position, that is, the position where it is higher than its descendants or it is an endpoint. The value 3 will finish in position 13. The contents of the triplet 3,6,7 will be 14,9,7. The contents of triplet 6,12,13 will be 9,8,3.

At the end of the first pass, the left subtree of the tree is arranged so that its root node (the left descendant of the root node of the whole tree) has the highest element to the left, and the right subtree is arranged so that its root node (the right descendant of the root node of the tree) has the highest element to the right. The higher of the elements in nodes 2 and 3 must be the highest on the list unless the highest element is already in the root node.

The second pass begins by placing the contents of the root node in working storage. This frees the root node position for the winner of the first comparison of the pass. Initial comparisons are (1) that of the left and right descendants of the root node with each other and (2) the higher with the initial contents of the root node now in working storage. The winner is highest on the list and becomes the new contents of the root node.

If working storage is not highest, a position for its contents must be found in such a way that the triplet-ordering conditions which were true at the beginning of the second pass are true at the beginning of the third. The process of finding a position for working storage is undertaken on the subtree which contributed the winner. The search begins by comparing the descendants of the winning node with each other to determine the higher, which is then compared with the contents in working storage. If working storage is higher, its contents are placed in the position just freed by the transfer of the subtree winner to the root node. If the contents of working storage are lower, then the highest descendant is transferred to the free position and another level of the tree must be inspected to find a position for working storage. This continues until working storage is found high and becomes a triplet vertex, or until it descends to be an endpoint.

The final action of a pass is to exchange the contents of the root node with the last current node. When the exchange is made, the last current node becomes last location minus 1. From pass to pass an ordered list grows from the bottom, each pass contributing the next highest element of the list until the $(N - 1)$ highest have been found. If for any pass the contents of the root node are initially higher than the contents of its descendants, the exchange of root node and last current naturally occurs immediately.

The basic comparison mechanisms of the second and subsequent passes may be the same as that of the first pass. The process of finding a position for the displaced root node is the same as the process of ordering the subtrees on the first pass. If the contents of each vertex on the first pass are put into temporary storage, then the coding may be identical.

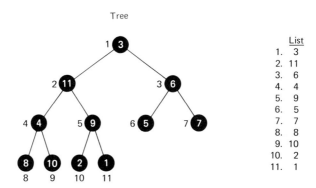

Tree

	List
1.	3
2.	11
3.	6
4.	4
5.	9
6.	5
7.	7
8.	8
9.	10
10.	2
11.	1

Fig. 5.17 Minimum-storage tournament, initial tree.

5.3.2   Example

The usual numbers are organized into the tree in Fig. 5.17. The first pass causes the list and the tree to be transformed to those given in Fig. 5.18; the tree has the property that the vertex of every triplet has the highest key of the triplet. Figure 5.18 shows the comparisons and exchanges of the first pass. The compare sequence works up the list, across the tree from level to level. At the time of the 5 : 4 comparison the contents of node 4 have already become key 10.

The second pass is represented by Fig. 5.19. The contents of the root node (key 3) are placed in temporary storage. The comparisons begin with the determination of the highest on the list. This is accomplished by finding the highest of TEMP, node 2, and node 3. The key of node 2 (key 11) is highest and is moved into the root node. Node 2 is now vacant. The element to fill this node will be the highest of node 4, node 5, and TEMP. Node 4 with key 10 is highest. The movement of the element in node 4 to node 2 leaves node 4 vacant. The element to fill this node will be the highest

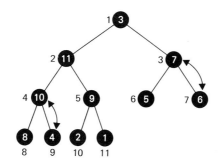

Position	Begin pass 1	End	Position compares	Position exchanges
1.	3	3	11 : 10	
2.	11	11	10 : 5	
3.	6	7	9 : 8	
4.	4	10	9 : 4	4 ↔ 9
5.	9	9	7 : 6	
6.	5	5	7 : 3	3 ↔ 7
7.	7	6	5 : 4	
8.	8	8	4 : 2	
9.	10	4		
10.	2	2	No root node comparison	
11.	1	1		

Arrows show exchanges

Fig. 5.18 Minimum-storage tournament, first pass.

of node 8, node 9, and TEMP. Node 8 containing key 8 is highest and moves into node 4. At this point no additional level of the tree is available for testing, and the contents of TEMP are placed in node 8. The exchange between root node and last current position (now node 11) is made, and the pass ends. Figure 5.20 shows the list and tree at the beginning of pass 3. The transfer from root node to last current has been made.

One detail which must be attended to as the method moves from pass to pass is the appropriate adjustment of the last current node. The reduction of the value of that variable is intended to provide an appropriate pointer to the last current position. But this is not the only use which can be made of that variable. As the list of unranked elements grows shorter and the tree loses nodes and levels, it is necessary to constrain sibling tests so that they do not include nodes which already hold ranked elements. We will see an example of this at the end of pass 3.

Pass 3 is pictured in Fig. 5.21, which shows the list at the beginning and at the end of the pass. The tree represents the end of the pass, with

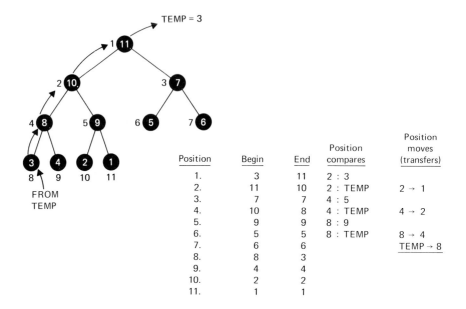

Position	Begin	End	Position compares	Position moves (transfers)
1.	3	11	2 : 3	
2.	11	10	2 : TEMP	2 → 1
3.	7	7	4 : 5	
4.	10	8	4 : TEMP	4 → 2
5.	9	9	8 : 9	
6.	5	5	8 : TEMP	8 → 4
7.	6	6		TEMP → 8
8.	8	3		
9.	4	4		
10.	2	2		
11.	1	1		

Fig. 5.19 Minimum-storage tournament, second pass.

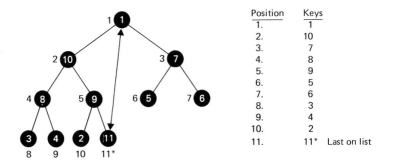

Position	Keys	
1.	1	
2.	10	
3.	7	
4.	8	
5.	9	
6.	5	
7.	6	
8.	3	
9.	4	
10.	2	
11.	11*	Last on list

Note: Asterisk shows ranked elements, Figs. 5.20, 5.21, and 5.22.

Fig. 5.20 Minimum-storage tournament, beginning of pass 3.

the current last position (position 10 for this pass) and root nodes having been exchanged. When the exchange occurred, node 1 had held key 10 and node 10 had held key 1. Node 10 will be excluded from further comparisons.

Pass 3 shows the suppression of the comparison 11 : 10. Since node 11 contains the highest element of the list, it must be excluded from entering

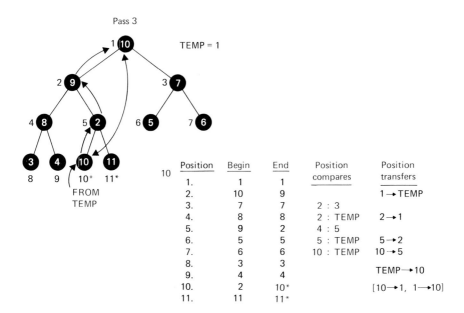

Position	Begin	End	Position compares	Position transfers
1.	1	1		
2.	10	9		1 → TEMP
3.	7	7	2 : 3	
4.	8	8	2 : TEMP	2 → 1
5.	9	2	4 : 5	
6.	5	5	5 : TEMP	5 → 2
7.	6	6	10 : TEMP	10 → 5
8.	3	3		
9.	4	4		TEMP → 10
10.	2	10*		[10 → 1, 1 → 10]
11.	11	11*		

Fig. 5.21 Minimum-storage tournament, pass 3.

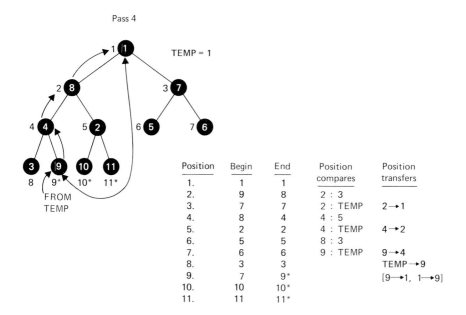

Position	Begin	End	Position compares	Position transfers
1.	1	1		
2.	9	8	2 : 3	
3.	7	7	2 : TEMP	2 → 1
4.	8	4	4 : 5	
5.	2	2	4 : TEMP	4 → 2
6.	5	5	8 : 3	
7.	6	6	9 : TEMP	9 → 4
8.	3	3		TEMP → 9
9.	7	9*		[9 → 1, 1 → 9]
10.	10	10*		
11.	11	11*		

Fig. 5.22 Minimum-storage tournament, pass 4.

into comparisons. Node 5 has at this time only a left descendant, node 10. The 10 : TEMP comparison occurs directly, and the contents of node 10 advance to node 5. Then TEMP is placed in node 10, the exchange of node 1 and node 10 (last current node) takes place, and the pass ends.

Figure 5.22 shows the fourth pass. The tree shows the list at the end of the pass after the final exchange is made. There are always N passes: $(N-1)$ to place $(N-1)$ highest elements and one to organize the tree.

5.3.3   Discussion: Number of Comparisons

The number of comparisons on the first pass is dependent on initial order in the data. The minimum number occurs when all triplets are in proper order. This number will be twice the number of vertices on the tree excluding the root node. N is a reasonable approximation for first-pass minimum comparing.

The maximum number of comparisons on the first pass will occur for a permutation which finds every vertex lower than its descendants and forced to descend to endpoints to place displaced vertices. The permutation which has this property is the ordered list. A satisfactory approximation for lists of more than 256 elements is $2N$. The maximum number of tests for small lists can be calculated reasonably directly from the tree. If one is willing to tolerate the error introduced by assuming perfect trees, a quick procedure for an estimate is as follows:

1. Calculate the number of vertices excluding the root and end points. To do this tabulate the number of vertices at each level of the tree from level 1 (2 vertices) to level $\lceil \log_2 N - 2 \rceil$. For $N = 127, \lceil \log_2 N \rceil = 7$, the table is

Level	Vertices
1	2
2	4
3	8
4	16
5	32

2. Calculate the number of comparisons at each level and sum. Two comparisons are required per vertex at the lowest level, $\lceil \log_2 N - 2 \rceil$. At each higher level twice as many comparisons are required. (See table at the top of page 81.)

The procedure works for any perfect N. Above $N = 256$, $2N$ is an adequate estimate. For $N = 2047$, the maximum number of comparisons is 4056; $2N$ is 4094.

The bulk of the comparisons will occur in the $(N-1)$ passes after the first pass. The minimum number is roughly 4 comparisons per pass, or $4(N-1)$ altogether. The derivation of this value is based on the following

Level	Vertices	Comparisons per Vertex	Comparisons per Level
1	2	10	20
2	4	8	32
3	8	6	48
4	16	4	64
5	32	2	64
		Total	228

considerations. As a result of the first pass, all endpoints are known to be lower than their parents, and this condition is maintained from pass to pass. Because of the end-pass exchange, the initial root node value may be the highest value only on the second pass. In all other passes the key in the root node has come from an exchange between an endpoint and the root node; therefore it cannot be the highest on the list. As a consequence, at least one set of comparisons must be undertaken to find a position for the displaced contents of the root node in each pass.

The total minimum number of comparisons for the sort is $N - 2 + 4(N - 1)$, or $5N - 6$. The maximum number of comparisons, which is achieved when each displaced root node sinks to the lowest level, is given by Floyd in the published algorithm as $2N(\log_2(N - 1))$. The expected number of comparisons is an analytically difficult number; the author knows of no available discussion of this in the literature.

5.3.4 Exchanges

Data movement for the method is best understood in terms of exchanges. If the parent is the highest in all triplets, there will be no exchanges. This occurs when the minimum of comparisons occurs. Each time a parent is lower, there will be an exchange. For each triplet two comparisons are required, but only one exchange may take place, so the maximum number of exchanges is half the number of comparisons. In general, two comparisons and one exchange order any triplet, and the number of exchanges approaches one-half the number of comparisons when the number of comparisons significantly exceeds the minimum.

In the algorithm shown, for all subsequent passes, each comparison causes a transfer. All comparisons which result in the movement of a sibling up the tree cause a transfer, and the comparison which finds the position for the contents of TEMP causes a transfer. The number of transfers is equal to the number of comparisons plus that of the final exchanges for all passes after the first. Since a transfer is roughly half an exchange, the number of exchanges is roughly half the number of comparisons.

5.4 DETAILS OF TOURNAMENT ORGANIZATION

The Tournament sort, in a form which makes use of additional space and which avoids the calculation of addresses, is currently a very widely used sort algorithm in sort/merge programs. Because of the wide interest in Tournament sorting and its replacement–selection variant, as well as in its purely internal versions, we will present here some further discussion of the organization of this sort. The example used in this section will pick lower values and rank from low to high rather than from high to low.

5.4.1 Organization of Space

The method is inherently a key sort. Given sixteen 100-character records with keys of five characters, the initial phase of the method might organize storage as shown in Fig. 5.23. This is essentially the pattern of *CACM* Algorithm 113, described in Section 5.2. The detached-table property of the sort is explicitly shown in Fig. 5.23, which represents the formed tree at the end of pass 1. The first $N/2$ locations (L through $L + 7$) represent the first level of the tree, the next $N/4$ locations ($L + 8$ through $L + 11$) the second level of the tree, the next $N/8$ locations ($L + 12$ and $L + 13$) the third level, etc., until the winner is represented in a sublist of length N/N. One difference between Fig. 5.23 and the discussion of *CACM* Algorithm 113 is that the location of the auxiliary space is chosen arbitrarily and is not necessarily calculated from an address on the main list of records. The initializing routine is provided with a base address, L, which it uses to reference the auxiliary list. The first $N/2$ spaces of this list are filled with tags representing the lower of the paired records at (0, 100), (200, 300), etc. Each boxed entry on Fig. 5.23 represents the location of a tag on a list beginning at location L.

The capability to arbitrarily choose auxiliary tree space can be generalized by the programmer's provision of a family of addresses, one for the beginning of each level of the tree. There will be as many base addresses as there are levels on the tree. The base addresses separate the levels of the tree, facilitating the use of a fragmented memory space, but more important, they reduce the number of divisions and multiplications, shifts, and other manipulations in the sort. As the sort finishes a level, it moves to a new index (L_i) until the index list is depleted.

One variation of the Tournament sort distributes initial tags to spaces between the records. Addressing on the list is to a location which is the address of the record plus its size plus 1. This technique enables the keys of these records to be effectively placed out of range without destroying the record, so that records may be written from the initial area. Out-of-range values are placed in the interspersed key spaces, effectively eliminating selected records from the tournament. Other variants use interrecord space

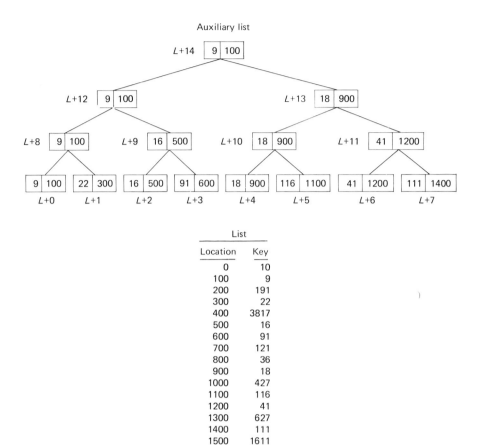

Auxiliary list

Location	Key
0	10
100	9
200	191
300	22
400	3817
500	16
600	91
700	121
800	36
900	18
1000	427
1100	116
1200	41
1300	627
1400	111
1500	1611

Fig. 5.23 Auxiliary list as a tree.

of this type to represent the entire tree, establishing conventions for first- and later-level referencing.

5.4.2   Path Finding

The program must determine a path through the tree after initialization. In Fig. 5.23 the initialization routine has selected the record starting in location 100 as the winner. Its sibling at location 000 is to enter the tournament. A problem develops. So far, there is no way of indicating the address on the tree to which the sibling tag is to be forwarded, no way of determining the address of the comparand at any level, no way of forwarding a tag to another level, and, once there, no way of determining a new comparand without address calculation.

The solutions to these problems are varied and diverse. They differ from machine to machine and from implementation to implementation. We will describe one method of organizing a tournament. It is based on a considerable expansion of the concept of the explicit pointers and causes a modification in our understanding of the use of the structure of a tree. The list of base address pointers is expanded to point to every node on the tree. Now the base address pointers form the tree, the structure of which becomes permanent and explicit. Part of the process of initialization is to associate with each pair of records the address of a location on the first level of the tree. Figure 5.24 shows a list of records with an associated pointer for each pair appended to record space. There are $N/2$ associated pointers to the $N/2$ first-level nodes. Figure 5.24 uses a location for each address, so that if pointer 1 is at location 1600, pointer 2 is at 1601, etc. The contents of locations 0–1500 are the records to be sorted (the list). Each record is 100 locations long with a key in its initial location. The "1600" locations are associated with records on the list. Location 1600 relates to records in 0 and 100; 1601 relates to records in 200 and 300; etc. For each pair of siblings on the list a "1600" location contains a pointer to a "comparison packet" which is a first-level node. The first-level nodes begin at location 2500. Each node is a three-word package that contains information about the tournament.

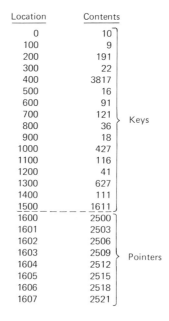

Fig. 5.24 Record list with associated pointers.

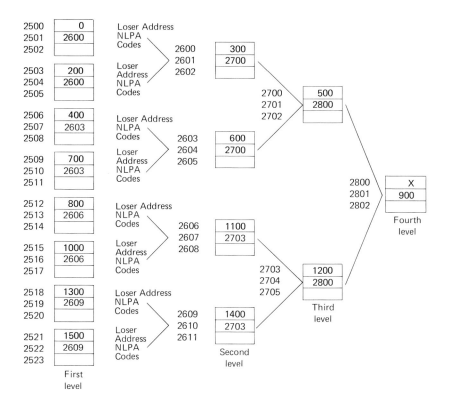

Fig. 5.25 Contents of nodes.

In Fig. 5.25 the contents of the first-level nodes beginning at 2500 are shown in terms of the example. Each first-level node contains the initial record list address of the loser, the address of a node at the next level (second-level packets), and a set of codes. Assumed are one-word addresses and a field of codes whose use is not discussed here. The set of first-level winners is implicitly defined by the first-level packets that record the losers' addresses. Notice that if we insert a pointer to the address of the loser on the original list, the method becomes a nondetached keysort. The addresses of the winners may be kept in packets if desired or assumed to be available in a register of the machine. The NLPA points to the next level in exactly the same manner that the "1600" pointers pointed to the first level. Figure 5.25 is a complete list of packets representing the tree. Thus 0 : 100 yields loser key 10, winner key 9. This result is recorded in node 2500 as (0, 2600), where 0 is the location of the loser, and 2600 is the next-level packet.

It is important to realize that there is no association between nodes at a level. That is, the structure of Fig. 5.25 does not describe the tournament when it is looked at vertically. We do not want to compare the element represented by the node at 2500 with the element represented by the node at 2503. We do not want to compare key 10 with key 191. A walk through the tree is worthwhile. At the end of the first pass the first-level packet associated with the winner contains the address of its sibling. Access to the packet comes through the pointer in location 1600 associated (Fig. 5.25) with the (0, 100) pair. At location 2500 the location of key 10 is indicated, and the key and its address may be brought to registers.

It is now required to find a proper comparand. The NLPA at 2501 points to it. This comparand is properly the key which lost to the previous pass winner at the second level. This was key 22 at list location 300. A comparison is undertaken. If the element described by the packet is smaller than the element in registers, then the list address in the packet is replaced and the old loser becomes a winner. If the contents of list location 0 were a key of 83, then the location 2600 would be modified to hold the address 0. However, the actual key at 0 is 10, and so key 22 loses again. Selecting its NLPA for a next comparison with key 10 advances key 10 in the tournament.

The next packet is at 2700. This packet points to list location 500. The content of 500 is key 16, which lost to key 9 in the previous round. Again, a comparison is undertaken; key 10 wins again and advances another level to to the comparison defined at 2500.

A comparison of Figs. 5.23 and 5.25 will confirm that the nonintuitive structure of Fig. 5.25 actually represents the simple tree of Fig. 5.23.

The general nature of this tournament organization should provide the reader with an appreciation of what tournament control may look like in detail. The serious user will wish to tailor his implementation to his machine.

6
Trees
in
Insertion

6.1 GENERAL

Insertion was described in Chapter 2 as a linear method in which a new element attempting to join an ordered list was compared with all members of the list until its proper position was found. At that time, room was made for the insertion of the new element by transferring members of the list.

The imposition of "structure" on this method reduces the number of comparisons made to find a position for the new element. Two widely known sorting methods are derived from the association of structure with insertion —Centered Insertion and Binary Insertion. Although simple in concept, both present some problems in the representation of the growing list in space. Binary Insertion, in addition, presents particular problems in controlling and generating compare sequences.

The use of these methods becomes attractive when lists of nontrivial size are collected dynamically and the growing list must be maintained in constant order.

6.2 CENTERED INSERTION: THE "TREE"

Consider that a list of numbers at some point in the sorting process looks like Fig. 6.1. The "tree" is not a binary tree. The two branches have as a parent a root node which holds the median (the element on an ordered list with as many above it as below it). One branch holds elements with key values below the median, the other with key values above it. Placing an assumed median in the central position leads to the method of Centered Insertion. In the next section we will assume that the median always stays in the central position. (Its *possible* movement is discussed in Section 6.2.2.)

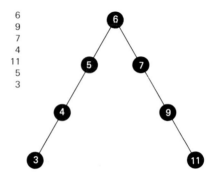

6
9
7
4
11
5
3

Fig. 6.1 A "weak" tree.

6.2.1   The Method of Centered Insertion

A position at the middle of a space large enough to hold the total number of elements to be sorted receives the first element. A counter representing the current length of the list is initialized to 1, and pointers to the occupied positions nearest the origin and the end of the list are established. They initially point to the central position.

Each subsequent call to the sort causes one item to be placed in position on the list. On the second call a new entry is compared with the value in the central position. If it is lower, it is placed on the origin side. The end pointer for the appropriate edge is advanced. A new value for length is developed by incrementing the counter by 1.

On each subsequent entry a new element is compared with the central position. If the new element is higher, the elements toward list end are searched consecutively until an element larger than the new element is found. When this occurs, that element and all elements lying beyond it are transferred one position toward the end of the list. The new element is inserted into the position freed by these transfers.

If the new element is lower than the contents of the central position, then all elements toward list origin are consecutively searched until an element on the list smaller than the new element is found. When this occurs, that element and all elements lying beyond it (in the direction of the origin of the list) are transferred one position toward the origin. The new element is placed in the position freed by these transfers.

If no element larger or smaller is found, the new element is added to the *end* of the appropriate edge of the list. For each new element placed, a modification is made to the appropriate pointer. These pointers control the length of a search in either direction and indicate end-of-list conditions.

An interesting variation involves using a good approximation for the median of the list, if one is known, as the midpoint value. This approximation need not occur on the list.

6.2.2 Centering

The downside and upside pointers are used to determine whether there is sufficient room for a new element at the proper side of the median. The space allocated to one "branch" may already be filled because of a one-sided run, which would offset the current list median from its central position in space. When space on one side is depleted, the list must be transferred in the reverse direction to find space for a new element. An alternative to reverse transfer, when space at the desired end is not available, is to track upside and downside pointers. Whenever space is becoming unbalanced, the list is moved in the direction of maximum space. This has the effect of making all transfers in the direction which will move the median toward the center location. The problem can be avoided by allocating up to $2N$ positions for the list.

6.2.3 Example

Figure 6.2 shows the passes for the Centered Insertion sort of the usual list of numbers. The status of the list is shown at the end of each pass. Initialization of unoccupied positions to 0 below the central position and 99 above it, as shown in the figure, is not required. For each pass, several values are shown. LENGTH is the length of the list at the end of the pass. HILIMIT is the position of the pointer at the end of the right branch; LOWLIMIT is the position of the pointer at the end of the left branch.

COMPARES and MOVES represent the number of comparisons and data transfers in each pass. The circled element is the element that joins the list in that pass. Arrows represent the direction of moves. The number of moves could be reduced in this example if movement were always made in a direction which balanced the number of elements around the center of the space.

6.2.4 Discussion: Number of Comparisons and Transfers

The minimum number of comparisons $(2N - 4)$ will occur in permutations for which there is exactly one comparison more than the median comparisons for $(N - 2)$ passes (after the initialization call). One permutation in which this will occur is shown in Fig. 6.3. In this permutation of the usual numbers, the median is the first element, all even-numbered positions form an ascending string, and all odd-numbered positions (except the first) form

Initial list	List at end of:				
	Pass 1	Pass 2	Pass 3	Pass 4	Pass 5
1. 3	1. 0	1. 0	1. 0	1. 0	1. 0
2. 11	2. 0	2. 0	2. 0	2. 0	2. 0
3. 6	3. 0	3. 0	3. 0	3. 0	3. 0
4. 4	4. 0	4. 0	4. 0	4. 0	4. 0
5. 9	5. 0	5. 0	5. 0	5. 0	5. 0
6. 5	6. ③	6. 3	6. 3	6. 3	6. 3
7. 7	7. 99	7. ⑪	7. ⑥ ↓	7. ④ ↓	7. 4
8. 8	8. 99	8. 99	8. 11	8. 6	8. 6
9. 10	9. 99	9. 99	9. 99	9. 11	9. ⑨ ↓
10. 2	10. 99	10. 99	10. 99	10. 99	10. 11
11. 1	11. 99	11. 99	11. 99	11. 99	11. 99
	Length=1	Length=2	Length=3	Length=4	Length=5
	HILIMIT=6	HILIMIT=7	HILIMIT=8	HILIMIT=9	HILIMIT=10
	LOLIMIT=6	LOLIMIT=6	LOLIMIT=6	LOLIMIT=6	LOLIMIT=6
	Compares=0	Compares=1	Compares=2	Compares=2	Compares=4
	Moves=0	Moves=0	Moves=1	Moves=2	Moves=1

Pass 6	Pass 7	Pass 8	Pass 9	Pass 10	Pass 11
1. 0	1. 0	1. 0	1. 0	1. 0	1. ①
2. 0	2. 0	2. 0	2. 0	2. ②	2. 2
3. 0	3. 0	3. 0	3. 3	3. 3	3. 3
4. 0	4. 0	4. 3	4. 4	4. 4	4. 4
5. 0	5. 3	5. 4	5. 5 ↑	5. 5	5. 5
6. 3	6. 4	6. 5 ↑	6. 6	6. 6	6. 6
7. 4	7. 5 ↑	7. 6	7. 7	7. 7	7. 7
8. ⑤ ↓	8. 6	8. 7	8. 8	8. 8	8. 8
9. 6	9. ⑦	9. ⑧	9. 9	9. 9	9. 9
10. 9	10. 9	10. 9	10. ⑩	10. 10	10. 10
11. 11	11. 11	11. 11	11. 11	11. 11	11. 11
Length=6	Length=7	Length=8	Length=9	Length=10	Length=11
HILIMIT=11	HILIMIT=11	HILIMIT=11	HILIMIT=11	HILIMIT=11	HILIMIT=11
LOLIMIT=6	LOLIMIT=5	LOLIMIT=4	LOLIMIT=3	LOLIMIT=2	LOLIMIT=1
Compares=3	Compares=5	Compares=5	Compares=6	Compares=4	Compares=5
Moves=3	Moves=4	Moves=5	Moves=7	Moves=0	Moves=0

Total comparisons   37
Total moves   23

Note: Circled numbers indicate element placed. Arrows indicate direction of element movement.

Fig. 6.2 Centered insertion.

a descending string. The effect is that the new element always lies between the median and its left or right successor.

The maximum number of comparisons occurs when each element must journey all the way from median to endpoint. The journey gets longer in each subsequent pass. On the first pass, one comparison is made (median); on the second, two are made; on the last pass, $N/2$ comparisons are made.

Position	Value
1.	6
2.	1
3.	11
4.	2
5.	10
6.	3
7.	9
8.	4
9.	8
10.	5
11.	7

```
                                          1     1
                                    1     1     2     2
                              1     1     2     2     3     3
                        1     1     2     2     3     3     4     4
                  1     2     2     3     3     4     4     5     5
                 (6)   (6)   (6)   (6)   (6)   (6)   (6)   (6)   (6)
                 11    11    10    10     9     9     8     8     7
                       11    11    10    10     9     9     8     8
                                   11    11    10    10     9     9
                                               11    11    10    10
                                                           11    11
```

Fig. 6.3 An optimum permutation.

The average size of a branch, which grows from 1 to $N/2$, is $N/4$. The search of a branch of average size $N/4$ occurs $(N-1)$ times. The maximum number of comparisons is $(N)(N-1)/4$ when all $N-1$ elements reach a branch of average size. The expected number of comparisons, $N(N-1)/8$, is based upon the assumption that $(N-1)$ elements will search half the half-lists. $N^2/8$ is an adequate approximation of expected compares.

6.3   BINARY INSERTION: THE INSERTION TREE

A structure which is a centered binary tree may be imposed on the list, as in Fig. 6.4. This is an ordered list of the numbers one to thirty-one. The tree has, in common with the tree in Chapter 4, the convention that the left successor of a node contains a smaller value and the right successor contains a higher value. The root node of this tree holds the median value on the list. Since the median value must be at the center of the list, the root node must be at the center of the space representing the list. Consequently, the root node of the ordered list of keys from 1 to 31 in memory positions 1 to 31 must be in position 16 and hold key 16.

The left and right successors of the root node are the positions which hold the first- and third-quartile key values, respectively. The left-successor node in Fig. 6.4 is node 8, containing the highest key which is lower than $\frac{3}{4}$ of the members of the list. The right successor node in Fig. 6.4 is node 24, containing the lowest key which is higher than $\frac{3}{4}$ of the members of the list. The left successor of node 8 is node 4, containing the value lower than $\frac{7}{8}$ of the list; the left successor of node 4 contains the element lower than $\frac{15}{16}$ of the list, etc. Naturally, the median is halfway down the list; the $\frac{1}{4}$ point is halfway between origin and midpoint; the $\frac{1}{8}$ point is halfway between the origin and the quarter point.

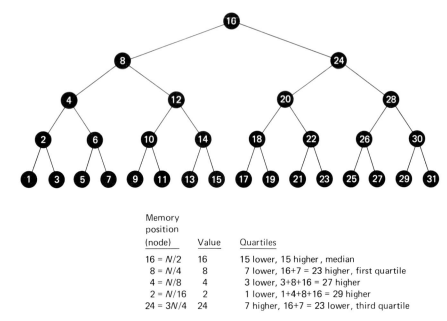

Memory position (node)	Value	Quartiles
16 = N/2	16	15 lower, 15 higher , median
8 = N/4	8	7 lower, 16+7 = 23 higher, first quartile
4 = N/8	4	3 lower, 3+8+16 = 27 higher
2 = N/16	2	1 lower, 1+4+8+16 = 29 higher
24 = 3N/4	24	7 higher, 16+7 = 23 lower, third quartile

Fig. 6.4 Centered binary tree.

6.3.1   Binary Insertion: Static Trees, with Example

The most direct way of implementing Binary Insertion is to consider that the tree which will represent the final list is already fully formed at the first pass. Consider Fig. 6.5. The list shows a space for 27 elements, of which 8 have arrived. The center of space is location 13, and this holds the assumed median (root node). The left quartile is location 6, the right quartile location 20, the octiles are at locations 2, 10, 16, 24. (Median and quartile positions are at 13, 6, 20, rather than 14, 7, 21, because the space has an origin address of 0 and not 1.) Any new element to be placed on this list will undergo five comparisons, one on each level of the tree. Because the list will occupy only partial space, some number of the comparisons may be between dummy low and dummy high values. These are represented by L and H on Fig. 6.5. Initially one half of space was filled with L values and one half was filled with H values. Every element attempting to join the list will be compared against location 13; then, depending on its value, with location 6 or 20 (quartile), etc. Figure 6.5 shows the comparisons for a set of keys. The significant features of this implementation are:

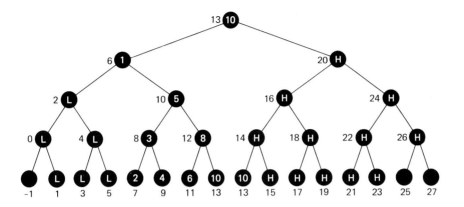

If next
arrival
is   Comparisons (positions)

If next arrival is	Comparisons (positions)
9	13, 6, 10, 12, 13
– 5	13, 6, 2, 4, 5
25	13, 20, 16, 14, 13
14	13, 20, 16, 14, 13

List

0.	L
1.	L
2.	L
3.	L
4.	L
5.	L
6.	1
7.	2
8.	3
9.	4
10.	5 ◄— Median
11.	6
12.	8
13.	10
14.	H
15.	H
16.	H
17.	H
18.	H
19.	H
20.	H
21.	H
22.	H
23.	H
24.	H
25.	H
26.	H

Fig. 6.5 Static-tree binary sort.

1. The locations used for comparisons remain the same from pass to pass. They may be precomputed, and address derivation may be avoided. The structure of the tree is fixed regardless of the shape of the growing list.

2. The number of comparisons is independent of the size of the current list and derives from the anticipated size of the entire list.

These two features are shown by Fig. 6.5. It is interesting to notice that the true median of the list of this figure is in location 10 and that the number of dummy tests can be very high. This is particularly true when the current list is small relative to its space. It is possible to prevent dummy comparisons by the use of limits, and it may be profitable to do so, on many machines. Note, however, as with −5 on Fig. 6.5, that an attempt to perform a dummy comparison does not terminate comparison.

Figure 6.5 has not attempted to balance the list. As with Centered Insertion, the median may be kept closer to the center and open space evenly distributed on both sides. Except for comparing at tree vertices rather than linearly, after the first comparison, the method is identical to Centered Insertion. It begins by placing an element in the center of space, etc.

6.3.2   Binary Insertion: Dynamic Trees

It is possible to reduce the number of comparisons by undertaking to impose tree structure on the actual list occupying list space. As with the static tree approach, the list is grown from the center of space and occupies, at any time, a portion of that space. At any particular pass the actual median and n-tile points of the existing list are used in comparison, rather than those points in total space.

Note Fig. 6.6. A list of eight elements, identical to those in Fig. 6.5, is represented in the same space. Figure 6.6 differs from Fig. 6.5 in that the list is centered in space, with the median at the center position. This could have been achieved in Fig. 6.5 with space balancing. The dynamic tree, however, not only balances space but reduces comparisons by restricting comparison space to actual elements. Doing this requires a good deal of additional work. The nature of this work is implied in the location address of Fig. 6.6. The number preceding the slash is the space-relative address of the element, its actual location relative to the start of the work space. The number following the slash is the list-relative address, the relative position of the element on the actual current list. For example, the element with key 3 occupies space-relative position 11, list-relative position 2. The list-relative position is the space-relative position minus the origin of the list—in this example, position 9. Each element added to the list defines a new tree. To sequence the comparisons of a pass, one must develop proper relative list positions and also relative space positions for each pass. The procedure for addressing a tree structure follows.

1. Compute the median address relative to the origin of the list. The address of the median in space is the sum of the list-relative address of the median and the origin address of the list. The list-relative address of the median is the integer half of list length; e.g., the current list of Fig. 6.6 is eight elements long—one half of eight plus the list origin is $4 + 9 = 13$.

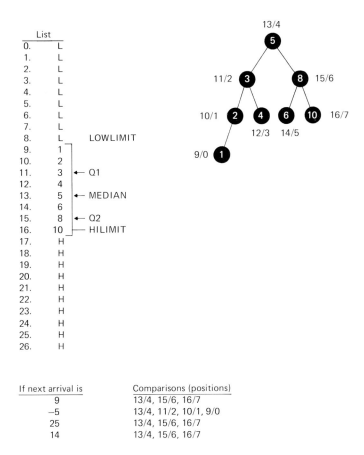

List	
0.	L
1.	L
2.	L
3.	L
4.	L
5.	L
6.	L
7.	L
8.	L LOWLIMIT
9.	1
10.	2
11.	3 ← Q1
12.	4
13.	5 ← MEDIAN
14.	6
15.	8 ← Q2
16.	10 ── HILIMIT
17.	H
18.	H
19.	H
20.	H
21.	H
22.	H
23.	H
24.	H
25.	H
26.	H

If next arrival is	Comparisons (positions)
9	13/4, 15/6, 16/7
−5	13/4, 11/2, 10/1, 9/0
25	13/4, 15/6, 16/7
14	13/4, 15/6, 16/7

Fig. 6.6 Binary insertion; dynamic tree.

2. To move from the median to another node, and subsequently from node to node in general, develop two variables, CURRENT and DISTANCE. The initial value of CURRENT is the median relative address. The initial value of DISTANCE is the integer half of the median relative address plus one. To move from median to left successor, subtract DISTANCE from CURRENT. To move from median to right successor, add DISTANCE to CURRENT. Then, add the list origin address to form the space address. To move from node to node, the value of DISTANCE is calculated as the integer half of DISTANCE + 1, each time. The distance from a quartile point to an octile is one-half the distance from quartile point to median. The new value of DISTANCE is added or subtracted from CURRENT to form the list-relative address. Using CURRENT = 4, as in Fig. 6.6, the initial

value of DISTANCE is 2. The quartile points, therefore, are at list positions 2 and 6. The actual space locations are formed by adding 9 (the address of the list origin) to obtain 13, 11, 15.

As the list grows, more levels are formed on the tree and more comparisons are required. The full current tree is traversed when a calculated value or distance is zero and/ or when an address is *developed beyond the limits of the list.*

6.3.3   Comments on Fixed vs. Static

The programmer must consider the burden of address formulation required to reduce the number of comparisons by eliminating dummy comparisons. Section 6.4 provides formulas for estimating comparisons. Some form of space control is probably desirable in order to minimize data movement, which uses a lot of time. Whether one address calculation per comparison is a penalty worth paying to obtain space control and minimize comparisons depends upon N and the machine. It is possible to compromise by using this dynamic tree until it reaches a certain size and then using the static tree. With $N = 1023$, for example, the final tree will have 10 levels, and 512 elements will be endpoints. Depending on machine architecture, it might be profitable to constrain comparisons to the actual list while the size of the list is under 128. At this point the dynamic tree of the list has at most 7 levels. When the list exceeds 128 and only 3 comparisons per element might be saved, the static tree could be used.

It is interesting, also, to consider the possibility that off-list values might be used in the method. If good approximations for median and n-tile points are known, they may be used as comparisons.

6.4   DISCUSSION: COMPARISONS AND TRANSFERS

The number of comparisons for the static tree is $N\lceil \log_2 N \rceil$. The number of comparisons with dynamic trees is the sum of $\log_2 K$ for each pass. K is the integer power of 2 such that the number of elements on the list lies between 2^{K-1} and $2^K - 1$. In general the expected number of comparisons is $N \log_2 (N/e)$, where $e \cong 2.72$. The number of transfers may be expected to be $(N)(N - 1)/8$.

7
Quicksort

7.1 GENERAL

The method of Quicksort was first described by C. A. R. Hoare in an article in the *Computer Journal* in 1962 [16] and by Algorithms 63, 64, and 65 in the *Communications of the ACM* (see Appendix A). Since its introduction, many variations have been published, some reflecting speculations and suggestions made in the original article. Like Shellsort, Quicksort is a generic name for a number of algorithms, which reflect variations in the approaches to the development of a critical parameter that affects the performance of the method.

Quicksort is a tree-based method, although the relationship between a binary tree and the sort algorithm is not so explicit as it is in Tournament or Binary sorting.

7.2 THE METHOD

An element of the list is selected as a "pivot." All elements lower than the pivot are placed in positions beginning at the origin of the list; all elements higher are placed in positions at the end of the list. Elements are compared against the pivot and change position when they are higher and earlier on the list, or lower and later on the list. This arrangement constitutes a phase of the sort. At the end of a phase, a position is available on the list for the placement of the pivot. This position is the proper rank position for the pivot because its relative location on the list is properly determined by the number of elements above and below it.

A phase defines two partitions. If the value of the pivot is a good approximation of the median of the list, the two partitions will be about equal in size. If the value of the pivot is the worst possible approximation of the median (the lowest or highest element of the list), then the partitions will be of sizes 0 and $(K - 1)$, where K is the size of the initial list.

The technique for building the partitions is essentially the same in all versions of the algorithm. Two pointers are established, one at the beginning and one at the end of the list. After the pivot is selected, a search for a lower value starts with the end pointer moving up the list toward the origin. Comparisons continue until an element lower than the pivot is found or the list is exhausted.

When an element lower than the pivot is found, it must be moved to a position which holds an element higher than the pivot. This position is found by comparisons using the beginning pointer. Comparisons are made down the list until a higher element is indicated by the beginning pointer. Elements higher and lower than the pivot have thus been identified, and they are now exchanged.

Variations occur in the exact method of moving items. One technique is to remove the pivot to temporary storage and transfer the contents of the top of the list into the free position. This creates a free origin position into which an element can be transferred. After a transfer, the free position shifts to a location near the end of the list. The top-down search looks for an element to transfer to this free position. As the method progresses, the free position alternates between top and bottom positions until the pointers converge.

After an exchange or double transfer, the search for another element lower than the pivot resumes. If one is found, then another search for a high element is undertaken with the beginning pointer. This process continues until the two pointers meet. The relative size of the partitions is determined by the position on the list at which the beginning and end pointers coincide. In some versions of the method it is necessary to compare the relative positions of the pointers after each movement up or down the list. However, it is possible to avoid doing so by using a technique of data bounding, which will be described in a later section.

The first phase generates two partitions. One is used as input to the second phase, and the other is placed in a LIFO stack. The common convention is to process the smaller partition first and remember the boundaries of the larger on the stack. The second phase generates two partitions. The smaller is processed in the third phase, and the larger is stacked on top of the larger partition generated by phase 1. The process of generating partitions, stacking the larger, and partitioning the smaller may continue until all partitions are reduced to one element. When a one-element partition is generated, no new partition is developed for the stack. Waiting partitions are unstacked and processed. The sort is complete when the stack is empty.

Processing the smaller of two partitions ensures that there will be no more than $\log_2 N$ waiting partitions on the stack. The sequence of partition processing is really quite arbitrary, and it can be altered by the programmer if he chooses.

Quicksort may be used in a combined algorithm to reduce partitions only to a predetermined size, at which point they are ordered by another method more efficient for small lists.

7.3   DETAILS OF A PHASE

Figure 7.1 contains a different permutation of the usual list of numbers. Some mechanism for selecting a pivot (explained in Section 7.5) will choose key 6 from position 3 and place it in a working storage called PIVOT. The element at the head of the list is transferred to position 3. The first position is now free to receive a value from bottom-up comparisons. The column headed "Pick Pivot" in Fig. 7.1 shows the list after the pivot value has been removed and the contents of the first position have been transferred. Pointer

List		Pick pivot	First exchange	Second exchange	Third exchange	Fourth exchange
1.	3	⊖ ← BEG	⑤	5	5	5
2.	11	11	— ← BEG	①	1	1
3.	6	③	3	3	3	3
4.	9	9	9	— ← BEG	②	2
5.	7	7	7	7	— ← BEG	④
6.	8	8	8	8	8	⑥ ← BEG, END
7.	4	4	4	4	4 ← END	⑧
8.	2	2	2	2	⑦	7
9.	10	10	10	10 ← END	10	10
10.	1	1	1 ← END	⑨	9	9
11.	5	5 ← END	⑪	11	11	11

Note: Circles indicate values moved.

PIVOT = 6

Comparisons and transfers		
POS 11 : PIVOT	POS 11 → POS 1	First exchange
POS 2 : PIVOT	POS 2 → POS 11	
POS 10 : PIVOT	POS 10 → POS 2	
POS 3 : PIVOT	—	Second exchange
POS 4 : PIVOT	POS 4 → POS 10	
POS 9 : PIVOT	—	
POS 8 : PIVOT	POS 8 → POS 4	Third exchange
POS 5 : PIVOT	POS 5 → POS 8	
POS 7 : PIVOT	POS 7 → POS 5	Fourth exchange
POS 6 : PIVOT	POS 6 → POS 7	

Fig. 7.1 Quicksort, first pass.

BEG is initialized to point to the first position on the list, pointer END to the last.

The initialization of these pointers defines the partition which a phase will process. Since the entire list is processed in the first phase, the pointers are initialized to the origin and the end of the list. On subsequent phases the pointers are intialized to the boundaries of the partition to be processed. This initialization occurs before pivot selection because it is necessary to constrain pivot selection to the partition currently "active."

The first comparison is between the element of the list (the partition) indicated by END and the selected pivot. The first entry in the list of comparisons and transfers is POS 11 : PIVOT. The element in position 11 (key 5) is lower than the pivot, and it is transferred to the free position at the head of the list. The free position is now position 11. A search to fill that position is undertaken from the top of the list using BEG. Since BEG is incremented before starting the search, the first comparison is POS 2 : PIVOT. The element in position 2 is higher than the pivot, so a transfer is made to the vacant position indicated by END. After this transfer, position 2 becomes the free position, and the pointers and the status of the list are as shown in the column headed "First exchange."

Bottom-up comparing now resumes. The first value tested (1 in position 10) is lower than the pivot, and it is transferred from position 10 to the free position 2. Position 10 becomes the free position, and the top-down comparisons are resumed. BEG travels to position 4 to find a value (key 9) that is larger than the pivot. The transfer to position 10 is made. The column headed "Second exchange" shows the list, pointers, and free position after this transfer and the adjustment of END in preparation for new bottom-up comparisons.

Figure 7.1 shows the remaining comparisons, transfers, and transformations of the list. The pass ends when BEG and END point to the same position. The means for discovering this condition lies in the fact that each time END is decremented it is compared to BEG, and each time BEG is advanced it is compared to END. When the pass ends, the pivot is put into the list in its rank position. The partitions which have been formed are the left and right branches of a tree whose root node is 6. This tree is shown in Fig. 7.2. The low partition runs from position 1 to position 5, the high partition from position 7 to position 11.

7.4   INTERPHASE OPERATIONS

The next phase will process the smaller of the two partitions just generated. The size of partitions is easily determined by subtracting starting and end position values. The partition selected will have its starting position placed

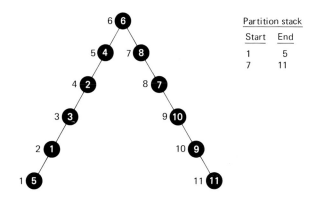

Fig. 7.2 "Tree" and partition stack.

at BEG and its ending position placed at END. The other partition will remain on the stack. When partitions are equal, the choice is arbitrary. Figure 7.3 shows the list with new values at BEG and END. Naturally, the pivot will be chosen from within the partition to be processed by the phase.

It is necessary to ensure that the partition about to be processed is valid. A valid partition in a pure application of Quicksort must have at least two elements. If BEG is not less than END, then a partition of one element has been developed. Another partition is obtained from the partition stack. If none is available, the sort ends.

Phase 2 partitions the elements 1–5 on the list into two partitions. If we assume the phase pivot to be key 3, these partitions are Positions 1–2 and Positions 4–5. Figure 7.4 shows the list at the end of phase 2, the compare sequence of the phase, the partitions generated, and the condition of the partition stack. The tree shows that the effect of the phase was to form a subtree on the left branch.

	List
1.	5 ← BEG
2.	1
3.	3
4.	2
5.	4 ← END
6.	6
7.	8
8.	7
9.	10
10.	9
11.	11

	Partition stack	
	Start	End
Top of stack →	7	11

Fig. 7.3 Quicksort, beginning of second phase.

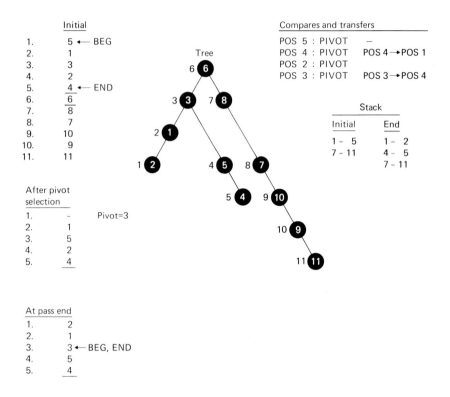

Initial

1.	5 ← BEG
2.	1
3.	3
4.	2
5.	4 ← END
6.	6
7.	8
8.	7
9.	10
10.	9
11.	11

Compares and transfers

POS 5 : PIVOT	—
POS 4 : PIVOT	POS 4 → POS 1
POS 2 : PIVOT	
POS 3 : PIVOT	POS 3 → POS 4

Tree

Stack

	Initial	End
	1 - 5	1 - 2
	7 - 11	4 - 5
		7 - 11

After pivot selection

1.	–	Pivot=3
2.	1	
3.	5	
4.	2	
5.	4	

At pass end

1.	2
2.	1
3.	3 ← BEG, END
4.	5
5.	4

Fig. 7.4 Quicksort, phase two.

Figure 7.5 shows the third phase. The development of a one-element partition is equivalent to reaching an end node of the tree. The selection of different pivots will form differently shaped trees with different values and a different number of associated comparisons.

7.5   THE PIVOT

Mechanisms for selecting the pivot of a phase have not been described. There are a number of techniques used in the published versions of the method. The goal is to pick a pivot close to the median of the partition, and to avoid making bad choices. A good mechanism must not be vulnerable to the possible occurrence of permutations in which it may too often pick the very low or very high elements as phase pivots. The impact of very bad pivot selection is to reduce the method to the effectiveness of Linear Selection with Exchange. Pivot selection determines the size of the partitions generated by a phase. The method is optimum when each phase pivot is

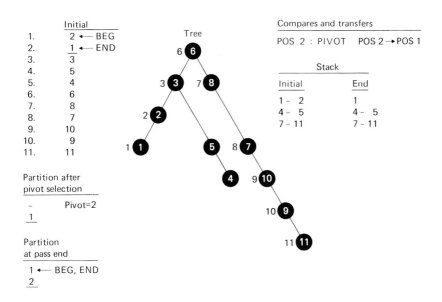

Initial
1.   2 ←— BEG
2.   1 ←— END
3.   3
4.   5
5.   4
6.   6
7.   8
8.   7
9.   10
10.   9
11.   11

Tree

Compares and transfers

POS 2 : PIVOT   POS 2 →POS 1

Stack

Initial   End

1 – 2   1
4 – 5   4 – 5
7 – 11   7 – 11

Partition after
pivot selection
 –   Pivot=2
 1

Partition
at pass end
 1 ←— BEG, END
 2

Fig. 7.5 Quicksort, phase three.

the median of the phase partition. Sophisticated pivot selection can reduce the expected number of comparisons for the method to better than $1.1N \log_2 N$ from $1.4N \log_2 N$.

If random numbers are assumed, then any permutation of the list of numbers is equally probable, and the median is equally likely to occur in any position. One pivot selection method, used in Hibbard's algorithm, is to take the element in the first position of the partition as the phase pivot. The danger here is that a strong tendency toward order in the list will probably result in the presence of a low element in the first location; therefore the partitions generated will be unbalanced. But if the first position supplies the pivot, then it automatically becomes the free position, and the transfer from first position to pivot position mentioned in the example is not necessary.

If a strong tendency toward order in the list is expected, then the middle position is a reasonable place from which to choose the pivot. If the list is indeed random, then selection of the middle position is not harmful—it is good as any—and if order does exist, that choice could be quite beneficial.

Singleton (Algorithm 347), following (perhaps) a comment in C. A. R. Hoare's initial article, attempts to qualify a median by selecting it from a sample of elements. In his algorithm he compares the first, last, and middle elements of the list on entry to a phase; and he uses the middle value of this sample as his pivot. (This procedure has a useful side effect, which we will discuss in the next section.) The general idea of doing a little work can be extended to taking a sample of any desired size to qualify a pivot. At the cost of three tests and transfers, the taking of the median of a random sample of a partition can reduce the multiplier to 1.188. Seven tests can reduce it to less than 1.1. Sample-sort by Frazer [9] and Van Emden's sort [28] use considerable sampling techniques to determine pivot points.

Hoare uses a random-number generator to generate a partition position for the pivot for each phase. The value of this lies not so much in qualifying a position as a good approximation of the median, as in protecting the method from unfortunate bias in the data. Randomly selecting a pivot for each new partition makes it highly improbable that the selected pivots will follow an undesirable pattern in the data.

If the programmer knows a lot about his data, further refinements are possible. "Distributive" sort overtones may be applied to comparative Quicksort by using off-list median points. It is not necessary that the pivot be an actual value of the partition. If the programmer knows the median, quartile points, etc., he may impose external pivot values. He may, by some inspection of a particular partition, be able to infer where its median lies and can then vary the algorithm for median selection for a given partition according to what he discovers.

7.6   ENDING A PHASE

Recognition of phase end can be accomplished in a number of ways. One method uses a comparison of pointer values at each exchange. In some machines this address compare could be quite expensive and might effectively almost double comparison time for the method. In others the address comparison can be more efficiently handled using indexing techniques.

A variation of this test-on-exchange technique is used by Quickersort. In this version of the method the search is reversed, starting from top-down at the second element of the partition following a vacated first position. If a larger element is never found, the top-down search finishes when it exhausts the list, and thus ends the phase. If a larger element is found by the top-down search, then the bottom-up search is undertaken. The bottom-up search is bounded by the value of the top-down pointer. When the pointers meet, the currently free position is set to the contents of the position one above where the pointers meet, and the pivot is placed into the position just freed. The pointers always meet at an element higher than the pivot; the position above the position at this element always contains a smaller element.

Hoare's first published ALGOL version compares a pointer with the limit of the list being processed by the phase. For the bottom-up pointer, the limit is the first position of the list; for the top-down pointer, the limit is the last position of the list. Each time a pointer is moved, it is compared with the limit. This limit comparison must be executed for each value comparison as a safeguard against running off the list.

The pointers are compared with each other only when an exchange is going to take place. If the top-down pointer (BEG) is not above the bottom-up pointer (END) on a comparison, a phase-end procedure is undertaken. As with Quickersort, the top-down search is undertaken first.

With Hoare's end-of-phase procedure, it is possible for the pointers to *pass each other* during a search (see Fig. 7.6). In the column "Third Exchange," the top-down search has discovered that key 7 in position 7 is higher than the pivot value. The bottom-up search begins at position 9 and continues until position 6, passing the top-down pointer and retesting a position tested by the top-down search. The pointer comparison, executed only when an exchange is to be made, discovers the crossing. The relative pointer locations are now determined. If the top-down pointer is above the

	Initial	First exchange	Second exchange	Third exchange
1.	3	3	3	3
2.	11	①	1	1
3.	6	6 ← BEG	6	⑤
4.	4	4	4	4
5.	9	9	②	2
6.	5	5	5 ← BEG	⑥ ← END
7.	7	7	7	7 ← BEG
8.	8	8	8	8
9.	10	10	10 ← END	10
10.	2	2 ← END	⑨	9
11.	1	⑪	11	11
PIVOT	6	Compares		
		1 : PIVOT	3 : PIVOT	6 : PIVOT
		2 : PIVOT	4 : PIVOT	7 : PIVOT
		11 : PIVOT	5 : PIVOT	9 : PIVOT
		BEG above END	10 : PIVOT	8 : PIVOT
		2 ⟷ 11	BEG above END	7 : PIVOT
			5 ⟷ 10	6 : PIVOT
				END above BEG
				BEG below PIVOT
				3 ⟷ 6

Note: BEG, END Pointers shown at end of comparison sequences.

Fig. 7.6 Quicksort, end of phase. PIVOT represents the temporary location holding the value 6.

pivot, an exchange between the values in the pivot position and the location of the top-down pointer places the pivot value properly. If the condition above does not hold, then an exchange between the pivot position and the bottom-up pointer will position the pivot. Thus, in Fig. 7.6, when END finds that key 5 in position 6 is lower than the pivot 6, it compares BEG to END and recognizes a crossing. It then sees that BEG is below the position of the pivot and exchanges PIVOT and END, positions 3 and 6.

An elegant approach to phase ending and control comes from Singleton's version of the method. It is a byproduct of his selection of the pivot. He places the smallest element of his three-element sample in the first position of the partition, the largest element in the last position. As a result, he does not need to perform any limit or pointer coincidence comparisons except when he has an exchange. The list is "data-bounded." There is a small key at the top of the partition and a large key at the end. A pointer moving from bottom to top will always force an exchange when it reaches the top of the list. There is no danger of searches going off the list. A similar effect could be obtained, independent of pivot selection, by placing at the edges of the list those values known to be out of list range.

A further side effect of Singleton's data-bounding is to limit the size of the partition that will be generated in the worse case to $(K - 2)$, since the small partition must contain at least the bound number (the low of the sample used for pivot).

7.7   QUICKSORT IN COMBINATION

The value of Quicksort can be enhanced by recognizing when a partition is too small to profitably support the overhead of further partition generation. That partition is then arranged by another method. The proper size of the minimum partition depends on the specific machine. As the reasonable candidate for the alternative technique, Singleton chose sifting. The process by which a given partition is arranged is irrelevant to other partitions, so there is no change in the method except for the addition of a test on partition size and a transfer to the alternative technique. In any case, there must be a partition-size test in some form, since it is necessary to check for valid partitions. The published algorithms other than Singleton's are all pure Quicksort, with the exception of Quickersort, which excludes partitions of fewer than three elements.

7.8   DISCUSSION

The expected number of comparisons ranges from $1.1N\lceil \log_2 N \rceil$ to $1.4N\lceil \log_2 N \rceil$, depending on pivot selection. The minimum number of comparisons occurs when the pivot selected is a perfect median every time. This minimum is less than $N\lceil \log_2 N \rceil$. The maximum number of comparisons will occur when the pivot is the worst choice in each phase. When this occurs only one valid partition is produced per phase. This partition will be only one element smaller than the phase input partition. The number of comparisons will be $(N-1)$, $(N-2)$, ..., $(N-N+1)$. The sum of this series is $N(N-1)/2$. Choosing a pivot from the first position of a partition when sorting the ordered list will produce the worst case.

The number of exchanges is difficult to calculate. There cannot be more exchanges than comparisons. An estimate of half the number of comparisons would provide a reasonable guide.

8
High-Order Selection Sorts

8.1 INTRODUCTION

The partitioning of lists into sublists is an inherently efficient approach to sorting, since it reduces the value of N (for example, in such expressions as N^2 and $N(N-1)/2$) to a value K less than N. It makes sort times additive rather than multiplicative in nature. In Linear Selection, the number of comparisons can be reduced from N^2 to $K_1^2 + K_2^2 + K_3^2 + X$ by partitioning into partitions of size K. The variable X represents the activity necessary to order ("merge") the ordered sublists relative to each other.

The methods in this chapter and the next differ from earlier tree-partitioning methods in two ways. First, they are not necessarily associated with a binary tree, although they can be visualized as tree-structure sorts. Second, the function of ensuring order in the entire list (between partitions) is explicit rather than implicit. Specific interpartition comparisons are performed to select the lowest element from candidates from each partition.

The literature has traditionally given attention to a family of sorts called "nth-degree selection," and it has given particular attention to *quadratic* selection. The reader should not assume, because of chapter order in this book, that the methods described in this section are as efficient as the $N \log_2 N$ sorts. It is useful to know the concepts and principles demonstrated in this chapter, however, before one approaches the internal merge.

8.2 QUADRATIC SELECTION

The method of Quadratic Selection derives its name and characteristics from the fact that the number of partitions into which the list is subdivided is the *square root* of the number of elements on the list. Linear Selection is used for the arrangement of each partition and for interpartition comparisons. The literature describes the method in two major variations. In one, the partitions are not sorted but are periodically scanned for the current lowest element; in the other, called Quadratic Selection with Presorting, the partitions are fully ordered before interpartition comparing is undertaken.

8.2.1   Quadratic Selection Without Presort: The Method

The list is divided into $\sqrt{N}$ partitions. If N is not a perfect square, the list is divided into $\sqrt{N'}$ partitions, where N' is the *next square higher* than N. Each sublist is subjected to a single pass of Linear Selection, which selects its lowest element. Each lowest selected element (or a tag representing it) is transferred to an auxiliary list. After all lowests have been selected (and replaced in their sublists by dummy high values), Linear Selection is applied to the auxiliary list to select the "lowest of lows," the lowest of the entire list, which is transferred to an output list. The size of the output list is a function of the relation between the sort and the successor process, as discussed in connection with Linear Selection.

The partition that contained the lowest element is scanned for a new lowest, which is placed on the auxiliary list. Another scan of the auxiliary list then chooses the new lowest, the element that is the second lowest of the total list. It is transferred to output. The process continues, alternating scans of the auxiliary list with partition scans, until all partitions are exhausted. This condition exists when all partitions send dummy values to the auxiliary list.

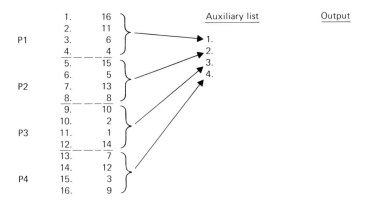

Fig. 8.1 Quadratic selection, partitions.

8.2.2 Example

Consider the list of numbers in Fig. 8.1. The 16 elements form four sublists, whose boundaries are positions 1–4, 5–8, 9–12, 13–16. The auxiliary list has four locations. Location 1 receives elements from partition 1, location 2 from partition 2, etc. The initial selection of a lowest from each partition produces Fig. 8.2. When the selection process applied to the auxiliary list determines that key 1 from partition 3 is the lowest, this key is transferred to output. Since the winning element came from the third location of the auxiliary list, partition 3, which contributed the element, will now be scanned for a replacement (Fig. 8.3).

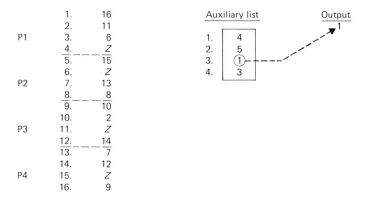

Fig. 8.2 Quadratic selection, initial selection.

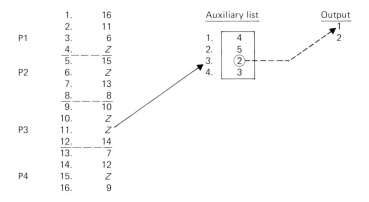

Fig. 8.3 Quadratic selection, replacement.

After a series of selections, the method produces the conditions of Fig. 8.4. Key 9 has just been transferred to output, and partition 4 will be scanned for a replacement. This scan will deplete partition 4. The selection of key 10 will empty partition 3 when it provides key 14, and then the selection of key 11 will empty partition 1.

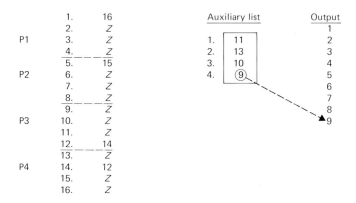

Fig. 8.4 Quadratic selection, partition depletion.

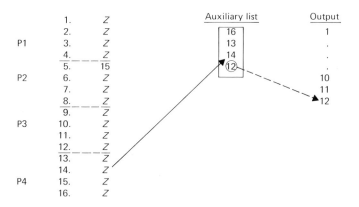

Fig. 8.5 Quadratic selection, scan completion.

Key 12 is selected for output in Fig. 8.5. The scan of partition 4 to replace key 12 will cause the first high dummy to enter the auxiliary list. Figure 8.6 represents the status when key 15 enters. All further selection from the auxiliary list will cause replenishment by dummies. The end of the sort is

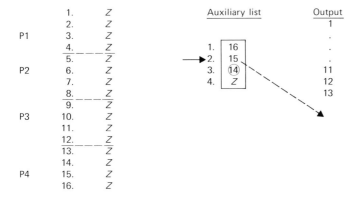

Fig. 8.6 Quadratic selection, emptying auxiliary list.

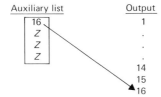

Fig. 8.7 Quadratic selection, end of sort.

shown in Fig. 8.7. Recognition of the end of the sort does not require dummies or depletions. It is sufficient to count the number of items transferred to output to bring the sort to an end.

In the example shown, it would be useless to attempt to reorganize the auxiliary list to suppress dummy comparisons. The high elements of the list are evenly distributed among the partitions, and therefore no list depletes very early. The first time a dummy value is involved in a comparison on the auxiliary list is for the selection of key 13. One dummy comparison is made for this selection. One dummy comparison is made to select key 14, two to select key 15, three to select key 16. Only seven dummy comparisons are made on the auxiliary list. No mechanism to avoid them would take less time than seven comparisons.

8.2.3   Depletion of Partitions

Various policies may be adopted regarding partition depletion. One is simply to ignore it. When a sublist is depleted, it will contribute a high dummy to the auxiliary list. Since the high dummy will never be selected as the lowest of the auxiliary, only that one "unnecessary" scan of the depleted partition will be performed.

It may be profitable to recognize depletions and to suppress unnecessary comparisons against high dummies on the auxiliary list. Normally the first partition will send its elements to the first position on the auxiliary list, the second partition to the second position, etc. When a partition sends a dummy, it is possible to compress the auxiliary list by suppressing references to both the dummy position and its corresponding partition. Adopting this procedure will eliminate dummy comparisons. If the programmer has a reason to suspect a heavy concentration of lower elements that will tend to cause a nontrivial number of partitions to deplete early, he should investigate the possibility of auxiliary list compression.

For example, suppose that on a list of 2500 elements partitioned into 50 sublists, there is reason to suspect that 15 of the partitions will deplete early. All the elements in these 15 partitions—a total of 750—might be, say, among the lowest 1200 elements. Thus, after the first 1200 elements have been selected and sent to output, these partitions will be empty. Each time the auxiliary list is searched, after depletion of these partitions, there will be a cost of 15 profitless comparisons. Since this search will occur 1300 times, there will be a total of 19,500 needless comaprisons. The following procedure will protect against them.

1. Compare for dummy value each time a transfer to the auxiliary list is made. This step will involve up to 2500 additional comparisons.

2. Count each time a depletion occurs. This step will involve up to 50 counts.

3. Test depletion count against a depletion limit. This step will involve up to 50 tests.

4. Reorganize the auxiliary list to modify destination addresses for transfers from a partition and to provide contiguous real key values.

If this procedure requires less time than that saved by the elimination of 19,500 comparisons, then it is worthwhile.

If the programmer has no particular reason to suspect unbalanced replenishment, he is well off leaving refinements alone. Instead of improving the Quadratic Sort slightly, he should evaluate altogether different sorts.

8.2.4   Discussion: Partitioning

The number of partitions into which the list is divided is roughly $\sqrt{N}$. Each partition and the auxiliary list will contain the same number of elements. This condition keeps the number of comparisons to a minimum.

To attempt to minimize comparisons in the auxiliary list by having fewer partitions would cost additional comparisons in the consequently larger partitions. On the other hand, defining more partitions would lead to more comparisons in the auxiliary list.

The task of defining partitions is performed only once, but it is not a trivial job. The process of partitioning is not thoroughly discussed in the literature. In our discussion here, we will explain one procedure and, hopefully, stimulate thought about others.

The partitioning routine must discover whether or not the number of elements on the list is a square. If it is, then the number of elements in each partition is equal to the number of partitions, and the size of the auxiliary list is equal to the number of partitions. The partitioning routine can deposit the starting address of each partition on a partition index, by starting with an origin address and adding the square root of N for the second partition, adding the same quantity again for the third partition, etc.

If the number of elements on the list is *not* a square, then the number of partitions will be the integer square root of the actual number of elements. To develop the number of elements in each partition, multiply the number of partitions by the square root of the next square (N'). If this product ($\lfloor \sqrt{N} \rfloor \times \sqrt{N'}$) equals N, then the number of elements in each partition is the integer square root of N', and all partitions are of equal size. Figure 8.8(a) shows four values of N for which this condition holds.

If the product $\lfloor \sqrt{N} \rfloor \times \sqrt{N'}$ is larger than N, the number of elements in a normal partition will be $\sqrt{N'}$, and the last partition will be small. Figure 8.8(b) shows four examples of this condition. The presence of a last partition smaller than other partitions will not disturb the sort since it will deplete naturally. If the last partition is very small, it may be absorbed into the penultimate partition, or boundaries may be adjusted to redistribute its few members.

A similar condition obtains when $\lfloor \sqrt{N} \rfloor \times \sqrt{N'}$ is less than N. In this case the number of elements in each partition is also $\sqrt{N'}$ with the remainder placed in a small partition. The number of partitions is increased to $\sqrt{N'}$ as well. Figure 8.8(c) shows this case for four examples.

8.2.5   Discussion: Number of Comparisons and Data Movements

The selection of the first element of the output list involves a scan of all partitions and a scan of the auxiliary list. If there are K partitions and K' elements in each partition, then the selection of the first element involves $K(K' - 1) + (K - 1)$ comparisons. This expression reduces to $N - 1$ when N is a perfect square. For the selection of each of $N - 1$ remaining elements, there will be a scan of one partition and a scan of the auxiliary list. This

Elements (N)	Next square (N')	Number of partitions ($\sqrt{N}$)	Elements per partition ($\sqrt{N'}$)
12	16	3	4
20	25	4	5
56	64	7	8
2450	2500	49	50

a) $\lfloor \sqrt{N} \rfloor \times \sqrt{N'} = N$

Elements (N)	Next square (N')	Number of partitions ($\sqrt{N}$)	Elements per partition Normal	Last
11	16	3	4	3
18	25	4	5	3
50	64	7	8	2
2402	2500	49	50	2

b) $\lfloor \sqrt{N} \rfloor \times \sqrt{N'} > N$

Elements (N)	Next square (N')	Number of partitions ($\sqrt{N'}$)	Elements per partition Normal	Last
15	16	4	4	3
23	25	5	5	3
60	64	8	8	4
2470	2500	50	50	20

c) $\lfloor \sqrt{N} \rfloor \times \sqrt{N'} < N$

Fig. 8.8 Partitions for several values of N.

scan pair will involve $(K' - 1) + (K - 1)$ comparisons. If N is a perfect square, this becomes $2\sqrt{N} - 2$. The scan pair occurs $N - 1$ times for a total of $(N - 1)(2\sqrt{N} - 2)$. The total number of comparisons for the case of the perfect square is therefore $N - 1 + (N - 1)(2\sqrt{N} - 2)$. A usable approximation, of course, is $2\sqrt{N}*N$, which differs by less than 10% in the range 100 to 1000.

If N is *not* a square, then K and K' will not be equal and the last (incomplete) partition is dummy-filled and treated as a full partition; the number of comparisons can still be reasonably approximated by $N - 1 + (N - 1)(2\sqrt{N'} - 2)$, or approximately $2\sqrt{N'}*N$.

The data movement involved in the method consists of the transfer of a record from the list to the auxiliary list, the transfer of a record from the auxiliary list to the output list, and the movement of key values for the Linear Selection process. Since each record is moved from the original list to the output list through the auxiliary list, there are basically $2N$ record transfers. In addition, each partition contributes a dummy record to the auxiliary list for each depletion except the last. There are therefore $\sqrt{N-1}$ additional transfers. Total record transfers are approximately $2N + \sqrt{N-1}$. In addition, there are N dummy fills and at least one key movement for each Linear Selection process. Use $2N + \sqrt{N}$ as an approximation for transfers.

8.2.6   Tag Variant

The movement of records to the auxiliary list can be replaced by the movement of keys; it is usually profitable to do so. Linear Selection involves the development of an address for the placement of a dummy value in the location of a winner. This address is available for forwarding to the auxiliary list, which then becomes a detached catalogue of current lows from each partition. Roughly half of the transfers can be reduced to key/address or address movements. Upon selection of a lowest from the auxiliary list, the address in the auxiliary list is used to effect a transfer of the record from the input list to the output list. The tag approach can be extended, of course, so that only tags are forwarded to the output area if physical ordering of the input list is not a requirement.

8.2.7   Presorting

The sublists of Quadratic Selection may be presorted—entirely arranged before the auxiliary list is formed or scanned. As a consequence, the number of comparisons can be reduced by applying a method more efficient than Linear Selection to the partitions.

   The number of comparisons may decrease because of the use of a more efficient algorithm on the partitions. The number of data movements may increase because of the movement inherent in actually arranging partitions. A more careful analysis would account for the fact that presorting eliminates filling with dummies and that a certain number of transfers can be reduced to tag movements, etc.

   Any combination of algorithms may be used in a combined sort of this type. The concept of presorting sublists and selecting the lows from elements off the top is a basic concept of the *merge*, which is described in the next chapter.

8.3   HIGH-ORDER SELECTION

It is possible to organize lists into structures of higher order, to have cubic or nth-degree selection, with or without presorting. The concepts involved are very much the same, but the mechanisms become more elaborate. Figure 8.9(a) shows sorting of the second, third, fourth, fifth, and sixth orders for various N's. Figure 8.9(b) shows how, for the same large file, different sort orders affect the process. When these sorts are represented as trees, the root node in each tree is the location in the output area, the endpoints are the original list, and intervening levels of nodes are the auxiliary lists.

		Number of elements in partition	Number of list partitions	Number of levels (aux. lists)	Number of auxiliary lists at each level				
					Level 1	Level 2	Level 3	Level 4	Level 5
Order (P)	N								
2	16	4	4	1	1	X	X	X	X
3	27	3	9	2	3	1	X	X	X
4	16	2	8	3	4	2	1	X	X
5	1024	4	256	4	64	16	4	1	X
6	4096	4	1024	5	256	64	16	4	1

Quadratic

a) Sorting lists of various size for orders up to 6.

2	65536	256	256	1	1	X	X	X	X
4	65536	16	4096	3	256	16	1	X	X

b) Sorting a list of 65536 elements at two different orders.

Fig. 8.9 Organization of various high-order sorts.

9
Internal
Merging

9.1 INTRODUCTION: BASIC MERGE PROCESS

Basic to the process of a merge is the fundamental notion of arranging data by interleaving the elements of two or more ordered lists. Figure 9.1 shows two ordered partitions of a list of ten elements. The ordered union of these two partitions represents a final ordered list of ten elements. The process of achieving this order is called a two-way merge—"two-way" because there are two input lists. The sequence of comparisons is shown in Fig. 9.1, as is the ordered list showing the initial positions of elements. The data space required includes, in addition to the space for S1 and S2, space for the output list.

The comparison sequence of Fig. 9.1 shows that the low-valued elements of each partition are compared and the lower is advanced to output. The process of comparing elements, advancing the lower to output, and selecting a linear successor in the "winning" partition continues until one of the partitions is depleted. The depletion of a partition causes all remaining elements in the other partition to be forwarded to output.

Partition depletion may be recognized in a number of ways. The way in which recognition is accomplished implies some important characteristics of the merge algorithm. If each partition is of a known (or predetermined) size, then partition depletion may be discovered by any technique of loop control or element counting. A merge process in which each partition is fixed in size is called *straight merging*. There is an alternative, called *natural merging*, in which partition size is not fixed, and the recognition of partition depletion is based on discovery of attributes of the data. We will first discuss

	Position	Key	Output
	1	1	1 (S1,1)
	2	2	2 (S1,2)
S1	3	5	3 (S2,1)
	4	7	4 (S2,2)
	5	8	5 (S1,3)
			6 (S2,3)
	1	3	7 (S1,4)
	2	4	8 (S1,5)
S2	3	6	9 (S2,4)
	4	9	10 (S2,5)
	5	10	

Comparison sequence

Positions	Keys	Transfers	
S1,1 : S2,1	(1 : 3)	S1,1	Output 1
S1,2 : S2,1	(2 : 3)	S1,2	Output 2
S1,3 : S2,1	(5 : 3)	S2,1	Output 3
S1,3 : S2,2	(5 : 4)	S2,2	Output 4
S1,3 : S2,3	(5 : 6)	S1,3	Output 5
S1,4 : S2,3	(7 : 6)	S2,3	Output 6
S1,4 : S2,4	(7 : 9)	S1,4	Output 7
S1,5 : S2,4	(8 : 9)	S1,5	Output 8
—		S2,4	Output 9
—		S2,5	Output 10

S = partition

Fig. 9.1 Merging.

straight merging and then natural merging. The reader will notice the similarity between merging of either type method and Quadratic Selection with presorting.

9.2   THE MULTIPASS TWO-WAY STRAIGHT MERGE

In the subsections that follow, we will discuss the organization of the straight two-way merge, describing string movement and performance.

9.2.1   Organizing the Merge Inputs

Figure 9.1 showed a straight two-way merge of one pass. Usually a merge will have more than one pass. The concept of the merge involves a multipass process, in which consecutive passes create larger and larger partitions until a final pass develops a partition containing all elements of the list to be sorted. The number of passes required to do this depends on the number of the initial partitions and the *order* of the merge.

		List	Output areas	Source
A	1.	3	3	1 ∪ 2
B	2.	15	15	
A	3.	12	11	3 ∪ 4
B	4.	11	12	
A	5.	6	4	5 ∪ 6
B	6.	4	6	
A	7.	16	5	7 ∪ 8
B	8.	5	16	
A	9.	13	7	9 ∪ 10
B	10.	7	13	
A	11.	8	8	11 ∪ 12
B	12.	10	10	
A	13.	2	1	13 ∪ 14
B	14.	1	2	
A	15.	14	9	15 ∪ 16
B	16.	9	14	

Fig. 9.2 Merge, first pass.

The size of the initial partition may be 1. Figure 9.2 has a list of 16 elements. The merging process is entered with 16 partitions of size 1. The first pass will merge the initial 16 partitions into eight partitions of two elements each, using a two-way merge.

There are a number of ways to associate partitions with each other in a merge. Figure 9.2 merges partitions 1 and 2, 3 and 4, 5 and 6, etc. Alternatively, the initial list may be split into two sublists, so that one is the first half of the list, the other the second. Figure 9.3 shows sublist A containing

		List		Output area	String no.	Source
A	1.	3	C	3	21	1 ∪ 9
	2.	15		13		
	3.	12		8	23	3 ∪ 11
	4.	11		12		
	5.	6		2	25	5 ∪ 13
	6.	4		6		
	7.	16		14	27	7 ∪ 15
	8.	5		16		
B	9.	13	D	7	22	2 ∪ 10
	10.	7		15		
	11.	8		10	24	4 ∪ 12
	12.	10		11		
	13.	2		1	26	6 ∪ 14
	14.	1		4		
	15.	14		5	28	8 ∪ 16
	16.	9		9		

Fig. 9.3 Merge, string formation.

single-element partitions 1 through 8 and sublist *B* containing single-element partitions 9 through 16. The eight resulting partitions are the results of the merge of partitions 1 and 9, 2 and 10, 3 and 11, etc.

The term commonly used to describe an ordered portion of a list is *string*. A string is a collection of elements known to be in order. In Fig. 9.3, area *C* has four strings and area *D* has four strings, each of two elements.

9.2.2 Dispersing Strings

In Fig. 9.3 the second comparison between *A* and *B* merges elements 2 and 10 and places the result in the first two positions of the second half of an output area, called area *D*. The reasons for this placement lie in an aspect of merging technique known as dispersion.

In Fig. 9.3 the output area is divided into halves, *C* and *D*, just as the initial input area was divided into halves, *A* and *B*. There should be an equal number of strings in *C* and *D*. The efficiency of a merge is always greatest when the number of strings is the same in each input sublist. One can achieve this result by a technique called balanced dispersion, which places one merged string on one output sublist and its successor on another. In a two-way merge this procedure is equivalent to alternating between output areas. In Fig. 9.3, because of the alternation, strings 21, 23, 25, and 27 are in *C*, and 22, 24, 26, and 28 are in *D*. (The string designations consist of the number of the pass for which these strings will be input—in this case, pass 2—in the first digit and the order of their formation in the second digit.) The dispersion of strings of a fixed two-way merge is often trivial. In the general merge, the algorithm used to disperse strings is often not trivial and may have serious impact on the performance of a merge.

9.2.3 Subsequent Passes

In Fig. 9.3, pass 1 has reduced the number of strings from 16 to 8 and increased the length of strings from 1 to 2. The strings are in areas *C* and *D*. The next pass will merge *C* and *D*, using *A* and *B* as the output area.

The first merge of pass 2 will be of strings 21 and 22 into *A* as string 31. The next merge will be to *B* to form string 32. The pass will end when all strings in *C* and *D* have been merged. Figure 9.4 shows the status at the end of the second pass. There are four strings, each of length 4.

Pass 3 merges 31 and 32 to form 41, and 33 and 34 to form 42. Each string is eight elements long, and there are only two strings, one in *C* and one in *D*. This is the "last-pass" condition—one string in each area—and therefore a single merge pass will form a single ordered string. Pass 3 and the final pass, pass 4, are shown in Figs. 9.5 and 9.6, respectively.

input				Output			
Area	String no.	String		Area	String no.	String	Source
				A	31	3	21 ∪ 22
C	21	3				7	
		13				13	
						15	
	23	8			33	1	25 ∪ 26
		12				2	
						4	
	25	2				6	
		6					
	27	14					
		16					
				B	32	8	23 ∪ 24
D	22	7				10	
		15				11	
	24	10				12	
		11			34	5	27 ∪ 28
						9	
						14	
	26	1				16	
		4					
	28	5					
		9					

Fig. 9.4 Merge, pass 2.

Input				Output			
Area	String no.	String		Area	String no.	String	Source
A	31	3		C	41	3	31 ∪ 32
		7				7	
		13				8	
		15				10	
	33	1				11	
		2				12	
		4				13	
		6				15	
B	32	8		D	42	1	33 ∪ 34
		10				2	
		11				4	
		12				5	
	34	5				6	
		9				9	
		14				14	
		16				16	

Fig. 9.5 Pass 3.

Area	String no.	String	Final output	Source
C	41	3	1	
		7	2	
		8	3	
		10	4	
		11	5	41 ∪ 42
		12	6	
		13	7	
		15	8	
D	42	1	9	
		2	10	
		4	11	
		5	12	
		6	13	
		9	14	
		14	15	
		16	16	

Fig. 9.6 Last pass.

9.2.4 A Variant in Less Space

It is possible to reduce the required size of additional storage for a straight two-way merge to an area one-half the size of the original list, by performing the initial merges shown in Fig. 9.2 rather than those in Fig. 9.3. That is, by merging contiguous elements rather than noncontiguous elements, one may save space.

In Fig. 9.2, when string 7 is merged with string 8, the upper half of the initial input area is effectively free. Merged strings beginning with 9 ∪ 10 can be placed in this area. The end of the first pass is shown in Fig. 9.7. The second pass may proceed by merging 25 and 26, and 27 and 28, into

String no.	String	String no.	String
	7		3
25	13	21	15
	8		11
26	10	22	12
	1		4
27	2	23	6
	9		5
28	14	24	16
Free	{ – – – – –		

Fig. 9.7 Merge variant using 1.5N space.

free space. The space vacated by those strings is used for the merge of 21 and 22, and 23 and 24. The reduction in space is achieved by conceptually rearranging the two sublists so that they are disjoint and interleaved, as shown in Fig. 9.2, where even-numbered elements are considered A elements and odd-numbered elements are B elements.

9.2.5   Number of Passes, Movements, Comparisons

The number of two-way passes required to form a single string from S strings is the number of times S must be halved before it is 1. This is $\lceil \log_2 S \rceil$. In the example of 16 initial strings, $\log_2 16 = 4$ passes are needed to complete the merge. When N is initially equal to S, the number of passes can be stated as $\lceil \log_2 N \rceil$.

The number of merge passes can be reduced by reducing the number of initial strings, i.e., making S smaller than N. This is done by some efficient method of separately sorting individual partitions or by a string-definition algorithm of the type to be described in Section 9.3.1.

A reduction in the number of passes will cause a reduction in the number of comparisons for the merge operation. The number of comparisons for a straight two-way merge is often given in the literature as $N\lceil \log_2 N \rceil$. This formulation represents the phenomenon that there are $\lceil \log_2 N \rceil$ passes and that, during each pass, each element which joins a forming string is involved in a comparison. There are N movements each pass, but there will not be N comparisons. $N\lceil \log_2 N \rceil$, therefore, represents an unachievable upper bound for the number of compares.

In general, for any straight two-way merge involving strings of equal size, there will be at most $N - 2^i$ comparisons per pass, where i is a pass index equal to $\log_2 N - P$, where P is the pass number. Note that $i = 0$ for the last pass and increases by 1 for each earlier pass. Figure 9.8 shows

	No. of strings	String size	Max. no. of comparisons per merge	No. of 2-way merges	Max. no. of comparisons
Pass 6 (Last)	2	32	63	1	63
Pass 5	4	10	31	2	62
Pass 4	8	8	15	4	60
Pass 3	16	4	7	8	56
Pass 2	32	2	3	16	48
Pass 1	64	1	1	32	32
					321

$S = N = 64$ $N/2 = 32$
$\log_2 N = 6$ $\log_2 (N/2) = 5$
$N \log_2 N = 384$ $N \log_2 (N/2) = 320$

Fig. 9.8 Maximum number of companions per pass for a 64-element list.

the maximum number of comparisons in each pass for a 64-element list. For the last pass there are $N - 2^0$ (or $64 - 1 = 63$) comparisons; for the fifth pass, $N - 2^1 = 62$ comparisons; and for the other passes,

$$N - 2^2 = 60, \qquad N - 2^3 = 56, \qquad N - 2^4 = 48, \qquad N - 2^5 = 32.$$

Figure 9.8 shows that the maximum number of comparisons is not $N\lceil \log_2 N \rceil = 384$ but a number closer to $N\lceil \log_2 (N/2) \rceil$. As an estimate of the number of comparisons for a merge, $N\lceil \log_2 N/2 \rceil$ will be reasonable and safe when the merge is entered with strings of size 1. Estimating the number of comparisons when the merge is entered with larger strings is somewhat more complex. It is necessary to find $\lceil \log_2 S \rceil$ (the number of strings), which represents the number of passes, and to develop the sum of comparisons in all passes.

Woodrum [30] has shown that the worst case for the number of comparisons in a two-way merge and the theoretical minimum to sort N numbers vary by only trivial amounts. For example, with 100 items, the worst case of a two-way merge is 8967, and the theoretical minimum is 8520. Therefore the use of $N\lceil \log_2 (N/2) \rceil$ for the number of comparisons is reasonable.

9.3   THE NATURAL TWO-WAY MERGE

In the subsection that follows, we will discuss the organization of the *natural* two-way merge, describing string movement and performance.

9.3.1   String Recognition and String Definition

The natural merge (or Von Neumann merge) differs from the straight merge in that it attempts to take advantage of strings already existing in the data. The list in Fig. 9.9 consists of ten initial strings of lengths 1, 2, and 3. The string definition table shows their numbers, origins, and lengths. One method for recognizing strings is to undertake an initial pass over the data, comparing each element with its successor, recognizing "steps down" (a successor lower than a predecessor), and building a string-definition table of the type shown in Fig. 9.9. This process involves $(N - 1)$ comparisons and the creation of the definition table. No movement of input data is involved. The first pass of the merge is replaced by this string-recognition process. The comparand addresses of succeeding merge passes are provided by the string-definition table, which also provides string-length limits.

In Fig. 9.10 the string definition pass has been undertaken in a different fashion. Strings of length 2 are created by grouping pairs of consecutive elements in either forward or backward direction. Those strings which have their earlier element higher are D (descending) strings, and the others are A (ascending) strings. This string-recognition pass is preparatory to a

List

1.	3 ⎫	S1L2
2.	15 ⎭	
3.	12 }	S2L1
4.	11 }	S3L1
5.	6 }	S4L1
6.	4 ⎫	S5L2
7.	16 ⎭	
8.	5 ⎫	S6L2
9.	13 ⎭	
10.	7 ⎫	
11.	8 ⎬	S7L3
12.	10 ⎭	
13.	2 }	S8L1
14.	1 ⎫	S9L2
15.	14 ⎭	
16.	9 }	S10L1

String definition table

Number	Origin	Length
1	1	2
2	3	1
3	4	1
4	5	1
5	6	2
6	8	2
7	10	3
8	13	1
9	14	2
10	16	1

S = String
L = Length
S1L2 = String 1 of length 2

Fig. 9.9 Natural string definition.

straight merge. Strings of length 2 are formed prior to the merge by means of $N/2$ comparisons and $N/2$ entries into the string-definition table.

A comparison between Figs. 9.9 and 9.10 shows very directly the difference between a natural merge and a two-way merge. In a natural merge, the size of the initial strings can be determined but not predicted. Consequently, neither the number of strings nor the number of passes can be precisely predicted with initial data.

List

1.	3 ⎫	A
2.	15 ⎭	
3.	12 ⎫	D
4.	11 ⎭	
5.	6 ⎫	D
6.	4 ⎭	
7.	16 ⎫	D
8.	5 ⎭	
9.	13 ⎫	D
10.	7 ⎭	
11.	8 ⎫	A
12.	10 ⎭	
13.	2 ⎫	D
14.	1 ⎭	
15.	14 ⎫	D
16.	9 ⎭	

String definition table

A	1
D	4
D	6
D	8
D	10
A	11
D	14
D	16

Fig. 9.10 Ascending/descending string definition.

Area	Input	(from Fig. 9.9)		Area	Position	Output	String no.	Source	Size
A	3, 15	S1L2		C	1.	3	21	S1 ∪ S6	4
	12	S2L1			2.	5			
	11	S3L1			3.	13			
	6	S4L1			4.	15			
	4, 16	S5L2			5.	2	23	S3 ∪ S8	2
					6.	11			
B	5, 13	S6L2			7.	4	25	S5 ∪ S10	3
					8.	9			
	7, 8, 10	S7L3			9.	16			
				D	10.	1	24	S4 ∪ S9	3
	2	S8L1			11.	6			
	1, 14	S9L2			12.	14			
					13.	7	22	S2 ∪ S7	4
	9	S10L1			14.	8			
					15.	10			
					16.	12			

S = String
L = Length

String definition table

Number	Origin	Length
21	1	4
22	13	4
23	5	2
24	10	3
25	7	3

Fig. 9.11 Natural merge with string definition, pass 1.

Figures 9.11, 9.12, and 9.13 show passes of a possible merge of the strings of Fig. 9.9. Note that the new strings fill the merge area from both ends. The string-definition table makes possible the prediction of the sizes of newly formed strings and the area they will occupy. A string-dispersion algorithm is developed from the definition table. Space requirements for natural merging are the same as for straight merging.

9.3.2   Natural Merge Without String Definition

A natural merge need not rely on a string-definition table or a string-definition pass. Consider a process which initially undertook a merge on the list of Fig. 9.14 without string definition. Define the origin of the total list to be the first position of A and the last location of the list to be the first position of B. This endpoint division effectively separates the list into two sublists with an indeterminate number of strings. Partitioning in this particular manner in order to avoid a string-definition pass is not necessary, but it is a fast and reasonable way to form two inputs to a merge. Algorithm 207 "Stringsort," a natural two-way merge shown in Appendix A, uses this top-and-bottom technique.

Area	Input	String no.		Position	Output	String no.	Source	Size
C/D	3	21		1.	3	31	21 ∪ 22	8
	5			2.	5			
	13			3.	7			
	15			4.	8			
	2	23		5.	10			
	11			6.	12			
	4	25		7.	13			
	9			8.	15			
	16			9.	4	33	25	3
	1	24		10.	9			
	6			11.	16			
	14			12.	1	32	23 ∪ 24	5
	7	22		13.	2			
	8			14.	6			
	10			15.	11			
	12			16.	14			

String definition table

Number	Origin	Length
31	1	8
32	12	5
33	9	3

Fig. 9.12 Natural merge with string definition, pass 2.

Input, pass 3	String no.	Position	Output, pass 3/ Input, pass 4	String no.	Output, pass 4
3	31	1.	1	41	1
5		2.	2		2
7		3.	3		3
8		4.	5		4
10		5.	6		5
12		6.	7		6
13		7.	8		7
15		8.	10		8
4	33	9.	11		9
9		10.	12		10
16		11.	13		11
1	32	12.	14		12
2		13.	15		13
6		14.	4	42	14
11		15.	9		15
14		16.	16		16

String definition table

	Number	Origin	Length
Pass 3:	41	1	13
	42	14	3
Pass 4:	Final	1	16

Fig. 9.13 Natural merge with string definition, passes 3 and 4.

Position	Input	Pointers	Output	String no.
1.	3	← A	3	21
2.	15		9	
3.	12	← A′	14	
4.	11		15	
5.	6		8	23
6.	4		11	
7.	16		4	25
8.	5		5	
9.	13		16	
10.	7		13	24
11.	8		7	
12.	10		6	
13.	2		12	22
14.	1	← B′	10	
15.	14		2	
16.	9	← B	1	

Fig. 9.14 Natural merge without string definition, pass 1.

The process shown in Fig. 9.14 rests on the concept of single and double "step-down." Comparisons are made between *A* and *B* elements by appropriate advancing of the *A* and *B* pointers, and transfers are made to the output list. Whenever an element of *A* is smaller, it is transferred to output and the pointer is advanced.

The Output List of Fig. 9.14 shows the first four elements of the new string as 3, 9, 14, 15. When those elements are on the Output List, the *A* pointer is at position 3, key 12. The *B* pointer is at position 14, key 1. But neither key 1 nor key 12 can join a string whose last element is key 15 because, to join a string, the key must be not only the lower of two contenders but higher than the latest element on the string. Therefore, every element that succeeds in joining the string must have passed two tests; the first determined that the element was lower than the eligible element in the other sublist, and the second determined that it was not lower than the last element already on the string.

At the point where an element is discovered to be lower than the last element of a string, a single step-down condition is said to exist. The condition of a single step-down causes a run-out of all elements in the other sublist until a step-down is discovered there also. When this occurs, a double step-down condition exists. A double step-down condition means that there may be no additions to the current string since there are no eligible candidates. Whatever string-termination procedure is used with the algorithm is now undertaken, and a new string is started.

The merge shown in Fig. 9.14 places the second string (22) at the end of the output area. This string is not built down from some point computed from a string-definition table; it is built up from the end position. Locations

16 through 13 of the output area hold a string of length 4 that ascends in key values from location 16. This is the simplest way of placing this string whose length is unknown, and the placement of ascending strings at the head of the list and descending (upward ascending) strings at the tail makes for a simple merge algorithm for later passes.

9.3.3   Number of Passes, Movements, Comparisons

The goal of the natural merge is to eliminate passes. The attempt is made because of the possibility that long natural strings exist in the input data. Recognition of these natural strings may result in fewer data passes than there will be with a method which imposes a fixed progression of string sizes regardless of natural ordering.

With random data the saving of one pass can be expected. The expected number of passes is therefore $\lceil \log_2 N \rceil - 1$ with natural merging. However, data with very long natural strings will have considerably fewer passes.

Without the string-definition table, the number of key comparisons increases for the natural merge, since it is necessary to perform comparisons for the end of the current string. This comparison replaces the limit test for end-of-string used in straight merging. The key comparison might be significantly more expensive than the limits test. Some authors ascribe $2\lceil \log_2 (N-1) \rceil$ comparisons to natural merge because of the additional data test for entry to the output string. This is an unachievable upper bound.

There will be (roughly) $N\lceil \log_2 (N-1) \rceil$ string-entry comparisons since each element will be so tested in each pass. The number of normal element comparisons does not increase, however, and so the number of comparisons for the worst case is

$$N\lceil \log_2 (N-1) \rceil + N\lceil \log_2 (N/2) - 1 \rceil$$

if a pass is saved, and

$$N\lceil \log_2 N \rceil + N\lceil \log_2 (N/2) \rceil$$

if a pass is not saved.

In straight merging there will be roughly $N\lceil \log_2 N \rceil$ string-end tests to determine whether a string has ended. In any determination of the preferability of one approach over the other, a critical factor is the existence on the machine of a mechanism which renders the string-end test more efficient than the step-down comparisons.

The advantage of natural over straight merging does seem marginal for internal merging. The techniques are equivalent to implement; in the absence of any particular study of "run length," the programmer may decide between them by investigating whether the peculiar overheads and timings associated with his machine make one or the other procedurally faster.

9.4   FURTHER TOPICS IN MERGING

Some considerations relevant to the merge that have not yet been discussed
are covered in the subsections that follow.

9.4.1   Key/Tag Sorting

The merge provides an ideal situation for the use of key-sort techniques.
The area required for additional storage can be reduced to that required
by keys and addresses (or addresses only), and the significant data movement can be reduced to address or tag movement time. There are countless
coding variations and approaches to the implementation of a merge in both
its record and tag or address forms. Structures like the string-definition table
may be created to support the merge, in tag as well as record form.

9.4.2   Multiway Merging

It is possible to perform merges of an order higher than two. With a list
of 16 elements, there might be three-way merges, four-way merges, or
merges of any order up to 16. These merges might be straight or natural,
with or without presorting or specialized first-pass string definition.

Figure 9.15 shows a straight four-way merge on a list of 16 elements.
The output area (which need not be larger than for a two-way merge)
receives strings formed by the four-way merge of initial one-element strings
on the input list. A merge of the four strings created by pass 1 will form

Position	Input	Output	String no.
1.	3	3	21
2.	15	11	
3.	12	12	
4.	11	15	
5.	6	4	22
6.	4	5	
7.	16	6	
8.	5	16	
9.	13	7	23
10.	7	8	
11.	8	10	
12.	10	13	
13.	2	1	24
14.	1	2	
15.	14	14	
16.	19	19	

Fig. 9.15 Four-way merge. Input of 16 strings of length 1 produces output of
four strings of length 4.

a single ordered string. The complete list will therefore be ordered on two passes using a four-way merge.

The number of passes required to order a list using straight M-way merging, with all strings assumed to be of initial size 1, is $\lceil \log_M N \rceil$. The log of 16 to the base 4 is 2. This reduction in the number of passes is important because it in turn reduces data movement to $N \lceil \log_M N \rceil$ moves (some of which may be key or address movements only).

A problem in the use of higher-order merges comes from the fact that computers tend to be designed with comparison instructions for comparing two elements at a time. If devices existed which could perform fast M-way comparisons, then internal M-way merging would require only as many "compare" operations as a two-way merge, and the use of higher-order merges would be unquestionably attractive. Although various M-way comparison mechanisms have been proposed and M-way merging using such devices has been (whimsically) described in the literature, such devices do not, in fact, exist, and the implementation of M-way merges in core faces the requirement for additional program control mechanisms to undertake M-way comparisons. These involve not only comparison instructions but the organization of routines to sequence and define such comparison loops. To select a single lowest in an M-way merge requires $M - 1$ comparisons. The maximum number of comparisons for an M-way merge is $N(M - 1) \lceil \log_M N \rceil$.

The technique used to accomplish an M-way merge need not be so simple as an $M - 1$ technique. In merges of very high order it might prove profitable to use a form of Tournament sort or any other more efficient method for selecting the next element of a forming string. The constraint on order of merge must be that the complexity of organizing M-way comparing does not overbalance the savings found in reducing data movement.

9.4.3   Effective Order of Merge

From pass to pass of a merge of any order, it is possible for the order of the merge to change because of unbalanced string dispersion or sublist depletion. In Fig. 9.16, the effective order of this three-way merge is truly 3. Each pass has reduced the number of strings by a factor of 3. But the effective order of a merge is equal to its nominal order only when the initial number of elements (or strings) is a perfect power of the order of the merge. In Fig. 9.17, the effective order of the merge is not 2. Unbalanced situations develop, and the pass-by-pass reduction in strings is not the order of the merge.

When natural merging is undertaken and string sizes vary widely, it is difficult to evaluate what the effective merging power truly is.

Areas:	A	B	C	D	E	F	No. of strings
	9(1)	9(1)	9(1)	—	—	—	27
Pass 1	—	—	—	3(3)	3(3)	3(3)	9
Pass 2	1(9)	1(9)	1(9)	—	—	—	3
Pass 3	—	—	—	1(27)	—	—	1

Fig. 9.16 Perfect merge. Number preceding parentheses indicates number of strings in area; number in parens indicates length of strings. All strings are initially of nominal length 1 ($\log_3 27 = 3$).

Areas:	A	B	C	D	No. of strings
	6(1)	6(1)	—	—	12
Pass 1	—	—	3(2)	3(2)	6
Pass 2	2(4)	1(4)	—	—	3
Pass 3	—	—	1(8)	1(4)	2
Pass 4	1(12)	—	—	—	1

Fig. 9.17 Imperfect merge ($\lceil \log_2 12 \rceil = 4$).

9.4.4   Partial Passes

In those merges which have developed imbalances, up to this point we have assumed a copy from one area to another of the extra string. It is possible to organize merges so that this copy need not be undertaken. In internal merging this need not be difficult to do. An alternative to copying is the partial pass, where the "pass" concept is weakened and the sort moves from phase to phase instead of from pass to pass. The distinction lies in whether or not the entire list is examined before a new pass (or phase) is undertaken. Consider Fig. 9.12 again. String 33 is a copy of string 25, which is undertaken to end pass 2 and free the C/D area. But with other storage management tactics, the copy of this string might not be necessary. It might be left where it is without changing the nature of subsequent merges at all. Since it is equally accessible from any position, its movement is not an absolute requirement. If it is not moved, then the merges of Fig. 9.12 do not truly constitute a pass, since not all of the data has been processed. A partial pass or phase has been completed; and the next phase begins with the condition of Fig. 9.13, except that the sequence 4, 9, 16 is in another location.

The partial-pass concept may also be made to work the other way, by working on more data than that constituting a pass. For example, the remaining string might be merged with a string formed during the phase. Now the entire concepts of pass and order of merge of a pass become imprecise. The neat image of the merge that has existed to this point becomes somewhat fuzzy. The designer of a merge need not be constrained to define merge "passes" in any rigorously restricted way; he is free to vary the order of a merge from phase to phase as he chooses, and to select strings for merging on any basis he likes. Even within the conceptual framework of a "pass," the design of merges is highly flexible.

9.4.5   Tree of a Merge

Further discussion of merge design requires the concept of a *tree*. The existence of such trees as those of Fig. 9.18 was mentioned earlier. These trees are perfectly valid, despite the fact that the number of descendants associated with a node may vary. Burge [4] provides an interpretation of trees which we will follow. In this section we will concentrate on optimizing data movement since there is one data movement per comparison (for a record sort). The reader will see that any technique which optimizes data movement optimizes comparisons as well.

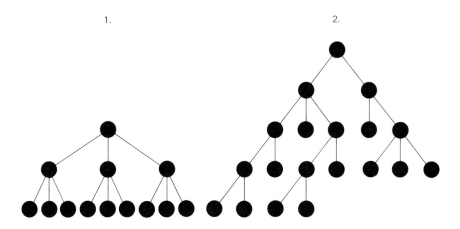

Fig. 9.18 Types of trees.

A merge can be represented as a tree in which a parent node represents a string formed by the merge of its descendants. Figure 9.19 is the tree of a straight two-way merge of 16 elements. The number of each node is the length of the string represented by that node. Each level in Fig. 9.19 represents a pass of the classical, multipass, internal merge.

Recall the concept of the value of a tree from Chapter 4. The levels of the tree of Fig. 9.19 are numbered from 0 to 4, starting from the root node, as we have previously numbered the levels of binary trees. Here we will use, for the value of a node, the length of the string represented by that node.

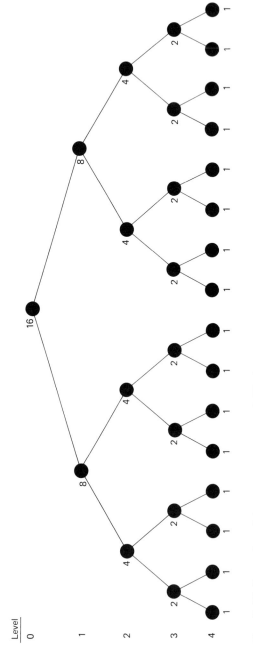

Fig. 9.19 Tree of straight two-way merge of 16 elements.

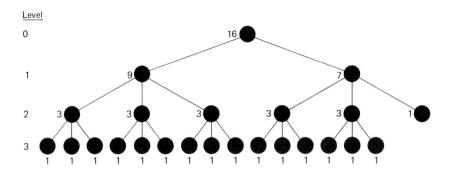

Fig. 9.20 Tree of three-way merge of 16 elements.

The highest level on the tree of Fig. 9.19 is level 4. The highest level number is the number of passes, the $\log_2$ of N or of N' (the next perfect power of 2). Figure 9.19 shows 16 endpoint nodes with values of 1. The sum of the values at level 4 is 16. This sum is the accumulated length of strings which are entering the merge at level 4. The product of the level number (number of passes) and the value of the nodes (length of strings entering the merge on the pass represented by the level number) is the number of element movements that are involved in the merge. The formulation for this is $N \log_2 N$, or $16 \times 4 = 64$. The tree shows us a graphic equivalent of the equation.

Figure 9.20 shows the same phenomenon with a three-way merge. There are 4 levels, with 3 the highest level number; $\lceil \log_3 16 \rceil = \log_3 27 = 3$. There are 15 strings which enter the merge at level 3. The sum of their lengths is 15. One string of length 1 enters the merge at level 2. The total number of string movements is

$$3 \times 15 + 2 \times 1 = 47.$$

In this case $N \lceil \log_3 N \rceil$ is $16 \times 3 = 48$. The difference is due to the elimination of a movement of the sixteenth string by delaying its entry into the merge until level 2. In general, the product of a level number and the sum of string lengths entering the merge at that level is the count of string movements and is the "value" of the tree for this interpretation.

9.4.6 Optimum Merges

Postulate a three-way merge for the strings of given length in Fig. 9.21. Use a merge design which combines the 18 strings in merges such that strings 1 through 6 constitute one sublist, 7 through 12 another, and 13 through 18 a third. Figure 9.22 shows the merge tree. Each string enters at the same

String	Length
1.	15
2.	3
3.	11
4.	6
5.	9
6.	41
7.	31
8.	27
9.	19
10.	61
11.	7
12.	8
13.	9
14.	21
15.	2
16.	14
17.	13
18.	17

Fig. 9.21 Strings in store.

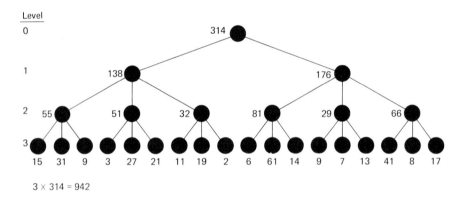

$3 \times 314 = 942$

Fig. 9.22 Standard three-way pass-oriented merge.

pass, and each is moved three times. There are 314 elements that are moved three times, and consequently the merge involves 942 movements to form one string.

For the merge of Fig. 9.23, strings have been selected for merging together in the order of their appearance on the list in Fig. 9.21.

The merges of Figs. 9.22 and 9.23 involve exactly the same number of data movements. String selection by size in a pass-oriented merge, in which all data is processed in each pass, does not impact the merge. The number of comparisons is similarly unaffected. Trees which can be transformed into each other by permuting endpoint nodes are equivalent trees.

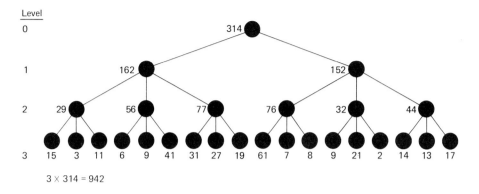

$3 \times 314 = 942$

Fig. 9.23 Another three-way pass-oriented merge.

But we may not conclude that string selection by size *never* impacts the merge. When some strings are not involved in each phase of a merge, the selection of both tree structure and individual strings at each node does affect performance. Consider again the merge of Fig. 9.23. The last pass is a two-way rather than a three-way merge. What would be the effect of redesigning the merge so that the last phase was always a merge of maximum order? In order to do this, delay the entry to the merging process of some strings. One form of the merge might be that given in Fig. 9.24. This merge follows the general principle that maximum merge order is used in later passes, so that the largest possible available strings can profit from the maximum merge. The resulting number of moves is reduced because not all strings were merged the same number of times. But now it becomes obvious that if some strings are going to be merged more often than others, it is desirable that the strings which are merged less often be the larger strings. It would be profitable to rearrange strings. The trees of a merge are not equivalent when a permutation of endpoint values causes a movement from level to level. Order the list of strings so that the smallest strings are the first ones merged, to get the tree in Fig. 9.25.

The merge in Fig. 9.25 has fewer than 800 data movements, following the principles of ensuring maximum order in later phases and introducing strings in order of size. By controlling "imbalances" and string size, one can accomplish a reduction of considerable size in the work of the merge. The cost of this reduction is certainly not trivial. A sort of string lengths is required, as is a merge design algorithm to develop the merge tree. It would be a massive internal merge indeed which could support such overhead.

The optimum merge (Fig. 9.26) is an extension of the merge of Fig. 9.25 in the following way. As each new string is generated, it is placed on

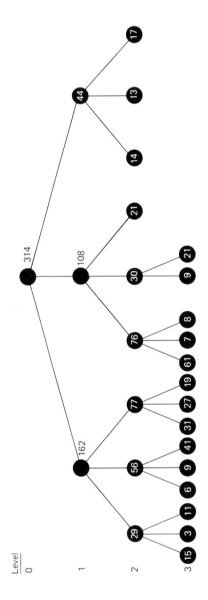

Fig. 9.24 Delayed-entry merge.

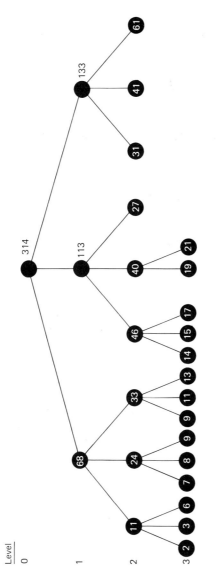

Fig. 9.25 Ordered strings as input to merge.

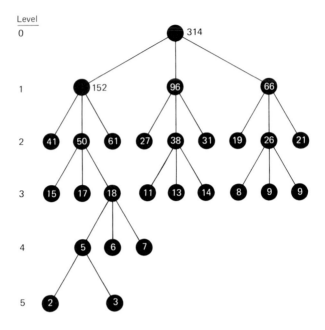

Fig. 9.26 Optimum merge tree.

a sorted list of strings to be merged. In Fig. 9.25, only initial strings were introduced into the merge by size. Only a single refinement remains, concerning a policy for strings of equal size when one is an initial and one is a generated string. A placement of the composite string ahead of an original string of the same size will have a tendency to minimize the number of levels on the tree.

When merging is done on tape or on disk, the selection of strings is not arbitrary. The realities of design on tape and disk, which constrain the considerations presented in this section, will be discussed when those devices are discussed. Merge optimization of this type is significant in multireel or multipack merging, where there are fewer devices than there are strings to merge. But these considerations are well shown in connection with core since it is only in core that all strings are equally accessible.

In the presence of very large, directly addressable core memories, such merge optimization might be useful. The number of strings involved in merges in these devices might be significant enough to warrant the overhead.

10
Distributive
Sorts

10.1 DISTRIBUTIVE SORTING

In Chapter 1, a distinction was made between *comparative* and *distributive* (or classification) methods for sorting. In this chapter, we will describe a number of distributive sorts and the problems and possibilities associated with them.

By their nature distributive sorts are not minimal-storage techniques. Since they "distribute" elements to receiving areas on the basis of some characteristic of the key, there must be an allocation of space for such receiving areas *other than* the space used by the initial list. The amount of space required for such areas and the penalty for misallocation are major considerations in weighing these methods. An exception to this general observation is the Binary Radix sort.

The algorithms of distributive sorting are equally applicable whether or not lists of input can be held in primary storage. Those algorithms which were initially intended to be internal algorithms are discussed as such in this text. Those introduced as external are discussed as external methods here. The logic of the procedures moves easily from one context to another.

A distinction to be made among distributive sorts which is not relevant to comparative sorts is that between block-sorting (range-sorting) methods and digit-sorting methods. Block sorts treat the entire key as a unit and distribute according to ranges which are defined to provide for a meaningful partitioning of the list. Digit sorts distribute elements into receiving areas based on the value of a specific digit of the key. The number of receiving areas and the distribution rules are functions of the number base of the keys.

The word "distribution" presents a problem. In statistics, "distribution" refers to the particular values which occur in a list, their frequency, and their range. In more general usage, it means the process of dispersing elements from one place to a number of places. Since both meanings have relevance in sorting, the subsequent text will use the word "dispersal" for the process of distributing elements.

10.2 BINARY RADIX EXCHANGE: THE METHOD

The basic algorithm of internal distribution sorting is Binary Radix Exchange [15]. The method is based upon the dispersal of keys according to the presence of a 1 or a 0 in bit positions of the key. The efficiency of this method depends on the ability of a machine to test bit positions easily and rapidly.

In Fig. 10.1 the usual list of numbers is given in both decimal and pure binary representation. Each bit position is scanned individually from the high-order bit (leftmost bit). The first pass develops two partitions, one in which all elements have 0 as the high-order bit, the other in which all elements have 1 as the high-order bit. The partitions are developed by searching from the bottom until an element with 0 as its high-order bit is found, then searching from the top until an element with 1 as its high-order bit is found, and exchanging the two. When the pointers meet, the pass is completed.

The size of the two partitions depends on the statistical distribution of values around 2^K, where K is the power of 2 represented by the first bit position (high-order bit) used in the scan. Values greater than or equal to 2^K will have a 1 bit in the K power position, values less than 2^K a 0 bit.

A second pass of the data scans the bit position to the right of the high-order bit, generating four partitions from two. The "zero" partition is transformed into two partitions whose leading bits are 00 and 01, respec-

Decimal	Binary
3	0011
11	1011
6	0110
4	0100
9	1001
5	0101
7	0111
8	1000
10	1010
2	0010
1	0001

Fig. 10.1 Decimal list with binary equivalents.

tively; the "one" partition is transformed into two partitions whose leading bits are 10 and 11, respectively. The process continues for a total of $(K + 1)$ passes; when the last bit position is scanned, the entire list is in order.

10.2.1   Example

Figure 10.2 shows the first pass of Binary Radix Exchange. Initially a mechanism must be established which will isolate the high-order bit of the key. Pure bit-addressable machines are rare, but many machines have devices which make efficient bit testing possible.

In addition to the bit-selection mechanism, the method requires beginning and ending list pointers. The first operation of the example tests the high-bit position of the key at location 11 of the list and finds that it is a 0. A search for an element to exchange is undertaken from the top of the list. The first element with a 1 in its high-bit position is in location 2. The contents of locations 2 and 11 are exchanged and the pointers advanced. For the status of the list after this first exchange, see Fig. 10.2, First Exchange column.

	List decimal	Binary		First exchange		Second exchange
1.	3	0011←BEG	3	0011	3	0011
2.	11	1011	①	0001	1	0001
3.	6	0110	6	0110←BEG	6	0110
4.	4	0100	4	0100	4	0100
5.	9	1001	9	1001	②	0010
6.	5	0101	5	0101	5	0101←BEG
7.	7	0111	7	0111	7	0111
8.	8	1000	8	1000	8	1000
9.	10	1010	10	1010	10	1010←END
10.	2	0010	2	0010←END	⑨	1001
11.	1	0001←END	⑪	1011	11	1011

		Pass end	Partitions*
1.	3	0011	1-7 (2)
2.	1	0001	8-11 (2)
3.	6	0110	
4.	4	0100	
5.	2	0010	
6.	5	0101	
7.	7	0111←END/BEG	
8.	8	1000	
9.	10	1010	
10.	9	1001	
11.	11	1011	

*Number in parentheses indicates the bit position to be scanned next.

Fig. 10.2 Binary Radix exchange, pass 1.

The bottom-up search now inspects the high-order bit of location 10 on the list and again finds a 0, so another exchange must be made. The top-down search for a high bit of 1 resumes. Location 5 has a 1 in the high-bit position, and an exchange is made, as shown under Second Exchange in Fig. 10.2.

The bottom-up testing resumes. The next element found to have a 0 in its high-order bit position is in location 7. However, no list location above it contains a 1 in its high-order bit. The search for an exchange value brings the top-down pointer to the position of the bottom-up pointer, ending the pass. Two partitions have been generated. One extends from location 1 to location 7 and contains all elements with high-order bits of 0. The other extends from location 8 to location 11 and contains all elements with high-order bits of 1. The effect of the pass has been to separate elements with keys of value less than 8 from elements with keys of value 8 or more. In effect, the value of 2^K is used as a partition pivot. The partition definitions are stacked.

The second pass will use the next bit position for its test and will reorder each partition into two further partitions.

Although the sequence of partition processing is logically arbitrary, there is some efficiency in selecting partitions so that all those which are to be scanned on the same bit are processed before those which are to be scanned on a bit to the right. The reason is basic: a reduction in the number of times the bit-selection mechanism must be reset. This can be accomplished by handling the developing partition stack as a FIFO list, not a LIFO list. Figure 10.3 shows the end of pass 2. Note that the method produces two partitions from every initial partition in the stack. Therefore, when all elements in an initial partition have the same value in the sorted bit, a null partition is created automatically.

		After processing of partitions 1–7		After processing of partitions 8–11		Partitions*
1.	3	0011		3	0011	1–3 (3)
2.	1	0001		1	0001	4–7 (3)
3.	2	0010		2	0010	8–11 (3)
4.	4	0100		4	0100	Null
5.	6	0110		6	0110	
6.	5	0101		5	0101	
7.	7	0111		7	0111	
8.	8	1000		8	1000	
9.	10	1010		10	1010	
10.	9	1001		9	1001	
11.	11	1011		11	1011	

*Number in parentheses indicates the bit position to be scanned next.

Fig. 10.3 Binary Radix exchange, pass 2.

		Initial			After pass 3	Partitions*
1.	3	0011	1		0001	
2.	1	0001	3		0011	1-1 (4)
3.	2	0010	2		0010	2-3 (4)
4.	4	0100	4		0100	4-5 (4)
5.	6	0110	5		0101	6-7 (4)
6.	5	0101	5		0110	8-9 (4)
7.	7	0111	7		0111	10-11 (4)
8.	8	1000	8		1000	
9.	10	1010	9		1001	
10.	9	1001	10		1010	
11.	11	1011	11		1011	

*Number in parentheses indicates bit position to be scanned next.

Fig. 10.4 Binary Radix exchange, pass 3.

A third pass will undertake to generate further partitions by scanning the next digit, digit 3, (Fig. 10.4). The scan of partition 1–3 produces partitions 1–1 and 2–3. The scan of partition 4–7 produces partitions 4–5 and 6–7. The scan of partition 8–11 produces partitions 8–9 and 10–11. The null partition of Fig. 10.3 is rejected.

The fourth pass is recognized as the last because the bit indicator has reached the low-order bit of the key. The stacking mechanism can be abandoned for the last pass. The final ordering of the list can be easily inferred from Fig. 10.4.

The amount of space required to hold partition definitions can grow quite large; in general, a maximum of $(2^K - 1)$ locations can be required. If partition definition space is not available in this quantity, it is possible to reduce the required amount by treating the partition stack as a LIFO stack and paying the price of resetting the bit selector. This technique reduces the storage space to roughly K locations.

10.2.2 Number of Tests and Data Movements

A pass constitutes the depletion of all partitions which are to be scanned on the same bit position. The number of tests is N for each pass, since there are N elements distributed across all the partitions.

The number of passes is the number of bit positions required to represent the highest key. In the example, since four bit positions are required to represent 11, there are four passes. This may be expressed as $(K + 1)$, or $\lceil \log_2 \text{MAX} \rceil + 1$, where MAX is the largest value in the list. The total number of tests is $N(K + 1)$, or $N(\lceil \log_2 \text{MAX} \rceil + 1)$.

The number of data movements depends upon permutation, because of the exchange feature of the method. The minimum number of exchanges is 0 and the maximum number is $N(K + 1)/2$. The expected number is half that, or $N(K + 1)/4$.

It is common in the literature to reduce the factor $(K + 1)$ to digits, D, or bits, b, and to express inspections as $N \cdot D$ and exchanges as $(N \cdot D)/4$.

10.2.3   Binary Radix Exchange and Quicksort

Binary Radix Exchange is a distributive version of Quicksort. The values of the bit positions being scanned are equivalent to Quicksort pivots. Is there, then, a reason for preferring Quicksort to Binary Radix Exchange, or vice versa?

The programmer must choose between $N \cdot D$ tests and roughly $1.4N \log_2 N$ comparisons. In the two methods, pass control and interpass control are so similar that they may be considered equal. Given 1024 elements to sort, Quicksort expects 14,336 comparisons ($1.4 \times 1024 \times 10$). What does Binary Radix Exchange expect? That depends on the size of the keys. If the highest key requires 36 bits (a UNIVAC 1108 word) for representation, then 31,744 bit inspections are required and Quicksort is preferred. If the highest key value requires only 13 bits, then the number of bit inspections is less than the expected number of comparisons with Quicksort.

When a key range is sufficiently small that the Binary Radix Exchange seems profitable, it is necessary to consider the relative cost of bit inspection versus comparisons. Certainly, some hand coding of the basic comparison loop and the bit-selection process for the particular machine is in order. Unless the machine can handle variable-length bit strings efficiently, the Radix Exchange may perform poorly.

The programmer may know enough about his data to wish to avoid the "pivots" used by Radix Exchange and to prefer the more sophisticated pivot selection available to Quicksort.

Binary Radix Exchange, like Quicksort, improves as a combined method. Exactly as with Quicksort, the use of sifting for very small partitions will speed the process [24].

10.3   DIGIT SORTING: BASIC METHOD

In its simplest form, digit sorting is a computer analog of mechanical punched card sorters. The output area is divided into A "bins" (also called "buckets" or "pockets"), where A represents the number of unique values which characters of a key may assume. If the characters of keys are known to be all numeric, base 10, then A equals 10. There must be a bin for every character; that is, A is the size of the alphabet used to form keys.

A pass of the method consists of dispersing elements into bins on the basis of the value of the character in the inspected character position, and then "collecting" elements for another pass. The dispersions are made, character by character, from low order to high so that, given D characters in a key, D dispersions and D collection passes are made, the final pass collecting elements in order.

List	Bins 0	1	2	3	4	5	6	7	8	9
314										
116	870	181	—	743	314	455	116	627	298	069
069	510			813						
743										
298										
455										
870										
181										
510										
627										
813										

Fig. 10.5 Digit sort, pass 1.

10.3.1 Example

There are 11 three-digit decimal numbers in Fig. 10.5; therefore $A = 10$, $D = 3$. Ten output bins, labeled 0–9, are set aside for dispersion. An initial pass on the units digit (rightmost digit) disperses the numbers as shown. At the end of the first pass, the input area is empty. A given bin will have from no numbers to N numbers in it, depending on the distribution of digits in the units position. Before the second pass, it is necessary either to collect the numbers from the bins and return them to the input area, or to have access to an alternative set of bins. If an alternative set is available, a dispersion can be undertaken immediately and a pass over the data is thus saved, but if not, then the numbers must be collected. The collection moves from the low-value bin to the high-value bin, placing entries from each bin in a continuous list ordered on the appropriate digit position. Figure 10.6 shows the list as collected after the first pass. When the collection is completed, the bins are empty and a new dispersion is undertaken.

The second dispersion is performed on the tens digit. Each number is dispersed to a bin in accordance with the value of its tens digit. The status after this dispersion is given in Fig. 10.6. The numbers are again collected

List	Bins									
	0	1	2	3	4	5	6	7	8	9
870	–	510	627	–	743	455	069	870	181	298
510		813								
181		314								
743		116								
813										
314										
455										
116										
627										
298										
069										

Fig. 10.6 Digit sort, collection and pass 2.

List	Bins									
	0	1	2	3	4	5	6	7	8	9
510	069	116	298	314	455	510	627	743	813	
813		181							870	
314										
116										
627										
743										
455										
069										
870										
181										
298										

Fig. 10.7 Digit sort, collection and pass 3.

and listed, as shown in Fig. 10.7. This list is in order by the units and tens position. The third and last dispersion is made on the hundreds digit. A collection from these bins will form the ordered list. The process has involved three dispersion passes and three collection passes.

10.3.2   Dispersion

The process of dispersion is the process of determining the proper location for an element and placing it in that location. The determination of precisely where an element should go involves an identification of the digit to determine its bin and a reference to an indicator showing the location of the next open position in the selected bin.

The mechanism for determining the identity of a digit can be a comparison or a calculation. A comparison mechanism compares the digit with the characters it might possibly be and, on the basis of a successful comparison, selects a proper bin. A calculation mechanism attempts to form a bin address by the direct manipulation of the digit. If a comparison mechanism is used, the number of comparisons becomes a consideration in the performance of the sort. The number of comparisons depends on A, the number of unique values which a digit position may take on, and the organization of the comparisons.

To avoid using comparisons, one can use some calculation technique to determine the identity of a character. Figure 10.8 shows basic address calculation. A Bin Location Table starting at location 1072 holds an address for each of the dispersion bins and a count of elements in the bin. Each digit is added to the table base to find the location holding the base address of the proper bin. A count of the number of elements already in the bin determines the position for the new element. The addend to bin base address would naturally be constructed so as to represent the real extent in address space. In Fig. 10.8, one-word records are assumed. If 10-word records were being dispersed, the count field might read 30 for convenience.

Adding digit values to table addresses or element counts to bin addresses need not depend on programmed additions when sorts run on machines that have index registers and/or indirect addressing.

The total number of digit inspections which must be made during a sort is a function of the number of digit positions required to represent the key. For D digits in a key, D dispersion passes are required. If a comparison technique is used, the number of comparisons per pass is $N \cdot D \cdot F$, where F is the number of comparisons expected in whatever comparison scheme is used to determine the proper bin. Alternatively, there would be $N \cdot D$ address calculations.

Bin locator table			Bin name
Location	Bin location	Bin count	
1072	10000	3	0
1073	12000	6	1
1074	14000	7	2
1075	16000	4	3
1076	18000	3	4
1077	20000	5	5
1078	22000	4	6
1079	24000	6	7
1080	26000	2	8
1081	28000	1	9

1. Add key digit to Locator Table Base;
   e.g., 1072 + 4 = 1076.
2. Contents of 1076 gives origin address of bin = 18000.
3. Count in 1076 shows first free position is 18003.
   Example assumes dispersion of one-word records.

Fig. 10.8 Calculating address of a bin and location within it for placement of an element.

10.3.3   Space

A punched-card sorter is designed for a small alphabet, $A = 10$ or 12, usually decimal digits. Alphabetic sorts require extra effort. Since a computer can have any number of bins, theoretically any alphabet can be handled. If the alphabet is large—for example, even the BCD alphanumeric character set can have as many as 64 characters—there may be a problem in allocating space for the bins. The punched-card sorter provides an essentially infinite amount of space, since it is possible to remove cards from pockets while the sort is in progress and to stack piles of indefinite size. A computer must impose constraints on the size of the dispersion area. If it is in core, then the constraint is the amount of available core. If the area is on disk or drum, the availability of space on these devices and its accessibility (in terms of speed) become considerations. Refinements to digit methods are basically attempts to reduce the amount of space required.

The most efficient dispersion algorithm requires space for the representation of $2(A \cdot N)$ elements. A digit position may assume A values. All N elements may have any of A values in a given position. Therefore, it is necessary to have bins of size N for each of the A values that a digit position may assume, in order to guarantee that no bin will overflow. But the provision of $A \cdot N$ spaces still requires a collection pass in order to clear the bins after each dispersion. This collection process contributes nothing to the ordering of the data. In order to avoid it, an alternative dispersion area, also of size $A \cdot N$ must be provided.

Given less core space than the nominal requirement, the programmer faces decisions as to how to allocate his space. If he has some knowledge of the probable distribution of his digit values, he may try to allot areas of various sizes to various values. For example, in a sort of proper names, he would set aside less space for X's, Z's, and Q's. The literature of data retrieval contains some material on the statistics of the occurrences of letters in various populations which might be useful for such designs. With this approach, it is necessary to provide some technique for bin overflow.

One obvious means of reducing required space is to disperse tags (or addresses) only. This not only reduces storage space to that required by the tags (or addresses), but saves time for the data movement involved in the dispersion. More sophisticated procedures are required when the amount of space available is nowhere near what is required. One of these procedures is counting, which we will describe in the next section.

10.3.4   Counting and Address Calculations

Counting is an attempt to reduce space requirements by determining the range and frequency of values in each digit position. The effect is to reduce additional space to that required to hold the entire list (or tags of the list) plus a counter location for each value. MATHSORT, an algorithm which takes this approach, is included in the appendix.

We use an output area of size N and a count area containing a counter for each symbol that may appear in a key. A pass over the input list inspects a digit position of each element key and counts the occurrences of each value. Figure 10.9 shows the status of counters after an initial counting pass. Indexing into the counter table is accomplished by adding the value of the digit position to the base address of the table.

List	Value	Count
314	0	2
116	1	1
069	2	1
743	3	2
298	4	2
455	5	2
870	6	2
181	7	1
510	8	1
627	9	2
813		
234		
169		
356		
492		
345		

Fig. 10.9 Status of counters, low-order digit.

Digit	Count	Cumulative count
0	2	16
1	1	14
2	1	13
3	2	12
4	2	10
5	2	8
6	2	6
7	1	4
8	1	3
9	2	2

Fig. 10.10 Cumulative count.

After counting, a cumulative count is made from the counter table (Fig. 10.10). There are sixteen digits equal to or greater than 0, fourteen digits equal to or greater than 1, ten digits equal to or greater than 4. The dispersion is controlled by values in the cumulative table. For instance, digit 4

TOSORT	CUM count		Sorted
314	0	16	1.
116	1	14	2.
069	2	13	3.
743	3	12	4.
298	4	10	5.
455	5	8	6.
870	6	6	7. 314
181	7	4	8.
510	8	3	9.
627	9	2	10.
813			11.
234			12.
169	$N + 1 - CUM(4) =$		13.
356	$17 - 10 = 7$		14.
492			15.
345			16.

Fig. 10.11 Dispersion using cumulative count.

is less than or equal to 10 numbers and greater than 6. In an output list with digit 0 at the top, a number ending in a 4 will be in location 7 or after, since there will be 6 numbers before it in the direction of the origin. Note that the position of the first entry ending in a 4 is $BASE + N + 1 - CUM(4)$, that is, the base address of the list plus the list length plus 1 minus the cumulative count for the digit 4. In Fig. 10.11, then, 314 ends up in position 7 (BASE is zero). This is the relative address on the output list of the first permissible location for digit 4. when the key is placed, the cumulative value is decremented. The next occurrence of digit 4 will develop $17 - 9 = 8$ as a relative address. Figure 10.12 shows the decremented value of $CUM(4)$ and the placement of the next digit (6 of 116). A pass is over when the entire list has been placed. At the completion of a pass the list is in order by the number of positions so far inspected. The dispersions are made from alternating areas, and therefore a collection pass is not necessary.

10.3.5 Bin Grouping

If an allocation of a bin for each value results in too many bins or in bins that are unacceptably small, the size can be changed by assigning multiple values to a bin. In a tape environment, the number of tapes required for the dispersal may be reduced by writing multiple values on the same tape. The penalty for grouping is additional passes. Each grouped bin must eventually be dispersed into individual bins. The breakdown of a grouped bin could be accomplished during a collection pass. However, there seems little to recommend this technique over the counting technique, since it does not provide as great a reduction in space or enough protection from bin overflow.

TOSORT	CUM count			Sorted	
116	0	16		1.	
069	1	14		2.	
743	2	13		3.	
298	3	12		4.	
455	4	9*		5.	
870	5	8		6.	
181	6	6	$(17 - 6 = 11)$	7.	314
510	7	4		8.	
627	8	3		9.	
813	9	2		10.	
234				11.	116
169				12.	
356				13.	
492				14.	
345				15.	
				16.	

*Decremented formula for next digit 4 is $17 - 9 = 8$.

Fig. 10.12 Adjusted cumulative count.

10.3.6 Transformations

Machines have a variety of number bases, a variety of codes for the representation of data, and a variety of ordering sequences based on these codes. If a digit sort is to be applied, it is necessary that the key be properly represented. For example, in Binary Radix sorting, some care must be taken that the pure binary bit stream represents the collating sequence of the machine and that the negative value conventions of the machine are such that a continuum from lowest to highest number is formed.

The programmer must determine whether conversion to or from various encodings is necessary or desirable. Key transformation is often used to change the numbering base of keys, in order to reduce the number of bins required by representing the digits in a system of lower base with consequently fewer unique digits. The cost of this is additional passes over an elongated key. Alternatively, the base may be raised in order to reduce the number of passes, at the cost of more bins.

10.3.7 Amphisbaenic Sorting

The nature of access time to external devices is such that the collection passes which may be tolerated internally are unacceptable when bins are on I/O devices. To avoid collection passes, one must have an alternative bin area. In a tape system, this means that $2A$ tape units are required for the sort of keys of radix A. Thus, for decimal keys, 20 tape units, or 20 areas on disk, are required. As a means of reducing the number of tapes or areas, two techniques may be employed. One, mentioned above, is transformation of the keys. The other is development of a method of dispersion and collection in which fewer than $2A$ units are required. The Amphisbaenic sort does both [26]. It differs little from an external radix sort with key transformation, but it eliminates the collection passes and uses only $(A + 1)$ tapes. The base of A may be changed to some value R to accommodate the actual number of tapes available. The method when tape-oriented depends on the ability to read tape backwards, a characteristic from which the method takes its name (two-headed).

The determination of the radix to be used for the dispersion is based on the number of available tape units (or bin areas when used internally). At any time there will be R $(R \leqslant A)$ units receiving output from a single input unit. If there are six tapes available, for example, then dispersion will be based on radix 5. Input values are translated to the dispersion base; i.e., decimal input becomes quinary input. Positional notation gives each digit d the value $(d \times R_p)$, where p is the position of the digit; counting from the rightmost digit, $p = 0$, and R_p is the multiplier of the digit in the pth position. Thus, 314 decimal is

$$3 \times 10^2 + 1 \times 10^1 + 4 \times 10^0.$$

The equivalent quinary number is

$$2 \times 5^3 + 2 \times 5^2 + 2 \times 5^1 + 4 \times 5^0, \qquad \text{or 2224.}$$

Initially, data is read from the input tape and dispersed to tapes by most significant digits. Figure 10.13 shows a list of decimal values with equivalents in base 5 representation. Figure 10.14 shows an initial dispersion of that list (and some other numbers). The arrows show the position of the read/write head of the tape drives after the dispersion.

After the initial dispersion, a series of further "splits" occurs. The group being split is always the one with the highest key values. The next dispersion will split T_4 (holding 4xxx) into smaller groups. Tape T_4 will become the

	Decimal				Quinary		
Position values:	100	10	1		125	25	5	1
	3	1	4		2	2	2	4
	0	6	9		0	2	3	4
	2	9	8		2	1	4	3
	4	5	5		3	3	1	0
	1	8	1		1	2	1	1
	5	1	0		4	0	2	0
	2	3	4		1	4	1	4
	1	6	9		1	1	3	4
	3	5	6		2	4	1	1
	4	9	2		3	4	3	2
	3	4	5		2	3	4	0

Fig. 10.13 Decimal numbers with quinary equivalents.

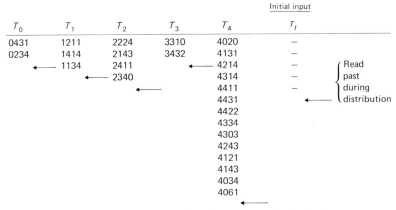

Fig. 10.14 Amphisbaenic sort, initial pass.

a) Dispersing second digit

T_0	T_1	T_2	T_3	T_4	T_I
0431	1211	2224	3310	—	—
0234	1414	2143	3432		—
4061	1134	2411	4303		4422
4034	4143	2340	4334		4431
4020	4121	4243	4314		4411
	4131	4214			

b) Dispersing third digit of 44xx

T_0	T_1	T_2	T_3	T_4	T_I
0431	1211	2224	3310	—	—
0234	1414	2143	3432		
4061	1134	2411	4303		
4034	4143	2340	4334		
4020	4121	4243	4314		
	4131	4214	4431		
	4411	4422			

Note: Arrows indicate direction. Underscore indicates inspected digit.

Fig. 10.15 Second pass of amphisbaenic sort. Arrows indicate tape direction.

input tape, and T_I (initial input tape) becomes a dispersion tape, receiving elements with T_0 through T_3. Reads and writes on T_4 and T_I are now in a backward direction. The dispersion of second digit develops groups of 40XX, 41XX, 42XX, 43XX, 44XX on tapes T_0, T_1, T_2, T_3, T_I. These groups are written behind the dispersed elements already on tape if there are such elements. The input tape, T_4, will be at load point when the dispersion is finished. Figure 10.15(a) shows this dispersion, with the underline indicating the digit position which caused the appearance of the element on its current tape.

The next dispersion will split 44XXs into 440X, 441X, 442X, 443X, 444X. This is shown in Fig. 10.15b. The following dispersion develops 4440, 4441, 4442, 4443, 4444. There is no further splitting that can occur with the 444X group. All members of the group are collected on T_I, written backward as a descending string of 4444, 4443, 4442, 4441, 4440. The splitting process begins again with the 443X group, which is split into 4434, 4433, 4432, 4431, 4430, and collected on T_I behind the 444X group. When this is accomplished, the 442X group is processed. The pattern of splitting a group from the left until the entire key is processed, before collecting and proceeding to the next largest group, continues until a full string is collected on T_I.

It is possible to abort the process by internal sorting when a group is small enough to fit into store. For example, the entire 44XX group might be ordered without further splitting, if it fits, after its dispersion to T_I.

The method, although described as a tape method, is even more applicable to disk because tapes that write backward are rare. The tape reversals are equivalent to scanning a list forward or backward. Information just written in a sublist is always read in the opposite direction. Writing continues in the same direction as preceding writes. The method may also be used internally, of course.

In tape or disc versions, the number of times a key is dispersed depends on the amount of internal sorting that is done. If there is no internal sorting, the key will be dispersed $(D + 1)$ times, where D is the number of digits required to represent keys in the radix R. If there is internal sorting, its effectiveness depends on how small a group must be if it is to be subject to internal sort. The internal sort reduces the number of passes over a key by eliminating dispersions on later digits.

10.4 RANGE SORTING

Range sorting is analogous to digit sorting, as Quicksort is analogous to Binary Radix. Dispersions are based on the value of an entire key within a specified range. The method is suitable for use as an internal or external technique with disk or tapes. It is particularly interesting in the external environment as a possible alternative to an external merge. The technique is to split the input data into a number of bins, each of which has a unique key range. The initial groups are then split into smaller and smaller groups until the list is ordered, or a comparative sort is performed to order the final small groups.

10.4.1   Determining the Bins

One would like as many bins (ranges) as possible, since the number of passes depends on the number of bins. A decision about "order of dispersion" (number of bins) depends on the number of drives or areas available. If there are six tapes, one will be an input tape and the dispersion will be five-way.

It is necessary to determine what definition of ranges will ensure an even dispersion; on this the efficiency of the method depends. If ranges were defined so that three of five bins never received elements, the effective dispersion would be two-way. To define ranges properly requires a knowledge of the range of values and the distribution of values of the list to be sorted. It is rare to have such detailed information about data. Given 100 elements with a possible key range of 1 to 1000, and five output bins, the ranges 0–200, 201–400, 401–600, 601–800, 801–1000 may or may not be reasonable and effective. It may well be that a particular list has 63 keys in the range 0–200, 250 in the range 201–400, 17 in the range 401–600, etc. In the absence of adequate information, a mechanism for acquiring knowledge throughout the sort process must be provided; otherwise, a distribution-sensitive technique should be avoided. One method of acquiring distribution data is to scan the input file on an initial pass, to create a range table. Counts are made for discrete ranges representing considerably more bins than are planned for the dispersion. At the end of the pass, ranges are combined to form new ranges whose limits span *approximately equal* numbers of elements.

It is possible to accept a bad dispersion on the first pass and then refine subsequent passes. Quite arbitrary ranges can be used for an initial dispersion and the range definition can be refined on the basis of collected data.

10.4.2   Example: Dispersion on Tape

Figure 10.16 shows three-digit numbers and their distribution in intervals of 200. An initial dispersal using these arbitrary ranges is shown in Fig. 10.17.

The next dispersion will be a partial pass, which splits the elements of one range into smaller ranges. Range limits for this dispersion may be selected on the basis of collected range tables or arbitrarily calculated as one-fifth the range of the first dispersion. If the latter, the new ranges for values on tape 5 are 801–840, 841–880, 881–920, 921–960, 961–1000. Tape 5 is read backwards, and the other tapes receive new elements behind the dispersed elements of pass 1. At the depletion of T_5, elements are distributed as in Fig. 10.18.

List					Distribution	
184	352	376	505	604	0–200	21
388	369	742	482	256	201–400	29
672	523	351	440	973	401–600	20
534	666	279	298	278	601–800	18
154	574	148	830	795	801–1000	12
116	104	469	293	267		
312	659	327	970	441		
265	201	163	507	243		
492	069	947	372	751		
570	194	032	759	889		
212	137	554	240	798		
198	393	199	570	078		
222	858	335	757	354		
842	637	212	412	619		
417	698	258	957	685		
921	127	637	177	537		
551	718	497	230	488		
328	397	248	783	024		
197	941	181	626	152		
878	065	868	707	003		

Fig. 10.16 A list of data.

The next dispersion will be based on intervals of one-fifth of 40. A dispersion of the highest elements (961–1000), those on T_4, is made with these limits. With internal sort, this dispersion may be avoided, and the string (973, 970) may be written directly on T_5. After the internal sort or dispersion, a final descending-ordered list begins on T_5. All groups are thus reduced and written onto T_5, as shown in Fig. 10.19. At the end of the process there is a descending string on T_5. The ascending string would have been built if secondary splitting went from low to high rather than from high to low. A low-to-high or high-to-low pattern is necessary because the final output tape doubles as a distribution tape. When $(R + 2)$ tapes are used, or when dispersion is effected in a disk environment, the pattern is quite arbitrary and can be optimized at will to suit machine features.

10.4.3   Passes and Comparisons

The number of passes depends on dispersal order, range effectiveness, and the size of groups that can be ordered by a final comparative sort. The number of passes is determined by successive division of N by the dispersal order. This division goes to 1 or to the size that can be handled by the comparative sort. If that sort will handle groups of 100 or less, and 50,000 items are to be sorted with a ten-way split, the first dispersion pass will produce 5000-element groups, the second pass 500-element groups, and the third pass 50-element groups. Three dispersion passes are required.

The number of comparisons in a dispersion pass depends on the method chosen for finding appropriate range. Since the method deals with entire

T_I	T_1 0–200	T_2 201–400	T_3 401–600	T_4 601–800	T_5 801–1000
←—	184	388	534	672	842
	154	312	492	666	921
	116	265	570	659	878
	198	212	417	637	858
	197	222	551	718	941
	104	328	523	742	947
	069	352	574	637	868
	194	369	598	759	830
	137	201	469	757	970
	127	393	554	783	957
	065	397	497	626	973
	148	376	505	707	889
	163	351	482	604	←—
	032	279	440	795	
	199	327	507	751	
	181	335	570	798	
	177	212	412	619	
	078	258	441	685	
	024	248	537	←—	
	152	298	488		
	003 ←—	293	←—		
		372			
		240			
		230			
		256			
		278			
		267			
		243			
		354			
		←—			

Note: Arrows indicate list pointer. On tape r/w head.

Fig. 10.17 Range sort, first dispersal.

	T_I	T_1	T_2	T_3	T_4	Input T_5
From pass 1		001 to 200	201 to 400	401 to 600	601 to 800 →	Reading 801 ↑ to 1000
	830 ↓ ←—	868	889 ↓ ←—	957	973 ↓	
		858		947	970 ↓ ←—	
		878 │		941 │		
		842 ↓		921 ↓		
		←—		←—		

Tape	Range
T_I	801–840
T_1	841–880
T_2	881–920
T_3	921–960
T_4	961–1000

Note: Vertical arrow indicates direction. Horizontal arrow indicates position.

Fig. 10.18 Redistribution of tape 5.

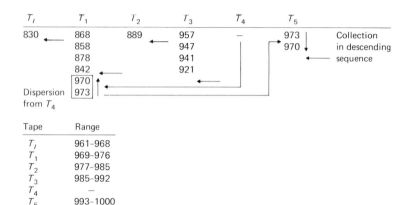

Tape	Range
T_I	961–968
T_1	969–976
T_2	977–985
T_3	985–992
T_4	–
T_5	993–1000

Only 801–1000 are shown; lower values lie "behind" toward the tape load points.

Fig. 10.19 Dispersion of 961–1000 and collection in descending string in T_5. Only 801–1000 are shown; lower values lie "behind" toward the tape load points.

keys, comparative methods will be more commonly used. The number of comparisons depends on the exact nature of the algorithm. For dispersions of low order, a simple $(N-1)$ technique will be commonly used. For dispersions of very high order, some tree-structure $\log_2 N$ technique may be effective.

10.5   ADDRESS CALCULATION

In Chapter 1 and again in this chapter, we have discussed the use of a mechanism to calculate a value which can be used as an address (or part of an address) on an output list. When there is a one-to-one mapping between a key value (or the value of a digit of a key) and part of the address space of a machine, then address calculation is trivial; each element is moved only once (from input to output), and only enough space is required to hold the final output list.

When address mapping is less direct, elements must often be moved more than once, and it is therefore necessary to provide more working space for the method. The nature of the relation between key and address values determines the complexity of the calculation used to develop an address and the probability of developing a directly usable address. The profitability of address calculations as a total method for sorting a list of elements depends on the range, representation, and distribution of key values and on the methods that are available for transforming those values into addresses [2,19].

10.5.1   Address Calculation Function

The heart of the address calculation is the function which transforms keys into addresses. Some methods treat the entire key as a unit and subject it to a series of manipulations. Division by a prime number and use of the remainder as the address is one such method. Multiplication (squaring) and truncation is another. Some methods develop each address digit by digit, producing a sum that represents an address of the machine. Conversion of the key into a radix system, in which it will be an address of the machine, is still another technique.

The programmer must be cautious in selecting a "hashing" scheme, since some do not preserve relative order between the keys they process. The major advantages of hashing are uniqueness of address and an even distribution through available addresses. Since retrieval of items may be accomplished by an identical hash, strict order is often not a requirement. In this regard, a programmer may well consider whether he really needs to sort his items, or merely to develop a way of retrieving them.

If the probability of developing unique addresses by means of hashing is low, range tables are used to support the sort function. The use of range tables concedes the need for some kind of search to find a proper location for an element. The extent of the search is partially dependent on the extent of the tables. In Fig. 10.20, there are 10,000 addresses allocated to 10 groups of keys. A key in each range would need to search within a group to find its proper position. Very often a hashing method will attempt only to produce a group address, from which a search is initiated.

When the sort function is very good, the efficiency of an address-calculation sort is equivalent to a one-pass distribution. As the function becomes less and less exact, however, the method approaches insertion. Address calculation attempts to pick a good starting point for a location search on the basis not of relative magnitude but of a location mapping.

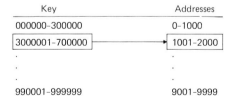

Key	Addresses
000000-300000	0-1000
3000001-700000	1001-2000
.	.
.	.
.	.
990001-999999	9001-9999

Key to be placed, 376172, will be somewhere between 1001 and 2000.

Fig. 10.20 Key to address conversion.

List	100	10	1	Value	Address*	Previously generated
MZJ	13	26	10	1570	8	no
VQT	22	17	20	2390	12	no
LZV	12	26	22	1482	7	no
CXR	3	24	18	558	3	no
QTP	17	20	16	1916	9	no
BAB	2	1	2	212	1	no
CMB	3	13	2	432	2	no
WTM	23	20	13	2513	12	yes
LVT	12	22	20	1440	7	yes
VWR	22	23	18	2438	12	yes
XLR	24	12	26	2546	12	yes
RVY	18	22	25	2045	10	no
TRV	20	18	22	2202	11	no
QMY	17	13	25	1855	9	yes
EXL	5	24	12	752	4	no
GQM	7	17	12	883	4	yes
HVO	8	22	15	1035	5	no
BAR	2	1	18	228	1	yes

*Address is calculated by rounding value to nearest 100, then dividing by 200 to obtain an integer quotient.

Fig. 10.21 Alphabetic key conversion.

10.5.2 Search

Figure 10.21 provides a list of 18 alphabetic keys. Each letter is transformed into a number equal to its position in the alphabet. The units position value is multiplied by 1, the tens position by 10, hundreds position by 100, etc. The values for each letter are summed. This total is rounded to a two-digit number (for example, 1570 becomes 16); the integer half of the result is taken as an address. The poor quality of the address calculation can be seen by the fact that, from a sample of 18 keys, it develops only 11 unique addresses. The performance of a hashing technique can be improved by providing additional space for the output list.

As each address is developed, an attempt is made to place the item. Keys MZJ through CMB may be directly placed, since each has an address not previously generated. (The procedure tests to see whether the location of the address generated is empty.) Figure 10.22, which shows the output list after CMB has been placed, has the property of being in order, though sparsely occupied. A list growing from an address calculation sort will always be in order as it grows. The generation of the address of WTM causes a problem, since location 12 is occupied. The "collision" forces a search for a location for WTM. The values of WTM and VQT (in location 12) are compared. Since WTM is larger than VQT, a search will be made for free space after location 12. An empty location is found immediately at 13. The same process occurs with LVT. Location 7 is filled; LVT is smaller than LZV and finds a place immediately before it at location 6.

1.	BAB
2.	CMB
3.	CXR
4.	—
5.	—
6.	—
7.	LZV
8.	MZJ
9.	QTP
10.	—
11.	—
12.	VQT
13.	—
14.	—
15.	—
16.	—
17.	—
18.	—

Fig. 10.22 List of Fig. 10.21 after placement of CMB.

	Before VWR	After VWR
1.	BAB	BAB
2.	CMB	CMB
3.	CXR	CXR
4.	—	—
5.	—	—
6.	LVT	LVT
7.	LZV	LZV
8.	MZJ	MZJ
9.	QTP	QTP
10.	—	—
11.	—	—
12.	VQT	VQT
13.	WTM	VWR
14.	—	WTM
15.	—	—
16.	—	—
17.	—	—
18.	—	—

Fig. 10.23 Placement of VWR.

Key VWR finds VQT at 12. It attempts to find a place after 12. However, 13 has not only an occupant but one that is higher in value than VWR. This key, WTM, must be moved so that VWR can be placed in 13. Thus, after an element has been placed, it may need to be moved to make room for another. The placement of WTM involves a search for a free address after that from which it must be moved. When an appropriate address is not free, its occupant and all elements between it and the next free position must be moved. Figure 10.23 shows the list before and after the placement of VWR.

The performance of address calculation is largely dependent on the frequency of this need to move data, which is controllable in some measure by the programmer [6,22]. For example, he can substantially improve performance by increasing allocations of space. Large areas of space increase the probability that a generated address will be unique and that if a collision does occur, there will be free space close to the collision location. Consider our example with 36 locations of space. The address can be the truncated values of VALUE, unhalved. There will be only two duplicated addresses. There are fewer collisions, and the need to move elements when collision does occur is reduced. It may be necessary to have a packing algorithm of some kind after the distribution is complete, depending on the requirements of processing after the sort. It is also possible to defer placing "collision" values until a subsequent merge [20].

The application of so dramatically tempting a method is worth studying if the relation between keys and addresses is almost obvious or easily derived, or if much time and care can be given to developing the sort function, testing its behavior, and undertaking a reasonable study of its properties. However, the casual user of sorting techniques will probably limit its use to that of an auxiliary technique (as in Radix sorting).

11
Comparison of Internal Sorts

11.1 INTRODUCTION

Figure 11.1 presents an index to discussions of the performance characteristics of the methods discussed in Chapters 3 through 10 of this text. In Hall [12] a table provides a guide for the comparison of sorting techniques. When an approximation of the running time of a sort is needed, the table can be helpful, but one should bear in mind the following limitations.

1. The expected number of comparisons or data movements may be off the mark by a considerable margin. Even slight bias in distribution and permutation can lead to important differences in performance.

Method	Chapter	Discussion of performance
Shellsort	3	3.4, 3.5, 11
Treesorts		
CACM 143	5	5.2.3, 5.2.4
CACM 245	5	5.3.3, 5.3.4
Centered Insertion	6	6.4
Binary Insertion	6	6.4
Quicksort	7	7.8
Quadratic Selection	8	8.2.5
Merging		
Two-way Straight	9	9.3.3
Two-way Natural	9	9.4.2
Natural	9	9.4.5
Binary Radix	10	10.2.2
Distribution	10	10.3.2, 10.3.3
Amphisbaenic	10	10.3.7
Address Calculation	10	10.5.2

Fig. 11.1 Index to performance tests.

2. The space estimates are not refined for the details of program environment or implementation. The use of detached or nondetached key sorts where they are optional, the nature of the use of data immediately following the sort, and the architecture of the CPU will all affect the amount of space required.

3. Various programming styles and data structures will affect the running time of the sort method. The details of pass control, loop control, subroutine linkage, key manipulation, and data-movement subroutines will determine how much actual machine time is spent in comparisons or movements.

The sort developer must avoid the temptation to accept any method discussed in this text as *the* method of sorting. It is always wise to investigate combined methods. Beyond the known combinations, such as Quicksort and sifting, there are many possibilities that may be useful in given situations. Any method that continually increases or reduces the size of partitions should be inspected for a proper "breakpoint" shift to another method with complementary characteristics. A sort that is very fast for small lists with some latent order can be used *in combination* with an efficient method for random data with large N. Beyond searching for combinations, the analyst or programmer should never completely overlook the possibility of *privately discovering* a new and better sorting method.

11.2 A SIMPLE STUDY

In order to provide some feeling for the performance of various sorts over different data, some of the sorts supplied in the appendix have been run with data lists of various sizes and diverse characteristics. The reader may be interested in the design and conduct of the effort as a miniature model of what might be undertaken to select a sorting method for various data.

11.2.1 Selection of Sorts

PL/I procedures have been developed for the following algorithms.

1. Stringsort (a natural two-way merge)
2. Quickersort
3. Singleton's sort
4. Floyd's treesort
5. Hibbard's Shellsort
6. Boothroyd's Shellsort
7. Shuttlesort

Each of these algorithms has appeared in the literature, and all but Hibbard's Shellsort have been printed in the *Communications of the Association for Computing Machinery (CACM).* The PL/I procedures, which appear in the appendix, follow published ALGOL versions.

These sort programs contain some obvious inefficiencies, which result from their being primarily designed for demonstration and exposition. Formal procedure calls, for example, are rather expensive in ALGOL and PL/I, and their use would not be normally encountered in an encoding of these sorts. This fact does not interfere with the experiments, however, since we are investigating the procedures representing the methods and not the methods themselves. This distinction is important when one is evaluating methods which have roughly the same characteristics as regards comparisons and data movements. An investigation of the methods themselves would require programming so that a uniform style would be used in all programs regarding subroutine usage, etc.

Of the seven selected sorts, there are only five distinct methods. Stringsort is a natural two-way merge defining initial strings from the opposite ends of a list. It is described in Chapter 9. Quickersort is a version of Quicksort, published as Algorithm 271, *CACM*. It uses the middle value of a partition as the pivot. Singleton's sort, published as Algorithm 347, *CACM,* is a composite sort which uses sifting for short partitions and Quicksort for longer partitions. The breakpoint used here is one suggested by the Singleton: ten elements. Experimentation with the breakpoint might be worthwhile. An important feature of the method is its initial sampling of a median to use as a partition pivot.

In the same spirit of observing variations of the same method, two versions of Shellsort are presented. They are much closer to each other than are the two versions of Quicksort, since they are identical in the handling of the critical parameter—the distance between elements during each pass. They differ only in the organization of list and pass control.

Floyd's treesort is not intended to represent an optimal form of Tournament sorting, since a tournament that does not use auxiliary address storage will not be so efficient as one that does; but it is an interesting minimal-storage algorithm. Shuttlesort is included in the list as the best of the "basic" methods, a version of sifting. The inclusion of this method will provide a standard by which to judge the significance of the improvements of the "new" techniques (especially Shellsort).

11.2.2   Data

The sorts were run against data lists of three sizes—100 elements, 1000 elements, and 5000 elements. In each case the unit of data was a "one-word" 32-bit record, in which the record and the key were identical.

For each sort, for each list, runs were made against data characterized as follows:

1. *Random:*   The list was generated by an encoding of RANDU, a standard IBM 360 random-number generator provided in the Scientific Subroutine package.
2. *Duplicate:*   A strong tendency for duplication existed in the data.
3. *Ordered:*   The list was already sorted.
4. *Interleaved:*   Even elements formed an ascending list; odd elements formed a descending list.
5. *Sorted in half:*   The first $N/2$ elements were sorted, and the last $N/2$ elements were sorted.

The duplicate test determines sensitivity to duplication, order, and run length. The interleaved and half tests determine sensitivity to run length and order. These are just two of the ways that nonrandomness can be expressed in a file. They are included here as examples of the kind of sensitivity for which one might search in testing a sort algorithm.

11.2.3   Tests

An IBM System 360/50 was used for the comparative runs of the selected algorithms. The system was run under OS/360 Release 21, using PL/I version 5–1. The runs were all made on a dedicated machine (i.e., with no multiprogramming), so that all differences between CPU time and elapsed time could be accounted for.

Each of the seven selected sorts was associated with a driver which performed the following functions in sequence:

1. Called a random-generator to generate 100 numbers.
2. Called the sort to order the generated numbers.
3. Called the sort to order the ordered list created by step 2.
4. Developed the "halved" data by rotating the position of the lower and upper halves of the ordered list. Called the sort to order the list.
5. Developed the interleaved data by inverting odd numbers. Called the sort to order the list.
6. Called the random-number generator to develop a list with significant duplication. The duplication was achieved by allowing the range of key values to be only one-half the values to be generated. Called the sort to order the list.

7. Repeated steps 1–6 for 1000 numbers.

8. Repeated steps 1–6 for 5000 numbers.

By means of the built-in TIME function of PL/I, the times of entry to and exit from the sort were recorded for each call. Elapsed time and CPU time differed only by the time required to handle time-slice interrupts occurring every 200 milliseconds. Since this is a *micro*second time which occurs five times a second, its effect is both minimal and easy to filter out of the results.

11.2.4   Results

The table in Fig. 11.2 shows the results of runs of the named algorithms. The reader is again cautioned that this table reflects only the relative performance of the coded algorithms with the data described. Some general conclusions about the inherent efficiency of the methods for sorting tags

	N	Random	Ordered	Half	Interleaved	Duplicated
Singleton[1]	100	1.0	0.7	0.9	0.7	1.0
(CACM 347)	1000	14.0	7.3	8.8	8.5	14.0
	5000	85.7	43.2	52.5	49.2	84.4
Quickersort[1]	100	1.3	0.8	1.3	0.8	1.3
(CACM 271)	1000	18.8	13.0	16.2	13.3	20.5
	5000	121.1	73.2	91.5	75.4	120.3
Treesort 3[2]	100	2.6	2.7	2.6	2.5	2.5
(CACM 245)	1000	34.3	37.9	36.2	36.5	36.3
	5000	220.8	230.8	222.5	221.9	220.7
Boothroyd/Shell[3]	100	2.0	1.0	1.7	1.5	1.8
(CACM 201)	1000	35.9	6.5	24.0	24.8	35.7
	5000	270.3	105.8	159.0	152.5	266.7
Hibbard/Shell[3]	100	2.2	1.3	2.6	2.5	2.5
(From CACM 6,	1000	42.3	20.8	36.2	36.5	36.3
May 1963)	5000	509.0	131.8	222.5	221.9	220.7
Stringsort[4]	100	4.7	2.7	2.8	4.7	4.7
(CACM 207)	1000	58.7	26.0	29.7	58.0	58.7
	5000	354.6	129.9	148.3	347.8	354.5
Shuttlesort[5]	100	7.0	0.3	7.3	7.3	6.7
(CACM 175)	1000	721.8	1.8	726.1	725.5	713.1
	5000	18,375.5	9.0	18,120.6	*	*

[1] Quicksort variant
[2] Tournament
[3] Shellsort variant
[4] Merge
[5] Sift

*Not collected

Fig. 11.2 Relative sort speeds.

may be indicated, but they must be qualified by consideration of the details of the program themselves. The use of a procedure call, as in Treesort 3, certainly extends the running time of a program and is only one example of a coding- and compiler-dependent detail.

Each entry in the table is normalized so that the basic time unit, 1.0, is the time taken by Singleton's version of Quicksort to order 100 random numbers. All other times shown are multiples of this value. For example, Singleton's algorithm requires 14 times as long to order 1000 random numbers as it does to order 100. Shuttlesort requires 721.8 times as long to order 1000 random numbers as Singleton requires to order 100.

Reading "across" a method for any value of N provides a measure of the relative efficiency of the method for various data conditions with a given number of elements. For example, for 100 elements Singleton seems to be insensitive to duplication, but it does respond to both the fully ordered and the interleaved list.

Reading "down" gives a measure of (1) the relative efficiency of each method for various sizes of N and (2) the relative efficiency of the methods. For example, it takes Singleton 85.7 times as long to sort a list of 5000 random elements as to sort a list of 100 random elements. Shuttlesort takes more than 1600 times as long (11837.5/7) to sort a list of 5000 random elements as to sort a list of 100 random elements.

Rank	Method	List size	Time per element
1.	Singleton	100	1.00
2.	Quickersort	100	1.30
3.	Singleton	1000	1.40
4.	Singleton	5000	1.71
5.	Quickersort	1000	1.88
6.	Boothroyd/Shell	100	2.00
7.	Hibbard/Shell	100	2.20
8.	Quickersort	5000	2.42
9.	Treesort 3	100	2.60
10.	Treesort 3	1000	3.43
11.	Boothroyd/Shell	1000	3.59
12.	Hibbard/Shell	1000	4.23
13.	Treesort 3	5000	4.41
14.	Stringsort	100	4.70
15.	Boothroyd/Shell	5000	5.40
16.	Stringsort	1000	5.87
17.	Shuttlesort	100	7.00
18.	Stringsort	5000	7.09
19.	Hibbard/Shell	5000	10.18
20.	Shuttlesort	1000	72.18
21.	Shuttlesort	5000	367.51

Fig. 11.3 Methods ranked by efficiency in random list sorting.

Figure 11.3 is a ranked list of methods in order of "efficiency," which is determined by dividing the normalized times for random-list sorting by N and multiplying by 100. The result gives an index of the normalized time per element required by each method. The list shows that every method is more efficient for smaller N ($N = 100$ and random) but that some methods are more efficient for $N = 1000$ than others are for $N = 100$.

Among the inferences which can be drawn from Figs. 11.2 and 11.3 are the following.

1. Never use simple sifting for very extensive random lists. Use that method only if there is reason to suspect nearly perfect initial ordering.

2. Methods may be coded to be insensitive to duplication. Only Hibbard's encoding of Shellsort shows a meaningful response to significant duplication, taking advantage of its occurrence with large N.

3. Different encodings of the same method will behave differently. Note the difference between Boothroyd's and Hibbard's Shellsort for 5000 random numbers.

4. Methods show different degrees of sensitivity to different biases in data. For example, Treesort 3 and Shuttlesort do not respond to the half list, but other methods do. On the other hand, Stringsort does not respond to the interleaved list.

The poor performances of Treesort 3 and Stringsort come as a surprise, as does the difference in performance between the Hibbard and Boothroyd (*CACM* 201) implementations of Shellsort. A possible explanation for the Stringsort performance is its selection of strings from the beginning and end of the list. None of the biases is good for this technique, since none develops long natural strings from the bottom.

The weakness in the performance of Treesort 3 may be due to the procedure call. The differences between the two forms of Shellsort may lie in the different expressions used to calculate distance, or in different sensitivity to order and duplication. Analysis of the algorithms themselves is not the intent of this chapter. The technique of the experiment, the results it generated, and the questions it raised should underscore the cautions (about comparing sorts) that were raised in Chapter 1 and at the beginning of this chapter.

Part 2
External
Sorting

12
The Sort Phase
of an
External
Sort

12.1 EXTERNAL SORTING

When the list of data to be sorted is too large to fit into primary memory, the concerns of a sort designer increase considerably. The problem changes from one of writing an excellent subroutine to one of determining the proper balance of an entire program with all the attendant considerations of I/O and CPU balance and utilization, blocking and buffer techniques, and data interfaces between program components. The efficiency of the sort process must be measured in terms of the total amount of time required to read, sort, and write all data elements, a time only partially determined by the ordering algorithm itself [4].

The most common form of external sorting is the sort/merge. The nature of this procedure forms the content of this and succeeding chapters. Distributive sorting using I/O devices is an alternative, as indicated in Chapter 10. The future may bring the end of explicitly external techniques by the use of "virtual memory," which is discussed in Chapter 20.

12.2 NATURE OF SORT/MERGE

The sort/merge process consists of two basic phases—the sort phase and the merge phase [18]. The sort phase usually operates first to disperse partially ordered records across a family of tapes or disk areas, which will subsequently provide input to the merge.

The sort (or dispersion) phase reads as much data as can be held in available memory at one time and, in some manner, develops ordered strings with the data in memory. These strings are written out according

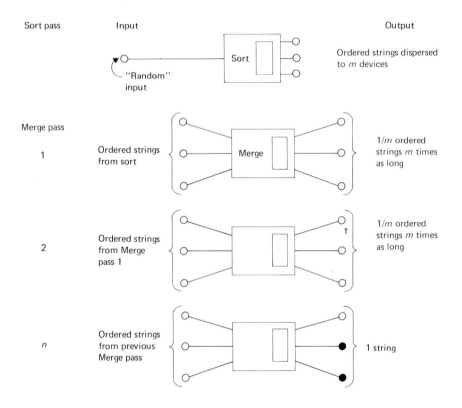

Fig. 12.1 Sort merge process.

to a dispersion algorithm tailored to fit the expected merge. For most algorithms, all of the data is passed by the sort before a merge is undertaken. At the beginning of the merge, there is a population of strings dispersed on devices. The merge then undertakes to form one ordered string. Figure 12.1 is a pictorial representation of the entire process.

The purpose of the sort phase is to reduce the number of merge passes. The sort phase relates to the external merge in the way that a prior internal sort relates to an internal merge.

The operation of the merge is affected by the following sort-related elements.

1. The number of strings produced by the sort phase
2. The size of strings and whether all are of the same size
3. The blocking characteristics of the data written by the sort
4. The dispersion of strings across devices, space on devices, channels, etc.

The goal of sort/merge design is to balance the times for the sort and merge phases so as to minimize the running time for the entire ordering procedure. This goal is not necessarily achieved by optimizing the sort phase or the merge phase. Only an analysis involving the complex interrelations between sort and merge characteristics and requirements will produce a balanced and efficient sort/merge.

Various secondary services may be performed by the sort phase. The repacking of keys and general rearrangement of records for the convenience of the merge are common functions of the sort phase. Functions that are not sort-related may be introduced. Various system considerations may make it desirable to abandon the concept of an efficient sort phase in favor of a program which produces a multiplicity of outputs as a result of its pass over the data. Among such nonsort functions may be initial data editing and error reporting, listing of data, listing with counts, or distributions on criteria unrelated to the sorting key. The inefficiency of the sort phase is compensated for by the ability to avoid additional passes over the data. The value of such functions depends on the nature of the application system; *application design decisions transcend considerations limited to sorting efficiency.*

12.3   ELEMENTS OF A SORT PHASE

There is a rich set of topics and relationships to be considered in developing a sort pass. Fundamental to development are the following.

1. Selection of a sort method
2. Determination of the dispersion algorithms
3. Determination of the method of communication between sort and output
4. Determination of the method of communication between sort and input
5. Determination of the allocation of storage for sort, input, and output
6. Determination of the impact on sort time of alternatives for steps 1–5
7. Determination of the impact on merge time of alternatives for steps 1–5

Steps 6 and 7 represent a continuing process of evaluation for steps 1–5. The issue of storage management, set forth as such in step 5, more or less directly affects all steps and is the point at which many considerations come together. Section 12.4 discusses step 5, Sections 12.5 and 12.6 discuss steps 3 and 4, Section 12.7 discusses step 2, and Section 12.8 discusses step 1. Because these considerations are interdependent, comments are made in all sections about the interrelations among the steps.

12.4 MANAGEMENT OF STORAGE: NUMBER AND SIZE OF STRINGS

Available storage must be split among input, output, and the sort. Contending for core are:

1. The sort procedure
2. The input procedure
3. The output procedure
4. The sort area
5. Input areas
6. Output areas

The critical factors in determining the desirable size of a sort area is the effect that size has on the length of string that can be produced, the speed of the sort phase, and the impact on the merge. To determine the impact of string size on a merge, it is necessary to know the order and "type" of the merge and the number of records in a file. Knowing this enables one to determine how long strings must be to achieve various merge levels. Figure 12.2 shows the number of strings of various sizes that are produced for a three-way merge of 50,000 elements. The conclusion to be drawn from this example is that it is not necessarily efficient to develop the largest possible strings. Because the number of merge passes is $\lceil \log_2 (N/s) \rceil$, where s is string size, strings of size 8 and strings of size 20 both have 8 merge passes. The storage area to produce strings of size 8 is smaller, and the space saved might be profitably used elsewhere as I/O buffers or as work space for a faster nonminimal storage method.

String size	No. of strings	No. of passes
2	25,000	10
4	12,500	9
8	6,250	8
10	5,000	8
12	4,166	8
14	3,570	8
16	3,125	8
18	2,777	8
20	2,500	8
25	2,000	7
30	1,666	7
40	1,250	7
50	1,000	7
68	729	6
80	625	6
100	500	6
150	333	6
200	250	6
250	200	5

Fig. 12.2 Size and number of strings and number of passes for a balanced three-way merge of 50,000 elements.

The exact number of elements to be sorted is often unknown, and the number of strings to be produced must be estimated. This estimate is basic to sort design. It is impossible to intelligently select a size for the sort area without making assumptions about merge characteristics which, in turn, are derived in part from assumptions about string population.

The sort phase may be designed to produce the largest possible strings or to run in the most efficient manner. The largest possible strings result from minimizing blocking and buffering to provide the largest sort area. Alternatively, the sort may be "balanced" (performance optimized) on an overlapped I/O machine by providing a smaller sort area, increasing block sizes, and extending buffering. Machine characteristics that are important to this design decision include the following.

1. Time required to produce strings of various sizes with various sort algorithms
2. Relative CPU–I/O speeds and overlap capability
3. Capability for chained reading of small blocks (Can short strings be linked later on?)
4. Maximum orders of merge conceivably available

In addition to expected merge characteristics, the size of the sort area depends on the following.

1. Record size (How many records will fit?—$8k$ of core for 100-character records is not $8k$ of core for 1000-character records.)
2. Requirements of sorting algorithm (minimal- or nonminimal-storage technique)
3. String-length property of sort (Can longer strings be produced in less space?)

Some sort algorithms may produce a string that is larger than the size of the sort area. The exact size of the string is not predictable; it depends on natural order in the data. If it is not necessary to know the exact size of a string, then considerable advantage can be gained by using a sort algorithm which may achieve string lengths nearly twice the number of records in the area with only a small increase in required space [7].

12.4.1   Storage Management: Input and Output

The determination of the storage area required to support the input and output activities is made concurrently, of course, with the selection of a sort and the determination of a sort area. Certain considerations must be included.

1. The blocking factors at the input and output sides
2. The time relation between the sort and I/O in terms of the possibilities of effective overlap with various buffering depths and techniques
3. The time saved in the optimizing of overlap in a dispersion pass, as against time lost in a merge because of shorter strings

The input block size is often beyond the control of the designer of the sort. The size of the input block determines the amount of space which must be used to hold input and affects the feasibility of using buffers for input backup.

The size of output blocks affects the use of core in much the same way that the size of input blocks affects it. The size of the block determines the feasibility or profitability of buffering output, and the number of records determines when writes can be undertaken.

The output block length is determined by considerations in the design of the merge. The block produced by the sort phase should be the block that is most efficient for the merge. The block length most efficient for the merge is that which provides for the best balance of overlap of operations during the merge phase, within the constraints of reasonable I/O device performance.

The space to be allocated to I/O depends on an appreciation of the relative speed of the compute, read, and write activities of a dispersion phase and on characteristics of the machine. If there are no overlap possibilities on the system being used for the sort, all space can be given over to the sort. The pattern of the program is to fill memory with input, then sort, and then write the complete string, using a block size good for the merge.

If there is I/O–computer overlap (we will henceforth assume that we are dealing with systems with at least a read–write–compute capability), then the problem of buffering levels must be faced. The speed of the sort relative to I/O determines what might be a proper buffering level for the sort. The establishment of a proper buffer level depends on whether or not the adverse affect on string length will hurt the merge. Since space will be taken from the sort area for buffers, strings will be smaller and more numerous. This will hurt the merge if additional passes are required because of the larger string population.

Consider one way of handling the sort phase of a 50,000-record file, shown in Fig. 12.3. Input blocks of 20 records and output blocks of 10 records are planned. Various sort times and an estimate of allocatable storage are given. Figure 12.4 has allocated 8000 locations to the sort, effectively disconnecting I/O overlap for the system. The total time of the dispersion phase will be the sum of read, write, and compute times. The pattern of the run is to read four blocks to fill storage, sort, and then write

Input

Number of records	50,000
Size of record	100 characters
Input blocking	20
Number of blocks	2,500
Block size	2,000 characters
Block read time	50 ms

Output

Output blocking	10
Number of blocks	5,000
Block size	1,000 characters
Block write time	30 ms

Internal processing time

Sort of 80 records	100 ms
Sort of 30 records	35 ms
Sort of 20 records	15 ms

Size of storage available

Total	10,000
Procedure	2,000
Allocatable	8,000

Fig. 12.3 A configuration for sorting 50,000 records.

Assume 8000 characters (80 records) available for minimal-storage sort algorithm.

Read time:	2500 blocks at 50 ms	= 125 s
Write time:	5000 blocks at 30 ms	= 150 s
Sort time:	625 strings of 80 records each at 100 ms	= 62.5 s
		337.5 s

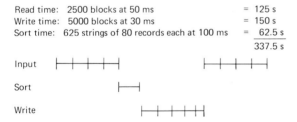

No I/O overlap; 625 strings of size 80 produced in 337.5 seconds.

Fig. 12.4 Sorting 50,000 records in 8,000 locations.

eight of the shorter merge blocks. The elapsed time of the run is 337.5 seconds to produce 625 strings.

An alternative design, shown in Fig. 12.5, would separate core into four areas of 2000 characters each. At any time the sort would be working on one area of this size. The other areas would be available for input or output depending on the rate of progress of each activity. Any particular buffer would serve at one time as the sort area and at other times as one of the

Assume 2000 characters allocated to minimal-storage sort algorithm. Three other 2000-character areas reserved for I/O buffering.

```
Read time:  2500 blocks at 50 ms                              = 125 s
Write time: 5000 blocks at 30 ms                              = 150 s
Sort time:  2500 strings of 20 records at 15 ms
            20 records per string; 2500 strings at 15 ms =    37.5 s
Run time (if write time sustained)                            = 150 s
```

Perfect overlap

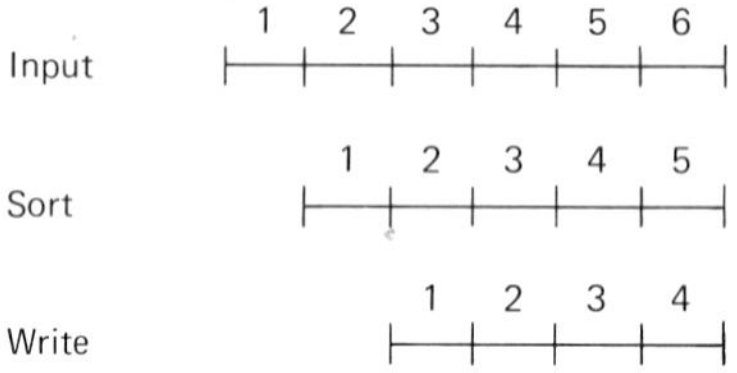

Numbers on line segments represent individual strings. String 1 is being written while string 2 is being sorted, etc. In 150 seconds 2500 strings 20 records long are produced.

Fig. 12.5 Sorting records in four 2,000-location areas.

I/O areas. The decision to pick a length of 2000 is due to the size of the input block. If perfect overlap is achieved, then the time of the run will be the longest of read, write, or compute. As Fig. 12.5 shows, write time is the longest. If the writing of all blocks can be accomplished so that there is no discontinuity—i.e., no period of time during which writing can *not* be undertaken—the run will take 150 seconds. If any writes are delayed, causing periods of inactivity on the write path, the run cannot achieve its minimum time.

Figures 12.4 and 12.5, represent extremes of design implied by read/write overlap goals. The former produces a minimum number of strings at the expense of overlap; the latter produces many more strings but achieves a balanced I/O–CPU sort phase. If the saving of 187 seconds on the design of Fig. 12.5 is not lost to a requirement of more than 187 seconds of additional merge time (because of extra strings), Fig. 12.5 is the better design. Figure 12.2 shows (for a three-way merge) 8 passes for 2500 strings and 6 passes for 625 strings. The time for each merge pass will determine whether or not two extra merge passes take 187 seconds.

12.4.2 Smoothing Performance With Buffering

The particular scheme of rotating buffer areas described in connection with Fig. 12.5 is a very flexible and common buffering technique. The ability of the buffering system to dynamically reassign space to functions on the

Buffer	1	2	3	4
(A)	R	R	S	W
(B)	R	S	W	W
(C)	R	S	S	W

R = Read
S = Sort
W = Write

Fig. 12.6 Various conceivable buffer patterns.

basis of their current need represents an important capability. In Fig. 12.6, the three rows show various conceivable buffer usage patterns at any point in time. Any permutation of A, B, C is equivalent to any other.

The true impact of buffering is apparent when there are variations in time for the particular functions. The time needed to read or write a block on tape will not usually vary; but it is, of course, quite common for the time needed to *sort* a list of records to vary. If the time required to sort elements is based on a method with an expected number of comparisons of $1.4N\lceil \log_2 N \rceil$ but a maximum of $N(N-1)/2$ (for example, an ordered list for Quicksort), sorting time can double for lists of even moderate size. Some blocks will take more time to sort than to read and write. If this is consistently true, then the sort phase becomes compute-limited.

If sorting a block (or blocks) requires a long time, the input of new blocks may be delayed by lack of a free input buffer. The reading of blocks will cease until the sort frees the buffer. The continuation of the sort will be delayed in turn because no blocks are available in memory for sorting. When a write is completed, the joint delays of reading and sorting may mean there is no block ready for writing, and the write must wait until one is available. If write is the limiting path, its waiting time is added to the total elapsed time of the run. Buffer policy is the determination of the number of buffers and their allocation in such a way as to minimize delays. For example, an added input buffer might avoid delays to sort and write and thus have an important effect on running time. Increasing the number of buffers must be justified, however, because it may reduce the block size and the sort area.

The scheme of rotating buffers is efficient with any number of buffers. There are alternatives which specify a particular number as input buffers, a particular number as output buffers, and the rest of space as work area. With this fixed-function buffering, elements are physically moved from and to buffer areas through work space. There is an increase in internal timing, since the time to sort records now involves their physical movement into positions in the sort area.

Whether rotating or fixed-function buffering is used, the question of the number of buffers is critical. A common standard is two buffers for input, two for output, and an "in use" area; however, this will by no means guarantee that there will be no delay, or that the sort phase will be "balanced."

12.5 RELATIONSHIP OF SORT TO OUTPUT

In the discussion of storage management, we have implied characteristics of the sort and the nature of data transfer between I/O and the sort area. In the discussion up to this point, the *block* has been the unit of communication between I/O and sort. Sort areas have been defined as blocks or multiples of blocks, and the sort has "released" entire sorted strings to output at the end of a sort. But the unit of transfer is a design decision, an aspect of the interface between the sort and I/O. Certain sorts have the property of growing an ordered list sequentially (selection sorts do; Quicksort does not). When a sort has this characteristic, there is a possibility of transferring data out of the sort area to an output area before the list is completely sorted. Whenever an element is ranked, it may be passed to output. The smoothing effect of buffered operation is enhanced because the delivery of records to output depends not on the rate at which an *entire block* or sort area (string) can be ranked but only on the rate at which *an element* can be ranked.

The use of a record (rather than a block or a buffer) as a unit of interchange is most naturally discussed in the environment of fixed buffer assignments. The physical transfer of a record from locations in the sort area to locations in the output area is a natural mechanism of interaction. However, it need not be the mechanism used. The architecture of the I/O subsystem may provide an alternative technique. Space allocated to I/O need not be continuous. A species of buffer rotation on a record level may be used. Ranked elements available for output may have their addresses recorded on a "gather write" list. An output block is formed when there are sufficient entries on the list. The output program gathers the records from diverse locations in memory under control of the write list and forms a continuous output block on a device. When a write is finished, the record locations are released to the sort.

12.6 RELATIONSHIP OF SORT TO INPUT

The unit of transfer between input and sort may also be a block, a number of blocks, or a single record. The interface between input and sort is particularly interesting, because it is related to string size and a technique of sorting

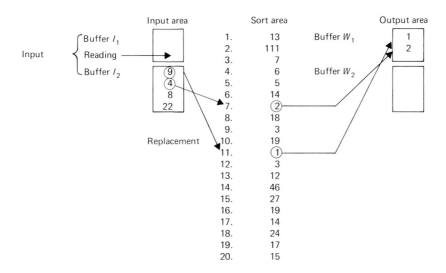

Fig. 12.7 Elements from input block replace winners as they are selected.

known as *replacement*. Selection sorts rank one element at a time in order. If, instead of replacing a "winner" with a dummy value or shortening the list, the program replaces the winner in the sort area with a new input element, the number of elements on a string can be larger than the number of elements in the sort area. The unit of transfer between I/O and sort necessarily becomes an element rather than a block.

Figure 12.7 shows a sort area that contains 20 records when it is first filled. Buffer area I_2 contains records of an input block; buffer area I_1 is receiving a new input block. Areas W_1 and W_2 are initially empty.

The sort selects key 1 in position 11 as the lowest on the list and transfers the record to W_1. The position vacated by the record with key 1 as its key is now filled with a record from an input buffer. The next available record from an input buffer, the first record in I_2, is transferred to the sort area position just vacated. The next cycle of the sort operates on the transformed list. It will select the record with key 2 for transfer to W_1. This record will be replaced in location 7 by the next record from the I buffers, the record with key 4.

The unit of *interface* between the sort and input has become a record. The pattern of developing string S while reading the blocks of string $(S + 1)$ is broken, and the development of a string and I/O for a string may go on concurrently. There is no long period of time during which the sort function is "closed" to communication. If the sort of Fig. 12.7 is well balanced, then, whenever a record is transferred to output, there is a replacement for it available in a buffer; and whenever a write is completed, there is a buffer full of records to be written.

The sizes of the blocks of input and output and the size of the sort area bear no relation to one another, since the unit of transfer is a record. Also, the size of the string, although related to the size of the sort area, is not the number of elements in the area because of replacement.

The physical transfer of records from an input buffer to the sort area can be suppressed by the use of "scatter read," the input analogue to "gather write." At any time, a list of locations free to receive input and a list of locations in the sort area dynamically define input and sort area spaces. When a write is completed, the write list can be made available as a read list.

As with the natural merge, there is need to ensure that an element too small for a string is not assigned to the string. It is possible for the lowest element in the sort area to be lower than the last element on an output string. One method is to associate with all keys a high-order bit, which can be set to eliminate them from the current string. This bit is set when the element enters the sort area if that element is lower than the element currently at the end of the string. All elements with the high-order bit set are precluded from being selected for the current string. The string-building process continues until all elements in the sort area are ineligible. The string is then terminated, high-order bits are reset, and a new string is started. The size of a string depends on the rate of arrival of elements that are lower than predecessor keys. The larger the sort area, the greater the tolerance for low keys. Thus string size is a function of sort-area size.

Strings produced by the application of replacement to a sort will obviously not be of the same size. There will be some distribution of string sizes around an expected size of twice the number of elements in the sort area. Tape merges and some disk merges will not be adversely affected by the variation in string size and may even work very well with larger strings. Some disk merges, however, cannot tolerate variations in size because of space placement problems (see discussion of Replacement Selection sort, subsection 12.8).

12.7 DISPERSION

One of the determinations which must be made by the sort phase is the pattern in which strings will be placed across devices. The nature of the dispersion algorithm is determined by whether the strings are of the same size, the nature and number of the devices being used by the merge, and the pathing (channel design) capabilities of the system.

The dispersion of strings on tape for any tape-merge design for a given order of merge involves the selection of I/O units for receipt of strings according to the set of goals associated with the merge. There are basically two types of algorithms, balanced and unbalanced. A balanced dispersion places an equal number of strings on each tape or disk/drum area. An unbalanced dispersion places strings to achieve an "ideal level." The concept of ideal levels will be described in association with each unbalanced merge type; in general, an ideal level is the dispersion of a specific number of strings which will allow an unbalanced merge to complete without a change to the merge algorithm from one pass to another.

Balanced dispersions are horizontal; unbalanced dispersions are either horizontal or vertical. A horizontal dispersion places the desired number of strings on each device by consecutive device selection, so that each device is brought gradually to its goal. A vertical dispersion brings one device to its goal and then brings others, one at a time, to their goals. Consecutive strings tend to be written on the same device. Preference for horizontal or vertical dispersion is partially a function of machine characteristics having to do with very specific details of I/O subsystem component configuration. For example, on a machine that can perform parallel writing, a horizontal dispersion may allow the writing of the first blocks of one string to overlap with the writing of the last blocks of another.

There are some systems on which consecutive writes to the same tape unit or to the same synchronizer involve a small delay; on such systems, tape switching may avert the delay. There are other systems, however, where consecutive writes are actually faster because interrecord stop/start is avoided; on such systems, increasing the number of blocks written consecutively on the same tape will be an advantage.

Of greater importance in regard to vertical or horizontal dispersions for the unbalanced tape merges is the degree to which the method is exposed to the risk of failing to achieve an ideal level. The penalty that must be paid when that level is not achieved will vary with the specific string dispersions associated with a vertical or horizontal dispersion pattern.

Dispersing onto disk or onto drums is a considerably more complex activity than dispersing onto tape because, in addition to selecting a device, the distribution algorithm must select a proper specific space for each string on the device.

12.7.1 String Variation and Extension

If strings of fixed size are produced, no special technique of marking string start or end is required. A block or record count, independent of the representation of the string itself, can be developed in the system to aid in the recognition of new strings.

If strings of differing sizes are produced, it is not uncommon—though not absolutely necessary—to write an "end sentinel" beyond the last record string or a "begin sentinel" at the beginning of a string. Alternatively, a new string from a tape can be recognized by comparing a record entering storage with a predecessor and taking whatever new string action is required when the new entry is less than its predecessor. Whether sentinels are used or not determines the feasibility of a technique of dispersion called *string extension*.

In an environment where strings of differing sizes are permissible, it is possible to extend the length of a string if the string has no end sentinel. At the beginning of each new string, the first record of the new string is compared with the last elements of other strings. If any string has a last element lower than the first element of the new string, that string can be extended, concatenated with the new string, to form one elongated string.

In Fig. 12.8 there are three tapes, each with three strings. The first and last key of each string is shown. The strings have been distributed horizontally, S_1, S_2, S_3, ... , S_7, S_8, S_9. The first element of S_2 is less than the last of S_1, the first of S_3 less than the last of S_1 and S_2; the first of S_4 less than the last of S_1, S_2, and S_3, etc. There has been no possibility of string extension for any strings developed thus far. The next string, however, will have a first element with a key of 57. Without string extension, S_{10} would be formed. However, a scan of the last elements of all available strings shows that S_7 has a final key that is lower than 57 and consequently can be extended, delaying the creation of string 10. The effectiveness of string extension is partly limited by the number of strings that are "available." In the merge of Fig. 12.8 three strings are available: S_7, S_8, and S_9. No other strings can be extended because they are not physically accessible. If there are multiple proper fits (for example, if the extending key were 67 instead of 57 in Fig. 12.8), it is possible to extend the first eligible string encountered or the string for which the difference between the last element and the first new element is smallest. Thus, if the first new element is 67 instead of 57, then S_8 rather than S_7 will be extended. In disk or drum environment, although a much greater range of available strings may exist in theory (all strings are equally available), space allocation policies may prevent the serious use of extension.

T_1	T_2	T_3
S_1 (9-481)	S_2 (2-517)	S_3 (41-119)
S_4 (6-211)	S_5 (53-71)	S_6 (16-374)
S_7 (19-55)	S_8 (23-61)	S_9 (8-141)

First element of new string: 57

Fig. 12.8 String extension.

12.8 CHARACTERISTICS OF THE SORT

Except in very small core sizes, the most commonly used internal sort method is Replacement Selection [20]. This method is a tournament of the type described in Chapter 5, to which replacement has been applied.

What are the properties that make this sort so attractive? A list of them will suggest the criteria that may be used in judging a sort for a premerge environment, while describing the properties of this method.

1. *Relation to I/O.* Since the method is one of selection, it ranks an element in every cycle. Consequently, the time before the delivery of an item to output is minimal, and the time before the freeing of a position in the input buffer is minimal. In addition, since the time for the selection of a record is reasonably fixed and predictable, an analysis of buffering levels can be made.

2. *Number of comparisons.* The number of comparisons to pick N winners (put all N members of the file on strings in a tournament of size F) is on the order of

$$N\lceil \log_2 F \rceil + N$$

after tournament initialization. Since there are $\lceil \log_2 F \rceil$ levels on the tree, the selection of a winner element will involve $\lceil \log_2 F \rceil$ comparisons plus the string inclusion comparison, or ($\lceil \log_2 F \rceil + 1$) total comparisons per winner selection. The time to deliver a winner is therefore minimal, not only in terms of when a delivery to output is made but also in terms of how long a "cycle" is.

The speed of the method of sorting is still important in premerge sorting. It is desirable that the time to create a string be less than the time to read or write a string, in order to keep the pass I/O-limited.

3. *Movements during the sort.* A record is moved only once during the sort; all other movements are of tags or addresses. However, there is some sensitivity to implementation here. A design may call for moving a record twice —from input buffer to sort area to output buffer. On the other hand, a design using gather-write and scatter-read does not move the records at all. If a rotating buffer scheme is used, the tournament area, the space used to represent the tree, lies in a private work space beyond the buffering domain.

4. *Use of space.* The method is inherently a key sort. There are various ways of representing the tree structure and of encoding comparisons for tree structure. Given a certain amount of space for the sort area, it is necessary to deduct space for tournament representation. For example, suppose that we are given 10,000 locations in a character-oriented machine and 100-character records. It is not possible to have a tournament of 100 records.

Suppose that nodes on the tree are represented by a two-address packet and each address is three characters long, so that six characters are required. (The addresses in the packet are the address of a packet address at the next level and the address of a loser at this level.) This leaves room for a tournament of only 94 records—a relatively small burden for the method, however. (See Chapter 5.4.1 for a discussion on tournament organization.)

5. *String size.* The size of a string is indeterminate and depends on order in the data. The first string produced by Replacement Selection will tend to be of length $1.7F$, later strings of length $1.9F$; very quickly an expected length of $2.0F$ is achieved. The long string length for a small number of comparisons is the feature of Replacement Selection which makes it so interesting and has given it much general popularity in the field [8,15,26].

12.8.1 Factors in Sort Selection

The discussion of the characteristics of Replacement Selection makes the desirable nature of the method apparent. However, it is not universally used. In situations with small core, alternative methods may be preferable. Methods often used instead of Replacement Selection are Binary Insertion, Sifting, and Internal Merging. To some extent, historical preference plays a role, but the possibility that pockets of resistance may have some rational basis deserves our attention.

Let us examine the issue of number of comparisons to form a string. For a string of length 20, Replacement Selection will perform 100 comparisons. Sifting, if we assume $(N^2 - N)/4$, will perform 95 comparisons. Twenty is a very small number of elements in a string but not for a machine with 16,000 allocatable characters. On small lists, the differences in number of comparisons are minimized, and very often the overhead associated with more sophisticated sorts (such as complex address calculations or pass controls) cannot be supported. For small memories, the use of Sifting as an alternative to Replacement Selection bears close investigation.

The issue of string length might encourage the use of Replacement Selection anyway, but there are considerations in this area which might make the method unprofitable. In a two-way merge, the long string capability of Replacement Selection may save a pass. This may be unimportant for very large files. Consider Fig. 12.9. With an assumption of 50,000 records

No. of strings	2-Way	3-Way	4-Way	5-Way	6-Way
1,024	10	7	5	5	4
2,048	11	7	6	5	5
4,096	12	8	6	6	5

Fig. 12.9 Effect of long strings.

and a sort area capable of holding 20 records, a nonreplacement technique will develop 2500 strings, a replacement technique roughly 1250. This difference will reduce the number of passes from 12 to 11, a saving of less than 10 percent in a two-way merge. If there is any added cost associated with dispersing larger strings, the advantage of long strings may well be marginal. Furthermore, as the order of merge increases, there is no longer any certainty that a pass will be saved. In this example, no pass would be saved for a four-way merge or a six-way merge. It is therefore not conclusive that Replacement Selection will save merge time.

There are situations in which the read/write/compute capability is not relevant. The irrelevancy may stem from the design of a small computer system, or it may be due to the fact that the resources assigned to the sort do not provide for parallel read/write paths. In such instances the smooth interfacing allowed by Replacement Selection is also not relevant.

The thrust of these remarks is to bring to the reader's attention the fact that the use of Replacement Selection should not be automatic, despite its excellent and important properties. In small machines a simpler method that may be faster (or as fast) on small lists without damaging the merge may well be worthwhile.

12.9   FINAL REMARKS

The usual reason for developing an external sort in a user environment is that the sorts provided by the manufacturer or a software product vendor are not applicable for some reason relating to the data or the configuration of a particular installation. A user is encouraged to develop a sort/merge if he observes that a particular set of his data has characteristics that are excluded by the features of the generalized sort package or that lend themselves to a specialized technique not provided in the package's repertoire of capabilities. Consequently, it may be worthwhile or necessary to develop a special sort.

Design is everything. There is often a tendency in program creation environments to underdesign, to rush into development of programs. This tendency is deadly in the development of sorts. Design must include not only a postulation of a system but a review of alternatives and a justification of choices.

In the design of a sort phase for a sort/merge, the more that can be known about the merge, the better. If merge specifications can be finalized before the sort phase is frozen, then the task of developing a good sort phase is considerably eased. The importance of knowing the nature of devices, the range of merge orders, block sizes, and merge patterns was indicated in our discussions of the string sizes and the performance that may be expected from certain storage management policies.

13
Tape
Merging

13.1 BALANCED TAPE MERGE

The first line of Fig. 13.1 represents the status of a sort/merge at the end
of the dispersion pass. Dispersion of nine strings each has been made from
tape 0 to half the tape units available to the sort. The number of tape units
is often called k in the literature, and balanced merging is called $(k/2)$-way
merging. The nominal order of the merge is $k/2$.

	T_0	T_1	T_2	T_3	T_4	T_5
	Input	—	—	9 (L)	9 (L)	9 (L)
Pass 1	3 $(3L)$	3 $(3L)$	3 $(3L)$	—	—	—
Pass 2	—	—	—	1 $(9L)$	1 $(9L)$	1 $(9L)$
Pass 3	—	—	1 $(27L)$	—	—	—

Fig. 13.1 Balanced merge summary.

The 27 strings created by the dispersion pass may be of different lengths
or of the same length. They are considered to be strings of length L, where
L is the nominal length of initial strings. The 27 initial strings represent
$27L$ elements to be merged. The final string produced by the merge will
be a string of length $27L$. The process of the merge is that of combining
strings into fewer and longer strings until one string containing all elements
is produced. The number of passes required to do this is $\lceil \log_m S \rceil$, where
m is the order of the merge and S is the number of initial strings. A three-way
merge of 27 strings requires $\log_3 27$, or 3 passes.

Pass 1 reduces the number of strings to nine. Since the length of each
newly formed string is $3L$, there are still $27L$ initial string lengths repre-

sented in nine merged strings. The number of strings at the end of a pass will be the next lower power of the "way" of the merge, and the string length will be the next higher power of the way of merge times L. Figure 13.1 shows 27, 9, 3, and 1 strings at various stages, and string lengths are L, $3L$, $9L$, and $27L$. When the number of initial strings is not a perfect power of the merge, these relations are approximate. The length of a string at any point in the merge is the length of all the strings which have combined to form it. In a three-way merge, three strings combine to form a string three times as long. When strings are not of the same length and/or the order of merge is not constant, it is necessary to sum the length of merged strings to determine the size of a string developed by the merge.

	T_0	T_1	T_2	T_3	T_4	T_5
	—	—	—	1	2	3
	—	—	—	4	5	6
	—	—	—	7	8	9
	—	—	—	10	11	12
	—	—	—	13	14	15
	—	—	—	16	17	18
	—	—	—	19	20	21
	—	—	—	22	23	24
	—	—	—	25	26	27
Pass 1	A(1,2,3)	B(4,5,6)	C(7,8,9)	—	—	—
	D(10,11,12)	E(13,14,15)	F(16,17,18)			
	G(19,20,21)	H(22,23,24)	I(25,26,27)			
Pass 2	—	—	—	J(A,B,C)	K(D,E,F)	L(G,H,I)
Pass 3	—	—	M(J,K,L)	—	—	—

Fig. 13.2 Balanced merge of Fig. 13.1, string by string.

Figure 13.2 shows the initial dispersion in considerably more detail. The strings are written across tapes 3, 4, and 5. Each number in Fig. 13.2 is an initial string number. The first merge pass combines strings 1, 2, and 3 to form string A of length $3L$ on tape 0. Similarly, strings 4, 5, and 6 are merged to form string B on tape 1, and strings 7, 8, and 9 are merged to form string C on tape 2. At the end of the first pass, the nine strings A through I, each of length $3L$, hold all records on tapes 0, 1, and 2. Tapes 3, 4, and 5 are empty.

The dispersion of new strings follows an algorithm identical to that for the dispersion of initial strings. Strings merged together are those "next" available (in a physical sense) to the read heads of the tape unit. String extension may be used, and advantage may be taken of natural order in the data. It is possible to produce one string in fewer than $\lceil \log S \rceil$ passes if maximum use is made of these techniques on highly ordered data. After each string number in Fig. 13.3, the values of the first and last keys of the

T_0	T_1	T_2	T_3	T_4	T_5
			1 (17-81)	2 (11-91)	3 (73-111)
			4 (86-121)	5 (93-171)	6 (136-183)
			7 (127-141)	8 (172-214)	9 (185-212)
			10 (149-183)	11 (215-282)	12 (217-281)
			13 (197-214)	14 (283-361)	15 (287-311)
			16 (219-279)	17 (362-380)	18 (371-416)
			19 (311-367)	20 (381-411)	21 (448-463)
			22 (373-411)	23 (417-511)	24 (469-510)
			25 (521-596)	26 (561-611)	27 (571-623)

Fig. 13.3 Natural merge can reduce in one pass.

string are given in parentheses. As an accident of string formation, each successive string on a tape starts with a key higher than the last key of the preceding string on that tape. Inspection reveals that each tape contains a single ordered sequence and that there are truly only three strings, not 27, on tape. A merge that recognizes step-downs and continues forming a string until there are no more candidates (every input tape starts with a key lower than the last member) can produce one string on a single pass.

13.1.1   Rewind

At the end of a pass, the input tapes are positioned so that the read/write heads are above the space immediately following the last block read. The output tapes are positioned so that the read/write heads are above the space immediately following the last block written. All tapes are "distended," and in systems which read only in a forward direction, it is necessary to rewind. The rewind time between merge passes will vary from sort to sort and from system to system. The total amount of time required to rewind depends on the following factors.

1. *The amount of data on the tape.* Different tape drives require different amounts of space to represent the same amount of data. The data rate is a function of packing density and the speed of tape movement. Blocking also affects the rate, since it determines how much space will be given over to interblock gaps. Space between blocks varies widely over the range of available tape units.

2. *Rewind rate.* Some tape units rewind at the same speed at which they read or write, and some can rewind at a much higher speed. Those tapes which have a high-speed rewind may brake at a certain point and move at read/write speeds from that point to the beginning of the tape.

3. *Parallel capability.* Given multiple tape units to rewind, systems will vary in the number that will be rewound at the same time. That number

is not necessarily the same as the number that can be read or written simultaneously. At worst, only one tape can be rewound at a time. This means that total interpass rewind time is the sum of all rewinds. The ability to rewind units in parallel is quite common in computer systems. The extent of parallel rewind varies from system to system according to the placement of the rewind controls. If controls are in the tape unit itself, each unit should be able to conduct an independent rewind. If controls are in the control unit and there is no special "dual" feature, one tape per control unit can be rewound. If controls are in the channel, one tape per channel can be rewound. If controls are in the CPU, only one tape per CPU can be rewound.

13.2 BACKWARD BALANCED MERGE

The necessity for rewinding tapes between merge passes can be eliminated if the tape units can be read backwards. The backward read is a standard capability, and balanced tape merging is now understood to imply its use. Figure 13.4 represents the distribution of 27 strings. Each of tapes T_3, T_4, and T_5, which received output strings, is positioned as indicated by the arrow. Tapes T_1 and T_2, unused during the dispersion, are assumed to be in initial position at load point. Tape T_0 is positioned at the end of the data.

The status of T_0, the original input tape, is a special problem. If the initial data is to be preserved (not a bad safeguard), T_0 must be removed and replaced unless the file to be sorted is very small relative to tape capacity. In that case, space on T_0 following the original file can be used safely (without danger of "overflow") for the remainder of the merge. If the tape on T_0 is to be used as a blank (abandoning original data), it must be rewound anyway. This is the only instance of rewind for the merge.

All the strings on T_3–T_5 are ascending strings; they are developed in such a way that the key elements increase in value from the beginning to the end of the string. It is possible to have an ordered descending string.

T_0	T_1	T_2	T_3	T_4	T_5
	LP	LP	1	2	3
			4	5	6
			7	8	9
			10	11	12
			13	14	15
			16	17	18
			19	20	21
			22	23	24
			25	26	27
			→	→	→

Note: LP = Load Point

Fig. 13.4 Position of tapes.

If we read ascending strings backwards, one element at a time, it is possible to form a descending string of merged values. The comparison mechanism associated with the internal operations of the merge selects highs in place of lows as winners. At the end of the merge of strings 25, 26, and 27, a string with descending values exists on T_2. The backward reading and forward writing continue for the merge of the next set of strings, 22, 23, and 24. At the end of the second pass, all strings are descending strings. Figure 13.5 shows the status of the system at the pass end. At that stage T_0, T_1, and T_2 will be distended, but T_3, T_4, and T_5 will have been returned to initial position. The backward reading has recovered the space used by the strings which had been written on T_3-T_5. Note that string A now consists of strings 25, 26, and 27, since these were the first available when backward reading began.

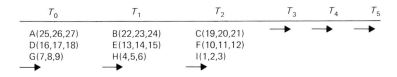

T_0	T_1	T_2	T_3	T_4	T_5
A(25,26,27)	B(22,23,24)	C(19,20,21)	→	→	→
D(16,17,18)	E(13,14,15)	F(10,11,12)			
G(7,8,9)	H(4,5,6)	I(1,2,3)			
→	→	→			

Fig. 13.5 First backward pass (descending strings).

Since tapes T_3-T_5 have been returned to initial position, they may be written on in a forward direction during the next pass. Tapes T_0, T_1, and T_2 are distended with descending strings. They will be read backwards during the next merge pass. The backward read of a string written in a forward direction as a descending string causes the string to be read as an ascending string; therefore, the next merge phase will form ascending strings. In general, if the initial strings are written as ascending strings by the dispersion phase, then odd-numbered merge passes will form descending strings, and even-numbered merge passes will form ascending strings. Figure 13.6 shows a merge, with the strings formed by each pass identified as ascending (A) or descending (D).

From Fig. 13.6 we can observe one problem that is inherent in the backward tape merge. The final pass has produced a descending string,

	T_0	T_1	T_2	T_3	T_4	T_5
	—	—	—	9 (A)	9 (A)	9 (A)
Pass 1	3 (D)	3 (D)	3 (D)	—	—	—
Pass 2	—	—	—	1 (A)	1 (A)	1 (A)
Pass 3	—	—	1 (D)	—	—	—

Fig. 13.6 Direction of strings.

which must be copied onto another tape. This requirement adds a complete pass to the merge phase. If the merges are I/O-bound, this pass will be as long as the others [10,29].

The need for a final direction-reversal pass may be anticipated by the merge phase. If the number of strings and the way of the merge are known, it is possible to calculate whether there will be an even or an odd number of passes. If an even number of passes is expected, the last pass will create an ascending string. If initial strings are known to be ascending and an odd number of passes is expected, then the last pass will create a descending string. When an odd number of passes is expected, a rewind before the first merge pass will ensure an ascending final string. As an alternative, the dispersion pass can reverse the direction of the strings it forms. If a final ascending string is desired for a three-way merge with 27 initial strings, the dispersion pass creates 27 descending strings. The first merge pass reduces the number to nine ascending strings, the second merge pass produces three descending strings, and the last pass creates a single ascending string. The success of this technique depends upon the ability to forecast the number of passes.

13.3   IMPERFECT BALANCED MERGE

There will be times when the number of strings developed by the dispersion pass will not be a power of the order of merge. In Fig. 13.7, with an initial 30 strings there are $\lceil \log_3 30 \rceil = 4$ passes. Since each pass moves each string, 120 string lengths are moved during the merge.

It is possible to improve the performance of a balanced merge somewhat by undertaking to recognize special situations and by undertaking partial passes. Let us first consider recognizing special situations. In Fig. 13.7, there is one of these during pass 3. After the merge of two strings of length $9L$ and a string of length $3L$ into a string of length $21L$, there remains on T_0 a string of length $9L$. As shown in the figure, this string is normally copied onto T_4. However, the copying may be avoided, since the string can just

End of pass	T_0	T_1	T_2	T_3	T_4	T_5	String lengths moved	Direction
0	10 (L)	10 (L)	10 (L)	—	—	—		A
1	—	—	—	4 (3L)	3 (3L)	3 (3L)	30	D
2	$2\begin{cases}1\,(9L)\\1\,(3L)\end{cases}$	1 (9L)	1 (9L)	—	—	—	30	A
3	—	—	—	1 (21L)	1 (9L)	—	30	D
4			1 (30L)	—	—	—	30	A

Fig. 13.7 Balanced merge, imperfect distribution ($3^3 < 30 < 3^4$ = 4 passes).

as well remain on T_0 for the final merge pass. The special situation recognized is that the number of existing strings at the time $21L$ is formed is less than the order of the merge, and a final merge can be undertaken immediately. The number of string movements is reduced to 111, and pass 3 is a partial pass.

Backward reading may complicate the partial-pass strategy. If strings to be "left in place" are inverted (as on Fig. 13.7), it is necessary to position read heads so that each string is in the same direction as all other strings. This can be done with an appropriate number of reverse-direction record reads or a rewind to the beginning of the string in the desired direction. These adjustments may make the strategy unprofitable. Note also that on a two-channel machine, serious channel interference develops when reading from T_0 and writing on T_2.

One may effect a more forceful partial-pass strategy by counting the number of strings left at any time and merging only as many as will generate a number of strings equal to the order of the merge. Figure 13.8 shows the status of the tapes at the end of the second pass of Fig. 13.7. A two-way merge of the string on T_2 with the string of $3L$ on T_0 reduces string population to three strings. This leads to the condition on the line labeled "Final," that is, the number of strings equals the merge order. A final three-way merge will develop one string. The total number of string lengths moved is 102. As before, if backward read is involved, the direction of the inverted strings must be changed. In terms of the string lengths moved, this approach will always be at least as good as that shown in Fig. 13.7.

End of pass	T_0	T_1	T_2	T_3	T_4	T_5	String lengths moved
2	$2\begin{cases}1\ (9L)\\1\ (3L)\end{cases}$	1 (9L)	1 (9L)	–	–	–	
"Final" (3 strings)	1 (9L)	1 (9L)	– 1 (30L)	1 (12L)	–	–	12 (Merge T_1/T_3) 30

Fig. 13.8 Final partial passes. Pass 2 is reduced to one-way merge of T_1/T_3.

An alternative to watching conditions at the end of the merge is undertaking an adjustment at the beginning. This involves copies, merges, and rewinds to move some of the strings to produce a perfect merge order before full merge passes are undertaken. This approach is, in general, the most powerful one. Consider Fig. 13.9, in which a partial-pass adjustment is shown. We do not claim that this adjustment is the best one for this situation; nevertheless we show it to illustrate the concept. The example

	Line number	T_0	T_1	T_2	T_3	T_4	T_5	String lengths moved	Cumulative string lengths moved
Partial pass 1	0.	10	10	10	–	–	–	0	0
	1.	9	9	9	1 (3)	–	–	3	3
	2.	9	9	8	0	1 (4)	–	4	7
	3.	8	8	8	0	0	1 (6)	6	13
Partial pass 2	4.	7	7	7	1 (3)	–	1 (6)	3	16
	5.	6	6	6	1 (3)	1 (3)	1 (6)	3	19
	6.	5	5	5	2 (3)	1 (3)	1 (6)	3	22
	7.	4	4	4	2 (3)	2 (3)	1 (6)	3	25
	8.	3	3	3	2 (3)	2 (3)	1 (6), 1 (3)	3	28
	9.	2	2	2	3 (3)	2 (3)	1 (6), 1 (3)	3	31
	10.	1	1	1	3 (3)	3 (3)	1 (6), 1 (3)	3	34
	11.	–	–	–	3 (3)	3 (3)	1 (6), 2 (3)	3	37
Pass 1	12.	–	–	–	3	3	3	–	37
	13.	1 (12)	–	–	2	2	2	12	49
	14.	1 (12)	1 (9)	–	1	1	1	9	58
	15.	1 (12)	1 (9)	1 (9)	–	–	–	9	67
Pass 2	16.				1 (30)			30	97

Fig. 13.9 Initial partial pass. Numbers in parentheses are string lengths.

assumes that the tapes read forward only. An initial merge of one string from each tape is performed. This develops the status of line 1 on Fig. 13.9. There are 28 remaining strings. A further merge of the string on T_3 and a string on T_2 creates a string on T_4 and reduces the number of strings to 27. The merge adjustment to develop these 27 strings has required 7 string movements. Line 3 shows the result of the resumption of three-way merging. The three-way merge of T_0, T_1, and T_4 to T_5 balances all normal input tapes. From this point on, normal three-way merging is performed. The string-by-string progress of the merge is shown on succeeding lines of Fig. 13.9. Only 97 string lengths are moved in the process of building a single string. To express the measure of merge efficiency in another way, each initial string is moved 3.23 times rather than 4 times throughout the merge; therefore the method is more efficient. Partial passes attempt to move the number of passes closer to $\log_2 M$ from $\lceil \log_2 M \rceil$. Notice that a backward read environment does not particularly complicate Fig. 13.9; here T_0, T_1, and T_2 are read backwards to form the string on T_3. This string may then be read backwards with T_2 to form the string on T_4. The new string on T_4 can be read backwards with T_0 and T_1 to form the string on T_5. All strings formed during partial-pass two are descending strings. No direction adjustment is required (line 12); however, the example produces the final string in descending sequence.

13.4   IMBALANCED BALANCED MERGE

It is possible to perform the Balanced Merge with an odd number of tapes. Dispersion occurs on $(k/2) + 1$ tapes, and the merge proceeds to merge $(k/2) + 1$ tapes onto $k/2$ tapes. The next pass will merge from $k/2$ to $(k/2) + 1$, and succeeding passes of the merge will alternate input and output in this manner.

13.5   CONSIDERATIONS IN MERGE DESIGN

The design of a merge is an even more complex problem than the design of a dispersion phase. The details of the characteristics of hardware and program design have such an impact on merge design that theoretically superior merges may run worse than do inferior merges. For example, a six-way merge that takes fewer passes and moves less data may actually run longer than a four-way merge over the same data on any given machine. The reason lies in the nature of I/O subsystems, the sizes of core, and the specific character of instruction sets. The art of designing a merge involves balancing a number of major considerations (which by their nature defy joint optimization) and a myriad of minor ones that usually have only trivial impact on performance but can provide some nasty surprises under certain conditions.

The major design decision areas follow.

1. *The merge network.* Here we have to choose the algorithm which will be used internally to combine multiple input strings into output strings. This process is basically a sorting process, and the techniques of internal sorting apply. In tape sort situations, where the order of merge is generally in the range of 2 to 20 (and more commonly 2 to 10), the basic sort algorithms are quite adequate, because there are so few elements in the network.

The time to select a member for the output string is important because it affects the balance of internal operations and I/O operations and the ability of the system to keep its tapes running at full speed. The order of the merge affects the speed of the internal operations by determining how many elements will be involved in the comparison to select the next member of the string.

2. *The order of merge, blocking, and buffering.* Theoretically, higher-order merges are preferable to lower-order merges; however, details of internal computer speeds, I/O subsystem organization, and available memory will create circumstances in which a lower-order merge is preferable.

The selection of merge order illustrates some of the interactions among design factors in a merge. The minimum running time for a merge pass is the write time of the output devices when they operate continuously at full speed. It may be possible to maintain a completely write-bound condition with a merge order of m but not with a merge order of $(m + 1)$ or $(m + 2)$. The extra work involved in handling extra input strings, including the increase in comparison time, may make a merge pass computer-bound and lead to periods of time when no write is in progress, delaying completion of the pass. The relation between merge order and merge internals centers on the ability of the CPU to keep merges of various orders write-bound. It may be that the longer merge times per pass are acceptable in a higher-order merge because they are offset by a reduced number of passes. Three passes of ten minutes each are preferable to four passes of eight minutes each. However, an increase in merge order may not reduce the number of passes. This is particularly true if the increase is small (1 or 2, for example). Figure 13.10 shows, for numbers of strings ranging from 100 to 10,000, the number of passes required to complete merges of order from 2 to 10. Note the diminishing return for merges of orders 6 through 10 for even small numbers of strings. A sort of 10,000 elements that may produce 100 strings takes the same number of merge passes for a merge order of 5, 6, 7, 8, or 9. The choice of merge order should be considered in the light of the number of expected strings and the time it takes for the CPU to support merges of different orders.

No. of strings \ Way	2	3	4	5	6	7	8	9	10
100	7	5	4	3	3	3	3	3	2
200	8	5	4	4	3	3	3	3	3
300	9	6	5	4	4	3	3	3	3
400	9	6	5	4	4	4	3	3	3
500	9	6	5	4	4	4	3	3	3
1,000	10	7	5	5	4	4	4	4	3
2,000	11	7	6	5	5	4	4	4	4
3,000	12	8	6	5	5	5	4	4	4
4,000	12	8	6	6	5	5	4	4	4
5,000	13	8	7	6	5	5	5	4	4
6,000	13	8	7	6	5	5	5	4	4
7,000	13	9	7	6	5	5	5	5	4
8,000	13	9	7	6	6	5	5	5	4
9,000	14	9	7	6	6	5	5	5	4
10,000	14	9	7	6	6	5	5	5	4

Number of passes = $\lceil \log_{way} S \rceil$

Examples

For $S = 100$, way = 2, passes = $\lceil \log_2 100 \rceil = 7$

For $S = 100$, way = 7, passes = $\lceil \log_7 100 \rceil = 3$

Fig. 13.10 Way of merge.

Merge order relates to blocking and buffering. Each input string must have associated with it an area in core large enough to hold at least one record from the string. The usual space associated with an input string is that required to hold at least one block. Given a certain amount of core, it may be necessary to restrict block size for a certain merge order, and vice versa. Populations of larger blocks can be read and written in less time. The order of merge must be so selected that the length of time it takes to write all blocks in each pass does not nullify savings in the number of merge passes.

Buffering is an absolute requirement if I/O and CPU overlap is to be maintained and a merge pass is to be write-bound. The order of merge affects the amount of space that will be available for buffers. The blocking factor, the order of merge, and buffering levels are the three most important design elements of a merge, once the major decision about merge type has been made. Buffering techniques and the impact of buffering deficiency will be discussed in the following sections.

3. *The merge algorithm.* The merge algorithm determines the particular pattern of merge reading and writing. The dispersion techniques, the selection of specific strings for combination with others, the adjustment of merge order during the course of a merge, the use of partial pass algorithms—all are considerations involved in this area. In tape merging, the algorithm may be one of a number of basic merge techniques: Balanced, Polyphase, Cascade, Oscillating. Once the major technique is determined, a partial-pass strategy is developed and the special situations to be recognized are defined.

An excellent list of the factors that determine the performance of a merge, first given in a letter by M. A. Goetz to the editor of the *CACM* in 1964 [28], is repeated here.

1. Number of tape units
2. Memory size
3. Fixed or variable size records, record length
4. Internal sort techniques selected
5. Independent data-channel reading and writing
6. Buffering
7. Expected volume of data
8. Initial string length
9. Distribution of strings
10. Processing time versus read-write time
11. Rewind time versus read-write time

12. Read-reverse time, write-after-read interlock time, fixed unit and channel times, switching time between two units on one channel

13. Read-backward capability

14. Computer processor characteristics (e.g., scatter-read, gather-write, indexing, indirect addressing, interrupts)

15. Interspace gap on tape, start–stop time

Goetz points out that for a given merge only one or two factors or all may be significant, and that "factors exhibit complex interactions, the evaluation of which is by no means simple. A final difficulty is that some of the factors are data-dependent."

13.6   I/O PATTERN OF MERGE AND SYSTEM CHARACTERISTICS

A pass of a merge usually reads and writes the same amount of data. The balanced merge writes strings cyclically across output tapes. Reads are performed when there is a need to replenish input strings. At any time during a pass there are $k/2$ input tapes and one output tape. The most efficient merge pass will write continuously and overlap all reading and internal operations with writing.

In $(k/2)$-way merging, each input tape contributes approximately the same amount of input data. If $m = k/2$, each input tape nominally contributes $1/m$ of an output string. This will not be true when certain local "runs" cause imbalances in the contribution of particular input strings to particular output strings, but it will tend to be true over the duration of a pass that each input tape contributes $1/m$ of output data. In order to guarantee that a required read may never be delayed and that a write will never interfere with a read, one must have a system with $(k/2) + 1$ independent available paths (channels) to tape units, since $(k/2) + 1$ tapes are potentially active at any time. It is possible to come very close to ideal performance with only two channels. Since each tape is reading $1/m$ of data, the time it takes to read input strings from all tapes is roughly the same as the time it takes to write all strings. Tape reading will be interleaved, each tape using the read channel $1/m$ of the time. A half-duplex two-channel system will allow efficient performance of the merge. However, because of the potentially uneven pattern of string contribution elements, it is possible that simultaneous reads will be necessary. In such cases, there will be read delay which may lead to delay of a subsequent write. This problem of read balance is discussed at length in connection with buffering.

The I/O activity imposes two burdens on the CPU. The first is control of the I/O subsystem, and the second is memory contention. The minimum amount of time a CPU may devote to I/O control in a single CPU system is the time it takes to send an I/O command and control list to a channel and the time it takes to respond to I/O interrupts. Smaller systems tend to require more of the CPU than larger systems for I/O support because data paths for I/O and CPU instructions are shared and CPU circuits are required for I/O functions.

When I/O is active, CPU operation is delayed because of contention for memory cycles. The transfer of characters between I/O and memory interferes with fetching of instructions and operands by the processor. CPU programs will execute more slowly when I/O is active than when it is not. This is a consideration in estimating CPU–I/O overlap and balance.

In systems that do not have two half-duplex channels, the balanced merge cannot run under optimum conditions. On a one-channel machine, each read will delay a write and each write will delay a read. Read/compute or write/compute is the best that can be achieved. The total time for the merge will not be some approximation of write time; it will be the *sum* of read and write times. If no overlap is available, then the time it takes to run a pass is the sum of read, write, and compute times.

On a two-channel system, care must be taken that the distribution of strings avoids read/write interference. For any pass, all input tapes should be on one channel and all output tapes on another. If this is not (or cannot be) done, merge performance will fall off because of read/write interference.

13.7   IMPACT OF BUFFERING ON MERGE

One can demonstrate the impact of buffering on merge performance by means of a microcosmic inspection of the merge. Assume a two-channel system and the three-way merge used earlier. At a particular point in the process, the storage of the machine is arranged as in Fig. 13.11. Twelve thousand words of storage are being used for the merge, and 8000 locations are given over to procedures. The blocking pattern is ten 100-word records per block. There are three input tapes, and 3000 words are set aside for input, beginning at location 8000. An additional 1000 words are set aside for the formation of the output block. The sort procedure maintains a pointer to the "current record" in each input block.

A current record is indicated in Fig. 13.11 by a circle around an element in the block. The process of sorting involves bringing a key from an input block into the sort area and comparing it with the keys already there. The lowest of the keys is selected as the winner, and the record having that key is transferred from its input block to the output block. The current-record pointer is advanced in the winning block, and a new key is brought from it into the sort area. When 10 records have collected in the output block area, a write gets under way. When 10 records have been selected from an input block, a read gets under way to bring in a new block.

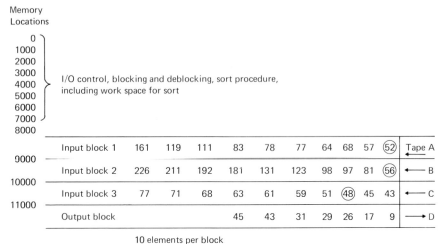

10 elements per block

Note: Circled elements in current comparisons.

Fig. 13.11 Impact of buffering on merge. Elements involved in current comparison are circled. No parallel operations.

At the time shown in the figure, seven records have collected in the output area. The current item in each of the first two input blocks is the first record; in input block 3, the third record. The sort is comparing the keys of these three records. The sort will select key 48 and transfer the third record of input block 3 to output. The record pointer for input block 3 will advance to key 51, and the process will continue.

The merge of Fig. 13.11 prevents significantly overlapped operations for the following reasons.

1. While an output block is being built, no reading can occur because there is no more input space. No writing can occur because there is no complete block to write.

2. While an output block is being written, no sorting can occur because there is no place to put a selected record, and no reading can occur since no space is available.

3. The only overlap possible occurs when an input buffer is depleted at exactly the same time that an output block is filled. Then reading and writing may occur together. The lack of alternate buffer areas has eliminated the read/write/compute overlap capability of the system. An additional output buffer will allow the possibility of overlap of writing and sorting. If space is not easily available for the additional buffer, then trade-offs have to be made between buffer and block size, buffer and order of merge, buffer and coding space.

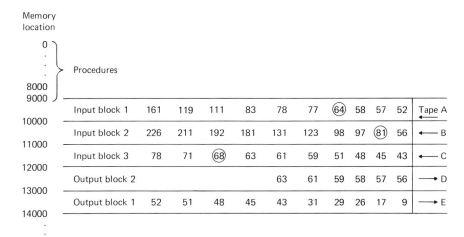

Fig. 13.12 Double-buffered output. Tape *E* is being written from block 1, as block 2 is filled. Write / compute overlap.

Figure 13.12 shows a block being written while another block is being filled. If output block 2 becomes full just as the writing of block 1 is completed, then there can be full overlap of write and compute. If output block 2 is filled before output block 1 is written, the CPU will idle until the writing is completed. The additional buffering, however, will have achieved its goal of maintaining continuous output. Whether there will be secondary effects depends on the tripartite relation of read/write/compute, which we will consider later. If output block 2 is not filled before the writing of block 1 is finished, there will be idle write time. If the time relations between CPU record selection and I/O are such that the process is inherently and constantly CPU-bound, then nothing can be done to sustain the write.

13.8   INPUT BUFFERING

A full understanding of the performance of the merge requires consideration of behavior on the read side. When an input block is depleted, a replacement block must be read, and while that reading is going on, the sort cannot continue. Idling the CPU during replacement reads may make it impossible to keep up with the writes. The write path will dry up because of the unavailability of blocks, and the merge time will be extended.

A usual solution to the problem of chronic read delay and potential write delay is to double-buffer every input, but there are two inherent problems. Double buffering may not be possible, because core does not exist to support doubling the number of input areas or core cannot be obtained for buffers without terrible penalty to block size or order of merge. Also double buffering may not be adequate because of the rate at which input blocks are depleted when particular patterns exist in the data. If the space is available, however, luck may well provide good results.

Figure 13.13 shows the added double buffers. The only change from Fig. 13.12 is that an additional block is available in storage for each input string. It is important to remember that, if tapes *A*, *B*, and *C* share a single path to core because of system architecture or system configuration, then only one can be read at a time. In the immediate future, the addition of records with keys 64, 68, 71, and 75 will complete output block 2. At that time, input block 4 will be depleted and will receive another block from tape *B*. During the reading from tape *B*, the sort can continue, using the alternative buffer (block 3) from *B*.

The goal of double buffering is to avoid delays in the sort and consequent delays in the writes. The means of avoiding delay is to ensure that each input string always has elements represented in core. This goal may not be achieved, however. The pattern of selection of elements from input strings may cause input blocks to be depleted unevenly. The pattern of input

Memory location											
0											
8000											
9000											
10000	*Input block 1	161	119	111	83	78	77	(64)	58	57	52	Tape A
	#Input block 2	320	315	310	271	239	237	214	197	181	171	←
11000											
12000	#Input block 3	226	211	192	181	131	123	98	97	81	56	Tape B
	*Input block 4	75	71	(68)	63	61	59	51	48	45	43	←
13000											
14000	*Input block 5	84	88	87	86	85	84	83	82	80	(78)	Tape C
	#Input block 6	116	115	110	100	96	94	93	92	91	90	←
15000											
16000	Output block 1	92	51	48	45	43	31	29	26	17	9	Tape D
	Output block 2					63	61	59	58	57	56	→
17000											

Notes: Circled elements in current comparisons.
*Active block input
#Backup block input

Fig. 13.13 Double-buffered input. Read/write/compute overlap.

block depletions and, consequently, the pattern of reads will be different if 20 consecutive records are chosen from the same input string than if the records selected come from all strings.

If one input string continues to be depleted while others are untouched, it may not be possible for the read operation to refill backup buffers fast enough, and the sort may be delayed. This problem may or may not affect the write depending on ratios of CPU and I/O time. If strings are depleted evenly, delays may still occur (on a two-channel machine), because it is impossible to read more than one tape at a time. The time for completing three reads may exceed the time the sort requires to deplete all backup blocks. Normally, however, variations in selection patterns are such that double buffering tends to achieve its goal over the course of a pass.

Friend [14] states that an absolute guarantee of avoiding "read hang-up" can be achieved only with $(I + 1)$ buffers for each of I input tapes. This configuration will ensure that reading can be sustained throughout the pass and cannot become a source of delay.

There are dynamic buffering systems which attempt to simulate double buffering with fewer buffers or which attempt to go beyond double buffering with fewer than the ideal $(I^2 + I)$ buffers. The basic technique is buffer-sharing and some form of *preselection,* or *forecasting* [11]. In basic preselection, a buffer is provided for each file, and a single additional floating buffer is shared by all. Whenever a block is read, the last element of the block is compared with all other last elements. The block with the smallest last element will be depleted first. In anticipation of this depletion, a read is executed to fill the floating buffer with the next block from the file whose current block will be depleted next. At any time there are $(k/2 + 1)$ input buffers; any one file has two buffers associated with it at any one time.

Under certain conditions, $(k/2 + 1)$ input buffering can simulate double buffers perfectly. However, the technique will be inefficient when there is a tendency for several buffers to be depleted simultaneously. In Fig. 13.14, it is determined by preselection that A will be depleted first, and a successor block has been read (or is being read) into the preselect area. Before the depletion of block $A(n)$, block $A(n + 1)$ will be in core and no read hang-up will occur. If the tendency is strong for one block to be depleted at a time, preselection will perform well. In Fig. 13.14, however, B and C will be depleted together. Even if a preselect block is read for B in time, there will

Block A(n)	Block B(n)	Block C(n)	Preselect block A(n+1)
2	5	3	8
4	6	7	9

Fig. 13.14 Preselection.

be an interval until the replacement block is brought in for C. This block would be present if standby blocks existed for all input strings.

The performance of preselection will potentially improve as the number of preselect blocks approaches the number of blocks which would be used for double buffering. Very much the same approach of buffer-sharing can be used to extend the power of double buffering. A common technique is to assign double buffers to each input string. Behind them a shared pool of $((k/2) - 1)$ buffers (or whatever can be afforded) helps to sustain continuous reading and thus avoid the read gaps that eventually lead to write gaps. If multiple-read capability exists in the system, it can contribute to performance by overlapping some percentage of the data read into the system and by providing faster response to multiple depletions.

13.9 MERGE EXERCISE

Figure 13.15 shows the characteristics of a file for which a merge is to be designed. This exercise will demonstrate a process by which various merge blocking factors and buffering levels may be chosen.

1. Determine a tentative merge order. The tentative merge order is the smallest one that results in the minimum number of passes for the expected number of strings. Figure 13.16 shows calculations for determining the number of passes required for a five-way merge, for a four-way merge, and for a three-way merge when the expected number of strings is 500. Since

Characteristics of file	
No. of tapes	10
Maximum order ($k/2$)	5
No. of items	50,000
Item size	100 words
Core	20,000 words
Work space	10,000 words
String length from sort phase	100 items

Fig. 13.15 Worksheet for designing a merge.

Step 1	Expected number of strings =No. of items; string length =		500 strings
Step 2	Maximum merge order will have $\log_5 500$ passes =		4 passes
Step 3	Reduced merge order will have $\log_4 500$ passes =		5 passes
Step 4	Reduced merge order will have $\log_3 500$ passes =		6 passes
Conclusion:	Consider 5-way and 4-way merge orders.		

Fig. 13.16 Tentative merge order.

each reduction in merge order requires another pass, both five- and four-way merges should be considered. If five- and four-way merges had used the same number of passes to reduce strings, 4 would have been used as the merge order.

2. Determine core allocation. For each order of merge under investigation, determine the balance of blocking and buffering for various buffering assumptions. Figures 13.17 and 13.18 show worksheets for four- and five-way merges, respectively. For various assumptions of buffering, the number of buffers, resulting block size, number of blocks, block level and pass level I/O time, and total merge time are calculated. The extended buffering scheme includes two output buffers, double buffering for each input tape, and additional floating buffers. A comparison of the two figures shows that for all buffering levels, the five-way merge is superior to the four-way merge. Note, however, that, for a block size of 1400 characters, the four-way merge with preselection outperforms the five-way merge with minimum buffering. Some systems in the past have had system-fixed block sizes. Some current devices have characteristics that make particular block sizes preferable. For example, track or half-track block sizes on disk devices.

The initial timing assumptions for the worksheets are that the merge is I/O-bound and that there is perfect overlap. For minimum buffering, this combination of circumstances cannot occur. A better timing assumption for

Work space = 10000

Buffering assumptions	Minimum $(k/2)+2$	Preselect $(k/2)+3$	Double $k+2$	Extended $3k/2$
Number of buffers	6	7	10	12
Block size, rounded[1]	1600	1400	1000	800
Number of blocks	3125	3572	5000	6250
Block tape time (ms)[2]	37	33	25	21
Pass time (sec)[3]	115.6	117.8	125.0	131.3
Merge time (sec)[4]	578.0	589.0	625.0	656.5
Merge time (sec), 5 passes[5]	1156.0	736.2	687.5	656.5

[1] Block sizes are rounded off to the nearest 100 words since they contain whole 100 word records. Block size=10000/Numbers of buffers.
[2] Includes 5 ms start time. Transfer rate is 50000 words per second.
[3] Assumes perfect R/W/C overlap. Pass time=Block tape time × Number of blocks.
[4] Assumes perfect R/W/C overlap. Time is Pass time × Number of passes.
[5] Assumes probable R/W/C overlap that can be achieved. Revised "fudge factor" judgment:
   1) Minimum buffer assumes run time is read + write time.
   2) Extended buffer assumes perfect overlap.
   3) Double buffer assumes 10% read delay.
   4) Preselect buffer assumes 25% read delay.

Fig. 13.17 Worksheet for a 4-way merge. $k/2 = 4$.

Work space = 10000

Buffering assumptions	Minimum $(k/2)+2$	Preselect $(k/2)+3$	Double $k+2$	Extended $3k/2$
Number of buffers	7	8	12	15
Block size, rounded[1]	1400	1200	800	600
Number of blocks	3572	4167	6250	8333
Block tape time (ms)[2]	33	29	21	17
Pass time (sec)[3]	115.6	120.8	131.3	141.6
Merge time (sec), 4 passes[4]	462.4	483.2	525.2	566.4
Revised merge time (sec), 4 passes[5]	924.8	604.0	577.7	566.4

[1]Block sizes are rounded off to the nearest 100 words since they contain whole 100 word records. Block size= 10000/Numbers of buffers.
[2]Includes 5 ms start time. Transfer rate is 50000 words per second.
[3]Assumes perfect R/W/C overlap. Pass time=Block tape time × Number of blocks.
[4]Assumes perfect R/W/C overlap. Run time is Pass time × Number of passes.
[5]Assumes probable R/W/C overlap that can be achieved. See Fig. 13.17.

Fig. 13.18 Worksheet for a 5-way merge. $k/2 = 5$.

minimum buffering is that there is no read/write overlap, a condition that doubles pass and merge time. Only extended buffering will come close to running at estimated times. The corrections for preselection and double buffering must be made by the designer, who uses his judgment and whatever knowledge he has of his data.

13.10　FINAL COMMENTS

There are some important merge considerations which we have not touched on. Among them are the following.

1. The distinction made between all other passes of the merge and the "last pass"
2. Checkpointing a merge
3. Predicting final output tape
4. Multireel sort/merge

13.10.1　Last Pass

When the number of remaining strings is reduced to a number equal to or less than the order of merge, a *last-pass* condition is achieved. The last pass of a merge is usually treated as a special phase of the sort/merge process. A common responsibility of the last pass is to interface with users and with subsequent processing programs. It must reblock and reorganize records; often it may transmit special reports, which can be effectively created during the development of the physically ordered file.

The need for record reblocking and reorganization occurs whenever block sizes are changed for the convenience of the merge. The optimum block size for the merge may not be the size in effect when the data entered the sort/merge, and it may not be the size the user expects at the end. It is not uncommon for records to be reformatted. When the record is about to leave the sort/merge process, it must be changed into whatever format the user wants.

13.10.2   Checkpoint

At the end of any merge pass, all the data exists in two places, on the new output tapes and on the tapes that were the original input to the pass. The input tapes are nominally depleted in that they have been completely read in order to form the merge, but of course they have not been erased. If they are dismounted and replaced with other tapes, they constitute a checkpoint for the merge. The merge can be resumed at the pass that used these tapes as input if they are remounted. The decision to dismount and save tapes is a function of the length of time a pass requires. The checkpoint procedure takes time, and the time it *saves* must be more than the time it *takes*.

In large systems with very flexible allocation schemes, it might not be necessary to physically dismount and remount tapes before the next pass of the merge can get under way. The input devices can be released by the sort, and premounted devices with fresh tapes can be dynamically acquired. While the fresh tapes are being written, the checkpoint tapes are being dismounted. The time for this process should be considerably less than for holding up the merge for tape mounting. If the operating system can be entered from the sort to effect the reallocation, the process should be very smooth. The profitability of this approach lies in a trade-off between using all possible available tapes for the merge and minimizing the cost of checkpoints.

If backward tapes are being used, it is probably desirable to take checkpoints after the input tapes have been read backwards and are near load point. When they are remounted, the strings will look like descending strings, but the restart pass can write descending strings as output and not suffer serious penalty.

Dismounting initial input tapes to save initial input data is a common requirement. Its treatment need not be different from that for other checkpoints.

13.10.3   Final Output Tape

Commonly the sort/merge tells the user where his ordered tape reel is. How much more can be done depends on how strongly the user insists that the final reel be on a particular tape drive. It is not difficult to arrange for the final output to be on either of two possible tapes. The dispersion algorithm for a pass can as easily be T_1, T_2, T_3 as T_3, T_2', T_1 or T_2, T_3, T_1, etc. On the last pass only one tape receives output, and it can be any of the available tapes. The difficulty is that it is not always possible to predict with absolute certainty whether there will be an odd or an even number of passes. If the number of passes can be predicted, then initial dispersion can be adjusted so that the proper set of $k/2$ output tapes can be selected, and within that set the selection of a given tape is easily accomplished. If the number of passes is not predictable, as is true when natural merging and string extension are used, then copying the final string onto the desired tape may be necessary.

13.10.4   Multireel Sort/Merge

There are two situations in which it is necessary to consider the nature of sorting for data that does not fit on one reel. One exists when the data exceeds one reel but fits on $k/2$ intermediate reels, and the multireel nature of the input and final output is accommodated by some simple technique for device allocation [6]. When the first reel is depleted, the second reel is mounted or, if an additional tape unit is available, the input unit address is switched to refer to the alternative drive. Since the data will not again be written on one tape until the final pass, the sort/merge proceeds normally until that pass.

The other situation exists when there is too much data to be accommodated by the sort/merge. Then separate sort/merges are undertaken, and the sorted subfiles are merged together. The design of these merges follows the principles discussed in the chapter on internal merging. The construction of the minimum tree for strings of various lengths and for orders of merge of various lengths applies perfectly to the multireel merge.

14
Polyphase Tape Merging

14.1 GENERAL

The Balanced Merge technique required k tapes to undertake a merge of $k/2$. Any time $(k/2) + 1$ tapes are active, $k/2$ are being read, one is being written, and $(k/2) - 1$ tapes are idle. Modification to the logic of the merge can increase the utilization of tapes. A family of "unbalanced" merges constitutes an attempt to achieve a merge power greater than $k/2$ by reading more than $k/2$ tapes at a time, using up to $k - 1$ inputs and a single output. Such unbalanced merges (which differ from one another in the range of merge orders actually used during a merge) form a continuum: At one end is Polyphase, which maintains a constant number of $(k - 1)$ input tapes. At the other is Cascade, which uses $k - 1$, $k - 2$, $k - 3$, $k - (k - 1)$ input tapes during various phases of a merge pass. Between these extremes are many "Compromise" merges distinguished by the range of inputs used during each pass.

The efficiency of the unbalanced merges relative to one another and to the Balanced Merge depends on k, on the number of strings, on architectural details of the running machine, and on many other factors. Comparative remarks are made during the description of the methods (Polyphase in this chapter, Cascade and Compromise in the next), and the methods are reviewed in Chapter 17.

14.2 FORWARD-READ POLYPHASE: THREE TAPES

If an internal dispersion pass has placed eight strings on tapes T_1 and five strings on T_2, the following merge can occur:

217

Phase	T_0	T_1	T_2
0	—	8	5
1	5(2L)	3	—
2	2(2L)	—	3(3L)
3	—	2(5L)	1(3L)
4	1(8L)	1(5L)	—
5	—	—	1(13L)

This is a phase-oriented partial-pass merge known as the "three-tape polyphase" [16]. It accomplishes a three-tape merge by shuttling strings onto a tape depleted in a prior phase until one of two current input tapes is depleted. It depends upon an uneven number of strings on each tape. For optimum performance, the pattern of strings must be a Fibonacci series.

14.2.1   Dispersion Pass: Fibonacci Series

In the example above, note that the number of strings $S = 13$. The dispersion placed eight strings on one tape and five strings on the other. These numbers are part of a Fibonacci series, in which each value is the sum of its two predecessors. The standard Fibonacci series is:

$$1, 1, 2, 3, 5, 8, 13, 21, 34, 55, 89, 144, ...$$

For any file, the dispersion pass attempts to make the number of strings on the two tapes receiving strings equal the two preceding Fibonacci numbers; that is, if $S = 13$, the dispersion will produce strings of eight and five, respectively. After dispersion every merge phase reduces the number of strings to the next two Fibonacci numbers,

$$(5,3), \quad (3,2), \quad (2,1), \quad (1,0),$$

until the records are ordered. Whenever the final numbers of strings on two dispersion tapes are consecutive Fibonacci numbers, an "ideal level" has been attained for the Polyphase merge.

If we know in advance that S is a Fibonacci number, we can calculate, in the beginning, exactly how many strings to place on each tape during dispersion. If we do not know S, we can still attempt to achieve a final Fibonacci sequence by generating intermediate Fibonacci sequences as long as the input lasts. This procedure builds the ideal levels for various values of S, as shown in Fig. 14.1. If S is not a Fibonacci number, the input will be depleted before the last ideal level is reached, and an adjustment to the Polyphase process, as discussed later, will be required. It makes no difference which tape has the larger number of strings, since any permutation of Fibonacci numbers is acceptable so long as an ideal level exists. Rules for computing strings to be placed on each tape are given in Section 14.3.

Input	Dispersion			
T_0	T_1	T_2	Fibonacci level	S
1	0	1	1	1
2	1	1	2	2
3	2	1	3	3
5	3	2	4	5
8	5	3	5	8
13	8	5	6	13
21	13	8	7	21
34	21	13	8	34
55	34	21	9	55

Fig. 14.1 Three-tape Polyphase merge, ideal Fibonacci levels

14.2.2   Merge Pass

The first line of Fig. 14.2 shows how 21 strings are distributed to two tapes. Since 21 is a Fibonacci number, the dispersion of 13 and 8 strings is an ideal level. The remaining lines show the phases of the merge, in which string lengths are extended by merging from two tapes onto the third, with ideal levels always maintained along the way. Since each step depletes only one of the merge tapes, each phase except the last is a partial pass.

Each phase of the Polyphase Merge empties the tape with the smaller number of strings. At the end of a phase, only the depleted tape and the newly filled output tape for the phase are rewound. This ensures that all tapes will be read in the forward direction on the next phase. The pattern of read/write overlap is such that a machine with read/write/compute can perform the merge efficiently (with a constraint to be shown later).

Since the merge does not move all strings in each phase, its effectiveness is not easily observed. Only the dispersion pass and the last merge phase move all strings in the file. Whereas in a Balanced Merge the string reduction power is $k/2$ and the number of merge passes is $\lceil \log_{k/2} S \rceil$, it is not so simple

Pass	T_0	T_1	T_2	No. of strings	Current length	String lengths moved
	—	13 (L)	8 (L)	21	1	—
1	8 $(2L)$	5 (L)	—	13	2	16
2	3 $(2L)$	—	5 $(3L)$	8	3	15
3	—	3 $(5L)$	2 $(3L)$	5	5	15
4	2 $(8L)$	1 $(5L)$	—	3	8	16
5	1 $(8L)$	—	1 $(13L)$	2	13	13
6	—	1 $(21L)$	—	1	21	21
						96

$$\frac{\text{String lengths moved}}{\text{Number of strings}} = \frac{96}{21} = 4.6 \text{ data passes, effective merge order.}$$

Fig. 14.2 Three-tape Polyphase merge, effective merge order.

to describe the Polyphase Merge. Much theoretical work has been done in computing the "effective order" of a Polyphase Merge. Gilstad [17] has calculated that with an infinite number of strings and for any number of tapes the maximum effective order of Polyphase is 4. This means that Polyphase will never reduce the number of strings to less than $\frac{1}{4}$ the number of input strings on any full pass. For the nontheoretically inclined, a very useful technique for computing the effectiveness of a merge is to divide the total number of strings passed during a merge by the number of initial strings. The quotient is the number of data passes. Figure 14.2 contains a calculation which shows that the merge of that figure requires 4.6 data passes. The required calculation is not cumbersome, even for many tapes and large populations of strings, and provides a figure of merit for Polyphase which can be used to compare it to other techniques.

14.3 POLYPHASE MERGE WITH MORE TAPES

The Polyphase Merge may be used with any number of tapes. The Fibonacci series must be generalized in order to determine what the ideal levels are with various values of S and k. The generalization applies to $k = 3$ as well.
    Given k tapes and S initial strings:

1. Distribute one string to all of $k - 1$ tapes. (The input reel is on the kth tape.) Distribute an additional string to all but one tape.

2. Identify the "augmentation" tape. This tape has the property of having the most strings. To form the next level add the number of strings on it to all other tapes but not to itself. When more than one tape has the same number of strings, select the augmentation tape arbitrarily.

3. Repeat step 2 until the input is depleted.

T_0	T_1	T_2	T_3	T_4	No. of strings	Level
I	0	0	0	1	1	0
I	1	1	1	1*	4	1
	+1	+1	+1	—		
I	2	2	2*	1	7	2
	+2	+2	—	+2		
I	4	4*	2	3	13	3
	+4	—	+4	+4		
I	8*	4	6	7	25·	4
		+8	+8	+8		
I	8	12	14	15	49	5

*Augmentation tape

Fig. 14.3 Polyphase dispersion ($k = 5$).

For programming convenience, it is normal to have a rule for breaking ties when locating the augmentation tape. We use a rule that associates the tapes with the columns of a worksheet and selects the rightmost tape when there are ties.

An example of the procedure (Fig. 14.3) shows how it works for $k = 5$ and $S = 49$. The data in Fig. 14.3 can be arranged in a generalized Fibonacci series, in which each sequence of four numbers is treated as a group:

(0,0,0,1), (1,1,1,1), (2,2,2,1), (4,4,2,3), (8,4,6,7), (8,12,14,15), ...

Each group of $k - 1$ elements is an ideal level. It is formed by adding the maximum element of the predecessor group to the $k - 2$ nonmaximum elements in the predecessor. This rule works for all $k > 2$; thus,

k	Fibonacci series	Reference
3	(0,1) (1,1), (2,1), (2,3), (5,3), ...	Fig. 14.1
4	(0,0,1), (1,1,1), (2,2,1), (4,2,3), (4,6,7), (11,13,7),	
	(24,13,20), ...	Fig. 14.7
5	(0,0,0,1), (1,1,1,1), (2,2,2,1), (4,4,2,3), ...	Fig. 14.3
6	(0,0,0,0,1), (1,1,1,1,1), (2,2,2,2,1), (4,4,4,2,3), ...	

Another way of calculating the value of the ith group in a Fibonacci series of $k - 1$ elements is to add the $k - 1$ predecessors. Thus, for $k = 5$ the fifth ideal level is

$$25 + 13 + 7 + 4 = 49.$$

The augmentation method is preferred because it isolates the current status of each tape, which is needed as a base for counting.

After the dispersion pass achieves an ideal level on $k - 1$ tapes, the merge pass steps back down through the generalized Fibonacci series until there is only one string left. Again, permutation of any ideal level is still an ideal level. (Compare the sequence for $k = 3$ above with Fig. 14.1. This fact may be useful when it is desirable, for any reason, to use a specific tape on a specific channel at a specific time.) Thus, in Fig. 14.3, any dispersion that produces 8, 12, 14, and 15 strings on the $(k - 1)$ merge tapes is acceptable. Note that the next ideal level for $k = 5$ is $S = 94$. Any file, $49 < S < 94$, will require an adjustment to make the Polyphase Merge effective. Figure 14.4 shows the merge phase for the dispersion of Fig. 14.3. Each phase of the merge reduces to the next "ideal" level below the starting level, and the number of strings decreases in a generalized Fibonacci series. In Fig. 14.4 there are five partial passes, equivalent to only 3.26 full data passes.

T_0	T_1	T_2	T_3	T_4	No. of strings	String lengths moved
—	8 (L)	12 (L)	14 (L)	15 (L)	49	—
*8 (4L)	—	4 (L)	6 (L)	7 (L)	25	32
4 (4L)	*4 (7L)	—	2 (L)	3 (L)	13	28
2 (4L)	2 (7L)	*2 (13L)	—	1 (L)	7	26
1 (4L)	1 (7L)	1 (13L)	*1 (25L)	—	4	25
—	—	—	—	*1 (49L)	1	49
						160

*Output tape for the phase

Data passes: 3.26

Fig. 14.4 Merge phase for strings dispersed in Fig. 14.3.

14.4 DISPERSION FOR FORWARD POLYPHASE

There are two techniques for dispersing initial strings for a Polyphase Merge: "vertical" and "horizontal." Vertical dispersion brings each tape to the desired level in turn; horizontal dispersion brings all tapes gradually toward the desired level, placing one string at a time across all tapes. There are variations in both dispersion methods that have implications for Polyphase performance. Achieving an ideal level is a relatively rare event, and the actual dispersion of strings at the time of the depletion of input is important. The actual number of string lengths moved during the merge depends on the specific pattern of strings that form an imperfect dispersion.

The dispersion algorithm for a Polyphase Merge must determine how many strings to place on each tape to achieve the next level. It must also keep a record of the dispersion of strings as it is actually accomplished. The number of strings to be placed on each tape to achieve the next level is independent of whether a horizontal or a vertical dispersion of strings is performed.

Two counters representing each tape in the merge are established during the dispersion pass. One (the "ideal" counter) represents the number of strings required on the tape at the next level; the other (the "actual" counter) represents the number of strings on the tape. Initially the ideal counters are set to 1. If input is depleted before the first perfect level (one string on each tape) is achieved, then the merge of all existing strings forms the ordered string on a single pass, which is effectively the "last pass." This situation is often referred to as "skipping the merge phase."

When the initial strings are laid out, an algorithm adjusts the ideal counters for each tape. Algorithms will always define a "next" level for all tapes, but the specific permutation of strings to be put out for the new level will vary. Factors affecting the algorithm include the desire to predict the final output tape and the ability to use backward reads in the merge passes. We will discuss these considerations further on.

The vertical dispersion described by Reynolds [37] has the characteristic that the tape with the highest number of strings at any ideal level is the tape that had the highest number of strings at the previous level. Further, the tape with the lowest number of strings at a level is the tape that had the lowest number of strings at the previous level. Figure 14.5 shows this relationship and also suggests a counter-manipulation mechanism. After the initial level of 1, 1, 1 has been established, a next level can be developed by adding T_1 to each of the other tapes in turn to form a new sum for T_1, T_2, ..., T_{k-1}. Tape T_1 is always the tape with the most strings at any level (and always the augmentation tape), and T_{k-1} always has the fewest strings. Once a level is formed as shown in Fig. 14.5, the dispersion to achieve it may be undertaken vertically or horizontally.

T_0	T_1	T_2	T_3
I	1	1	1
	2	2	1
	4	3	2
	7	6	4
	13	11	7
	24	20	13

Level $i + 1$:　$T_1 = T_1 + T_2$;　$T_2 = T_1 + T_3$;　$T_3 = T_1 + 0$;　$T_{k-1} = T_1$

Fig. 14.5 Reynolds dispersion.

T_0	T_1	T_2	T_3
I	1	1	1
	2	2	1
	4	2	3
	4	6	7
	11	13	7
	24	13	20

Fig. 14.6 Mendoza dispersion.

Another mechanism for forming ideal counters, described in Section 14.3, varies the augmentation tape from level to level (see Fig. 14.6). It is described in detail by Mendoza [35].

Figure 14.7 shows the step-by-step movement from a level of 13, 11, 7 to 24, 20, 13, by vertical and by horizontal dispersion. The alternative horizontal dispersion shows another new level, that is an equivalent permutation of the new level.

Vertical dispersion

T_0	T_1	T_2	T_3	No. of strings	Comments
I	13	11	7	31	
I	24	11	7	42	T_0 at next level
I	24	20	7	51	T_1, T_2 at next level
I	24	20	13	57	All at next level

Horizontal dispersion

T_0	T_1	T_2	T_3	No. of strings	Comments
I	13	11	7	31	
I	14	12	8	34	Each tape incremented by 1
I	19	17	13	49	T_3 full
I	22	20	13	55	T_2 full
I	24	20	13	57	T_3 full

Alternative horizontal dispersion

T_0	T_1	T_2	T_3	No. of strings
I	13	11	7	31
I	13	24	20	57

Fig. 14.7 Vertical and horizontal dispersion.

A critical difference between dispersions is the status of strings on tape if an ideal level cannot be achieved. An ideal level is not achieved when input is depleted with any of the string layouts in Fig. 14.7 between 13, 11, 7, and 24, 20, 13. If an ideal level is not achieved, some adjustment technique is necessary in order to undertake a Polyphase merge. The technique may be a partial pass or assumed dummies. Partial passes attempt to move and combine strings so as to get back to the previously achieved perfect level. Assumed dummies pretend that a perfect level has been achieved by "padding" with strings of length 0.

Mendoza and Reynolds describe partial-pass algorithms for various interlevel conditions in detail. In general, adjustments for horizontal conditions provide for more efficient merges. (This text does not describe such adjustments.) The usage of assumed dummies has become general because of its efficiency and simplicity.

14.5 DUMMIES

It is possible to pretend that an ideal level has been achieved. The assumed level is the one being sought at the time of depletion. The method is called the *method of assumed dummies* [37]. Dummies are strings that are represented only by internal counters. Almost any actual dispersion can be treated as an ideal level by use of internal counters. The specific treatment of dummy counters, however, will result in performance differences for the Polyphase merge. The dummy method will work with either horizontal or vertical dispersion, but its most effective use may lead to "mixed distributions," in which some of the strings are dispersed horizontally and some vertically.

T_0	T_1	T_2	T_3	T_4	T_5	
I	8	16	15	14	12	Ideal
—	8	12	10	9	7	Actual
	0	4	5	5	5	Dummy

Fig. 14.8 Dummy counters.

Figure 14.8 shows the completion of a dispersion and a set of associated dummies. Each tape has a dummy counter whose value is the difference between the number of strings actually on the tape and the number of strings required for the ideal level. These counters may be established by subtracting the actual counters from the ideal counters or by keeping a running count of strings yet to be placed on a tape.

The Polyphase Merge inspects the dummy counters for each tape when it wishes to include a string from that tape in the merge. If the counter is at 0, the tape is treated normally. If the counter is not a 0, then a 1 is subtracted from the counter and no I/O is undertaken for that tape. As counters reach 0, more and more tapes begin to contribute real strings to the merge. It is not necessary that a single pass clear all dummy counters.

In Fig. 14.9 an ideal level of (8,16,15,14,12) has not been achieved, and T_1 is initially the only tape with a zero dummy. A string is copied from T_1 to T_0 and all other counters are reduced by 1. Conceptually, this process merges strings of length 0 from T_2 through T_5 with a string of length (L)

Real						Dummy						String lengths moved
T_0	T_1	T_2	T_3	T_4	T_5	T_0	T_1	T_2	T_3	T_4	T_5	
0	8	12	10	9	7	—	0	4	5	5	5	
4 (L)	4	12	10	9	7	0	0	0	1	1	1	4
4 (L) 1 (2L) } 3		11	10	9	7	0	0	0	1	1	1	2
						0	0	0	0	0	0	
4 (L) 1 (2L) 3 (5L) } 0		8	7	6	4							15
												21

Fig. 14.9 Merge with dummies. (Ideal assuhl ws 8, 16, 15, 14, 12.)

from T_1. When this has occurred four times, there will be four strings of length $1L$ on T_0, and the dummy counters for T_2 through T_5 will be 0, 1, 1, 1, respectively. Since T_2 has a zero dummy counter, it will now begin to contribute real strings to the merge. When one string of length $2L$ is formed by the merge of strings from T_1 and T_2, all of the dummy counters become 0. At this time all tapes begin to contribute strings to the merge. Figure 14.9 shows that, when T_1 is depleted, a perfect real level (8,8,7,6,4) is achieved. Twenty-one string lengths have been moved.

The specific dispersion at depletion time may affect performance, and dummy counter strategy is not without effect. One method used to control dummies is called *balancing*.

The goal is to develop dummy values for all tapes which are as close to one another as possible. In order to achieve this goal, the dispersion algorithm is changed so that the tape with the smallest increment required for the next level receives no strings until other tape dummy counters are reduced to its value.

The effect of balancing is to minimize string copying and partial merging, replacing copies and partial merges with dummy movement. One reason for the preferability of one permutation of a level to another lies in the desire to control dummy dispersion from level to level. A further reason for choosing a particular permutation lies in the desire to pick the final output tape. The Polyphase Merge will always place the final ordered string on the tape that had an odd number of strings or an odd sum of strings and dummies at the end of dispersion. By controlling dispersion so that the desired output tape always has an odd number of strings from level to level, one can ensure the placement of ordered data on the desired tape. At every $(k - 1)$th level, however, all tapes contain an odd number of strings. In order to predict the final output tape, one must reject these "all-odd" levels as invalid, and establish the successor level as the next ideal.

14.6   BACKWARD POLYPHASE

The desire to avoid rewind and to use backward-read tapes to advantage has a number of profound effects on the Polyphase Merge. The dispersion algorithm must be expanded to include an entirely new requirement. The proper placement of dummies becomes considerably more subtle. The performance of the sort used for creating initial strings may be affected.

14.6.1   Dispersion

The first merge pass of Fig. 14.10 produces seven strings on T_0. If no rewind is undertaken after the initial dispersion, the first phase will read T_1, T_2, T_3 backwards until T_3 is depleted. The strings formed on T_0 will necessarily

T_0	T_1	T_2	T_3
—	13A	11A	7A
7D	6A	4A	0

Fig. 14.10 Backward-read.

T_0	T_1	T_2	T_3
	A01	D02	D03
	D04	A05	A06
	A07	D08	D09
	D010	A011	A012 ←
	A013	D014	
	D015	A016 ←	
	A017 ←		
	7	6	4

FIG. 14.11 Alternating direction of strings.

be strings in descending order. At the completion of the phase the strings on T_1 and T_2 will be initial strings of ascending order, and the merge will be unable to continue.

In order to effect a backward Polyphase, the sort and the dispersion algorithm must take cognizance of the direction of strings. To undertake the merge, it is necessary to achieve a level with only one tape holding an odd number of strings and to alternate the direction of strings on each tape. The first string placed should be a string of the desired final direction on the tape which will hold an odd number of strings. Figure 14.11 shows 7, 6, 4 dispersion which meets the requirements. Each A represents an ascending strings, each D a descending string. Numerals next to the As and Ds represent the order in which that string was placed. This dispersion assumes that T_1 was the desired output tape. The first strings merged will be A17, A16, A12, to form a descending string on T_0. Figure 14.12 shows the merge. The initial digit of each string shows the pass that produced it. The numbers in parentheses show the component strings. The merge is like the forward Polyphase in its passing characteristics, in its string growth, and in the reduction of strings. It differs only in the requirement for alternating string direction and in the absolute requirement for odd-even dispersion.

14.6.2   Dummy Placement

Since tapes will be read forward and backward, the clustering of long or short strings at the front or the back of a tape (important when reading forward) loses significance. It is possible to ignore this detail and handle dummies as though they were at the front end of the tapes; thus dummies

T_0	T_1	T_2	T_3	Comments
—	A01	D02	D03	
	D04	A05	A06	
	A07	D08	D09	
	D010	A011	A012	
	A013	D014		Initial
	D015	A016		
	A017			(7,6,4)

T_0	T_1	T_2	T_3	
D101 (17,16,12)	A01	D02	0	End phase 1
A102 (15,14,9)	D04	A05		
D103 (13,11,6)	A07			
A104 (10,8,3)				(4,3,2)

T_0	T_1	T_2	T_3	
D101	A01	0	D201 (104,7,5)	End phase 2
A102			A202 (103,4,2)	(2,2,1)
			D201	End phase 3
D101	0	D302 (102,202,01)		(1,1,1)
	A401 (101,301,201)			End merge

Fig. 14.12 Backward Polyphase merge.

are always considered to precede strings to be merged. But this approach will not provide for an optimum Polyphase Merge.

An approach to optimum dummy placement for backward-read Poly-phase would "scatter" dummies on tape so that they occupied string positions that are most often used, keeping real strings in those positions that are less frequently involved in the merge. Dispersion algorithms to provide for this approach are a current development topic among advanced workers in the field. The interested reader will find Shell's [40] article (listed in the bibliography) useful.

14.6.3   String Length

If a Replacement Selection internal sort is used to form initial strings, the need to change string direction may have a negative effect on string size and consequently produce more strings. The degree of impact of more strings is a function of the dispersion algorithm and ordering in the data. The production of a short string (closer to $1.5F$ than $2F$, where F is the number of elements in the tournament) occurs because the change in direction makes each string, effectively, an initial string. The dispersion algorithm will determine how often a change in direction occurs and consequently how often there is exposure to a short string. A strict vertical algorithm will alternate A and D strings so that every string changes direction. A horizontal algorithm will tend to have sequences of strings of the same direction. The sequence numbers in Fig. 14.12 show this effect.

14.7   INTERNAL EFFECT OF POLYPHASE

The nominal $(k - 1)$ merge has the following effect on the internal organization of the merge phase.

1. $k - 1$ rather than $k/2$ elements are involved in the comparison mechanism. More comparison time will be needed to select the next member of the output string.

2. More space will be needed for I/O support. A minimum of $k - 1$, rather than $k/2$, block spaces are required merely to operate; $2k - 2$, rather than k, buffer areas are needed to provide a backup block for each input. On the other hand, depletion rates are slower and a preselection technique may be adequate. If, because of the large block requirements of a $k - 1$ merge, there is a need to reduce buffering below effective levels or to reduce block size, the performance of a particular Polyphase Merge, even though it will move less data, will seriously degrade. With large tape populations and moderate-sized cores, there may be many cases in which the Balanced merge, even when moving more data, will outperform Polyphase because of its more modest core requirements.

14.8   I/O SUBSYSTEM AND POLYPHASE

The capabilities of the I/O subsystem have a serious bearing on the suitability of the Polyphase Merge for a system. It is necessary that the I/O subsystem be able to accommodate $(k - 1)$ tape input and one tape output. For single-channel machines, performance of the Polyphase Merge should present no particular problem, since read/write overlap is not possible in any case.

Two-channel systems will experience read/write interference delays while using a Polyphase Merge, because with such systems there can be no guarantee that reading an input tape will not conflict with writing. With $k - 1$ input tapes dispersed across two channels, the probability is very high that some reading will have to be done from a tape that is on the same channel with an output tape. Systems that spontaneously define free channel paths to devices, or that provide alternative paths, will overcome this difficulty.

14.9   FLEXIBILITY IN TAPE USE

The modifications to the dispersion of strings discussed in this section are not central to the topic of the Polyphase Merge. However, they may be useful to a programmer developing his own Polyphase Merge in a data-processing environment.

14.9.1   Piggybacking

In the interest of system efficiency, the user of the sort may wish to produce some reports during the dispersion phase. Given a six-tape system, he would have only four rather than five tapes available for the dispersion of strings. However, the user wishes to take full advantage of the six-tape system during the merge. Figure 14.13 shows this situation. Tape 0 is the input tape. Tape 1 is given over to "first pass own code" (FPOC), and is being used for some output unrelated to the sort.

T_1	T_2	T_3	T_4	T_5	T_6	
I	FPOC	0	0	0	0	
I	FPOC	2 (1,1)	1	1	1	5
	FPOC	2 (1,2)	2	2	2	9
	FPOC	5 (3,2)	4	4	4	17
	FPOC	10 (6,4)	7	8	8	33
	4	6	7	8	8	

Fig. 14.13 Piggyback.

The dispersion of strings is organized as if all five output tapes were available. One of the tapes, however, does double duty, holding the strings for itself and for the tape that is not available for dispersion. Unless there is a constraint, such as final output selection, the tape doing double duty should represent the tapes with the fewest strings. In this way the subsequent copy which occurs when the dispersion phase ends is minimized. When Tape 1 (Fig. 14.13) becomes available to the merge, the strings from T_2 are copied to T_1, and a proper five-way merge may be undertaken. With backward Polyphase, some care must be taken to ensure proper string direction. If an ideal level is not achieved, a copy is made and dummies are adjusted to represent whatever depletion pattern would have existed if all tapes had been taking strings.

14.9.2   Last Pass Release

A requirement analogous to that for the use of tapes on the first pass may exist for the last pass. The user may wish to take full advantage of all tapes throughout the dispersion and merge until the last pass, when he wishes to have extra tape available for "last pass own code" (LPOC). He may choose the tape he wishes to use during the last pass.

The dispersion is adjusted so that the tape which is to be empty at the start of the last pass will be depleted by that time. No string is placed on that tape when the first level of the dispersion is established. Throughout the merge, this tape remains short one string, and it will therefore be depleted by the last pass. Figure 14.14 shows the early release.

T_0	T_1	T_2	T_3	T_4	T_5	
0	1	1	1	0	1	
0	2	2	2	1	1	
	2	4	4	3	3	Dispersion
0	6	4	8	7	7	
4	2	0	4	3	3	
2	0	2	2	1	1	Merge
1	1	1	1	0	0	
0	0	0	0	X	1	Last pass

Fig. 14.14 Early release.

14.9.3   Rerun

Rerun points may be established after each phase by dismounting the tape which is depleted at phase end and replacing it with a new tape. If a malfunction requiring a rerun occurs in the next phase, a remounting of the saved tape and a repositioning of the other tapes will return the run to the condition at the beginning of the phase.

15

Cascade
and
Compromise
Tape Merges

15.1 GENERAL

The Cascade Merge is a nominal $(k - 1)$-way merge much like the Polyphase Merge. The differences between them, as a rule of thumb, make the Cascade more powerful for larger tape populations and the Polyphase more powerful for smaller tape populations. Although Cascade was initially conceived as a method for backward-read tapes, a forward-read version is feasible, and for the sake of consistency with the discussion of Polyphase, the description of the method will begin with a forward-read context.

15.2 FORWARD-READ CASCADE: DISPERSION AND MERGE

The ideal Cascade levels for a three-tape system are identical to the levels for the Polyphase described in Chapter 14. For larger tape populations the generalization of the Fibonacci series, which defines ideal levels for a Cascade Merge, differs from Polyphase.

Figure 15.1 shows an ideal level for a four-tape Cascade. The derivation of an ideal level will be described shortly. Phase 1 undertakes a three-way $(k - 1)$ merge of 6 strings. When T_3 is depleted, it is rewound and a two-way $(k - 2)$ merge of five strings is performed as phase 2. When T_2 is depleted, it is rewound and a one-way $(k - 3$ or copy phase) merge is made onto T_2. When this phase (3) is completed, a new ideal level has been achieved and the pass is finished. A Cascade "pass" is a succession of phases. Each phase has its own merge order. The first phase has an order of $(k - 1)$, and successive phases have orders of $(k - 2)$, $(k - 3)$, ... until the last phase, a

T_0	T_1	T_2	T_3	No. of strings	Phase	No. of strings moved
–	14	11	6	31	0	
6 $(3L)_R$	8	5	0_R	19	1	18
6 $(3L)$	3	0_R	5 $(2L)_R$	14	2	10
6 $(3L)$	0_R	3 $(L)_R$	5 $(2L)$	14	3	3
						31

Partial pass	Merge order	No. of strings
1	$k-1$	6
2	$k-2$	5
3	$k-3$	3

Fig. 15.1 One pass of Cascade (R = rewind).

copy (which may be omitted) concludes the pass. Each phase ends with a depleted input tape. After being rewound, the depleted tape is used as output for the next phase. The output tape of a phase is also rewound. There are $(k-1)$ phases per pass. Each phase requires two rewinds, so there are $2(k-1)$ rewinds per pass. The rewind of the depleted tape must be completed before the next phase starts, since it will be the output tape for the next phase. This rewind time is always additive to total merge time. The output tape of a phase may be rewound during the succeeding phase, since it is not used. In any case, it may overlap the additive rewind.

Despite its nominal power of $(k-1)$, the merge does not sustain an effective merge order of $(k-1)$. Where on a continuum between $k/2$ and $(k-1)$ the power of a Cascade Merge will lie depends on the number of tapes and the specific string population. The power depends on the percentage of strings involved in the $(k-1)$-way merge, the $(k-2)$ merge, etc. These percentages naturally depend on the specific level from which a pass starts. The method tends to be more powerful for large string populations. The string-reduction power is always superior to the power of a Balanced merge. The Cascade Merge will characteristically require fewer data passes. There will be string populations which require the same number of passes, but as the number of strings become significantly large, the Cascade Merge begins to save passes.

15.3 CASCADE MERGE LEVELS

To generate levels, disperse $(k-1)$ strings under column headings representing one input and $k-1$ output tape units. Call the first dispersion tape T_1. Each succeeding level is developed as a summation of strings.

1. $T_1(L + 1) = T_1L + T_2L + T_3L + ... + T_{k-1}L$
2. $T_2(L + 1) = T_1L + T_2L + T_3L + ... + T_{k-2}L$
3. $T_{k-1}(L + 1) = T_1L$

Here L is the current level, and $L + 1$ is the next desired level. Statement 1 means that the number of strings on T_1 at the next Cascade level is the sum of all strings on tapes at the achieved level. Statement 2 means that the number of strings on T_2 at the next level is the sum of strings on all tapes at the achieved level except the last tape. Each summation excludes an additional tape, until the last tape at the next level will be assigned the number of strings on T_1 at the prior level.

Figure 15.2(a) shows ideal levels for a four-tape Cascade. Figure 15.2(b) shows the string components of each tape according to statements 1, 2, and 3, above. Figure 15.3 shows how 190 strings would be merged by a six-tape Cascade. The ideal level of the dispersion is 55, 50, 41, 29, 15. Lower levels can be seen in the pass summary. The Cascade will pass through each of these ideal levels as it proceeds. At the end of any pass, the smallest strings are on the tape with the fewest strings.

Level	T_0	T_1	T_2	T_3	No. of strings
1	–	1	1	1	3
2		3	2	1	6
3		6	5	3	14
4		14	11	6	31
5		31	25	14	70
6		70	56	31	157

a) Ideal level

Level	T_1	T_2	T_3
2	1 + 1 + 1	1 + 1	1
3	3 + 2 + 1	3 + 2	3
4	6 + 5 + 3	6 + 5	6
5	14 + 11 + 6	14 + 11	14
6	31 + 25 + 14	31 + 25	31

b) Level formation

Fig. 15.2 Cascade levels.

The great string-passing capability of the method comes from the fact that the gap between levels is much wider than the gap between levels for Polyphase. More additional strings are required to define a new level, and any given number of strings will be merged in fewer passes.

T_0	T_1	T_2	T_3	T_4	T_5	No. of strings	
—	55	50	41	29	15	190	
15 (5L)	40	35	26	14	0		5-way
15 (5L)	26	21	12	0	14 (4L)		4-way
15 (5L)	14	9	0	12 (3L)	14 (4L)		3-way
15 (5L)	5	0	9 (2L)	12 (3L)	14L (4L)		2-way
15 (5L)	0	5 (L)	9 (2L)	12 (3L)	14 (4L)	55	Copy or rewind
10 (5L)	5 (15L)	0	4 (2L)	7 (3L)	9 (4L)		5-way
6 (5L)	5 (15L)	4 (14L)	0	3 (3L)	5 (4L)		4-way
3 (5L)	5 (15L)	4 (14L)	3 (12L)	0	2 (4L)		3-way
1 (5L)	5 (15L)	4 (14L)	3 (12L)	2 (9L)	0		2-way
0	5 (15L)	4 (14L)	3 (12L)	2 (9L)	1 (5L)	15	Copy or rewind
1 (55L)	4 (15L)	3 (14L)	2 (12L)	1 (9L)	0		5-way
1 (55L)	3 (15L)	2 (14L)	1 (12L)	0	1 (50L)		4-way
1 (55L)	2 (15L)	1 (14L)	0	1 (41L)	1 (50L)		3-way
1 (55L)	1 (15L)	0	1 (29L)	1 (41L)	1 (50L)		2-way
1 (55L)	0	1 (15L)	1 (29L)	1 (41L)	1 (50L)	5	Copy or rewind
0	1 (190L)	0	0	0	0	1	Last pass
—	55	50	41	29	15	190	
15	0	5	9	12	14	55	
0	5	4	3	2	1	15	
1	0	1	1	1	1	5	
0	1	0	0	0	0	1	

Note: Underlined tapes output or excluded from phase.

Fig. 15.3 Cascade merge of 190 strings.

15.4 BACKWARD CASCADE

The differences between backward-read and forward-read Cascade are minor. In place of rewinds, tape reversals are undertaken. No rewind is performed until the sort is complete. Depending on rewind and reversal characteristics, the increase in efficiency can be considerable. The penalty is the possibility of developing the final string in the wrong direction. This problem can be handled by doing the necessary copy or by guaranteeing that there will always be an even number of passes. Guaranteeing an even

T_0	T_1	T_2	T_3	T_4	T_5	Reversals
—	15A	14A	12A	9A	5A	All, T_0 rewound
5D	10A	9A	7A	4A	0	T_0, T_5
5D	6A	5A	3A	0	4D	T_5, T_4
5D	3A	2A	0	3D	4D	T_3, T_4
5D	1A	0	2D	3D	4D	T_2, T_3
5D	0	1D	2D	3D	4D	T_1, T_2

Fig. 15.4 Backward Cascade, one pass (A = ascending string; D = descending string).

number of passes (as is done in the SODA sort for UNIVAC III) may have a severe impact on dispersion technique.

Figure 15.4 shows a pass of backward-read Cascade indicating where reversals occur. Note that inverted strings are not reused during the pass. After the initial dispersion of strings, all tapes are reversed and read backward. Subsequently tapes are reversed as they are depleted and after they receive strings. The final copy phase is optional and depends on how much it costs on the hardware being used. The remaining string may be effectively reversed by a rewind.

15.5 STRING DISPERSION FOR CASCADE

The placement of strings for a Cascade Merge may be vertical, horizontal, or mixed; the interlevel adjustments may be partial-pass or dummy-controlled. The Cascade Merge commonly pays a higher penalty for failing to reach an ideal level than Polyphase pays.

15.5.1 Dummy Adjustments

The method of dummies may be associated with the Cascade Merge. The dummy adjustment moves all strings to reduce to an ideal level. For this reason the choice of horizontal or vertical dispersion and associated interlevel dispersions does not affect the performance of the merge. Figure 15.5 shows a vertical dispersion attempting to achieve 14, 11, 6 on three output tapes. Depletion occurs with 24 strings dispersed. The last level achieved was 6,5,3. The vertical dispersion has raised T_1 to the next level, but input is depleted after two strings have been written on tape 2. The dummy values, D_0 through D_3, are shown alongside the tape dispersions. In the example, $D_2 = 4$ and $D_3 = 3$ to pad the dispersion up to the desired level. The adjustment to reduce to the previous level of 6,5,3 proceeds in a manner identical to the Polyphase dummy method. Line 2 of Fig. 15.5 shows the movement of strings from T_1 (corresponding to the merge of T_1, D_2, D_3) to T_0. When three strings have been copied from this tape, the dummy counters of all other tapes have been reduced by 3, causing D_3 (corresponding to T_3) to become 0. Lines 3 and 4 show the status of strings following the merge of T_1, D_2, T_3 to T_0. This is a two-way merge, which involves only a single string since D_2 is depleted after the merge of one string. Lines 5, 6, and 7 show the three-way merge, which continues until T_3 is physically depleted. Lines 8 and 9 show the usual $(k-2)$ and $(k-3)$ phases of the process. At the end of the process, the level 6,5,3 is achieved.

When the dummy method is used, for any interlevel distribution, data movement involves one full data pass to reduce to a perfect level. Thus, if the distribution 6,5,3 is normally reduced to a single string in three passes, any number of strings between 14 and 31 will require four passes.

	Goal	14	11	6	No. of strings	No. of strings moved	D_0	D_1	D_2	D_3
	T_0	T_1	T_2	T_3						
1.	—	14	7	3	24	—	0	0	4	3
2.	3 (L)	11	7	3		3	0	0	1	0
3.	4 { 3 (L)									
4.	1 (2L)	10	7	2		2	0	0	0	0
5.	6 { 3 (L)									
6.	1 (2L)									
7.	2 (3L)	8	5	0		6				
8.	6	3	0	5 (2L)		10				
9.	6	0	3 (L)	5 (2L)		3 / 24				

Fig. 15.5 Dummy adjustment.

15.5.2 Partial-Pass, Pass-Pair Dispersion

An interesting and efficient, albeit somewhat complex, approach to Cascade dispersion was developed for the UNIVAC III SODA (sequential ordering of data) sorting package. It is included here because a partial-pass algorithm for Cascade has not previously been published in the general literature.

The method attempts to overcome the following two difficulties of the dummy method:

1. The dummy method is insensitive to string direction, and a final copy pass may be required to achieve a string in the order (ascending or descending) desired by the user.
2. The dummy method moves all strings in achieving an ideal level. The partial-pass approach may reduce the effect of an imperfect level.

The technique is based on horizontal dispersion and the definition of intermediate dispersion goals, achieved by string groups called "layers" [31].

15.5.2.1 Layer-oriented dispersion

A layer is a pattern of strings dispersed in such a way that a pair of partial merge passes will reduce the actual number of strings to a Cascade level. Figure 15.6 is an example of a layer-oriented dispersion. The process of defining a layer is discussed in the next section. The dispersion has skipped level 6,5,3 and has gone directly to 14,11,6 from 3,2,1. This pattern ensures that there will always be an even number of passes and eliminates the need for a final copy pass. The layers provide a mechanical way of bridging across alternate ideal levels.

	T_0	T_1	T_2	T_3	Layer applied	Layer
	−	3	2	1*		(1) 2, 2, 1
1.		5	4	2	1	(2) 2, 1, 1
2.		7	6	3	1	(3) 1, 1, 0
3.		9	8	4	1	
4.		11	9	5	2	
5.		13	10	6	2	
6.		14	11	6*	3	

*Ideal levels

Fig. 15.6 Layer-oriented dispersion.

Figure 15.6 shows the number of strings on each tape after dispersion of the defined layers. Layer 1 was dispersed three times (lines 1, 2, and 3), layer 2 twice (lines 4 and 5), and layer 3 once (line 6).

For a merge of maximum order $(k - 1)$, there are $(k - 1)$ unique layers. Thus, in Fig. 15.6 there are three layers:

$$2,2,1, \qquad 2,1,1, \qquad \text{and} \qquad 1,1,0.$$

From level to level, the number of times a layer is applied is determined by the number of strings on each tape at the previous level. Tape 1 (the tape with the highest number of strings) determines how many applications there are to be of layer 1; tape 2 determines the applications of layer 2; and tape $(k - 1)$ determines the application of layer $(k - 1)$. The number of layers successfully dispersed determines the number of strings which must be moved from each tape in a partial pass.

15.5.2.2   Calculation of Layers

To generate layers, let i represent layer numbers, j the tape units, and $l_{i,j}$ the increment to the jth tape in the ith layer. In Fig. 15.6, $l_1 = 2,2,1; l_{1,3} = 1$. That is, layer one is 2,2,1; the amount to be added to T_3 on layer one is 1. k is the number of tape units.

An algorithm to calculate layers is:

```
DO  i = 1, k − 1;
    DO  j = 1, k − 1;
        A = min (k − 1, k − i, k − j);
            IF i = j
                THEN  l (i,j) = A − 1;
                ELSE  l (i,j) = A;
    END;
END;
```

i	j	$(k-1)$	$(k-i)$	$(k-j)$	L_{ij}
1	1	4	4	4	3*
1	2	4	4	3	3
1	3	4	4	2	2
1	4	4	4	1	1
2	1	4	3	4	3
2	2	4	3	3	2*
2	3	4	3	2	2
2	4	4	3	1	1
3	1	4	2	4	2
3	2	4	2	3	2
3	3	4	2	2	1*
3	4	4	2	1	1
4	1	4	1	4	1
4	2	4	1	3	1
4	3	4	1	2	1
4	4	4	1	1	0*

$$L_{i,j} \begin{cases} = \min (k-1, k-i, k-j) \text{ when } i \neq j \\ = \min (k-1, k-i, k-j) - 1 \text{ when } i = j \end{cases}$$

Layer 1: 3, 3, 2, 1
Layer 2: 3, 2, 2, 1
Layer 3: 2, 2, 1, 1
Layer 4: 1, 1, 1, 0

*$i = j$

Fig. 15.7 Layer generation ($k = 5$).

The inner loop handles one layer for $(k - 1)$ tapes. The outer loop spans $(k - 1)$ layers.

The values of $l_{i,j}$ for $k = 5$ are shown in Fig. 15.7. These values are then applied to the achieved ideal level a number of times, indicated by the number of strings already on tapes 1 to $(k - 1)$. Figure 15.8 shows the results of the application of layers.

15.5.2.3 Partial-pass algorithm

At the completion of string dispersion, there will be "excess" strings on all tapes if a perfect level is not achieved. The total number of strings on a tape is the number of strings at the last achieved ideal level ("nonexcess" strings) plus the number of excess strings. For each layer i dispersed, a "nonexcess" string (one belonging to the previous ideal level) from tape i will be moved with all excess strings by the first partial-pass. Figure 15.9 shows the dispersion of input with 125 strings for a four-tape system. This is not an ideal population. It is more than 70 and less than 157 (see Fig. 15.2). The 40 excess strings on T_1 come from 14 applications of layer 1 and

	T_0	T_1	T_2	T_3	T_4	
	I	1	1	1	1	
		4	3	2	1	Basic level
$4 \times$ layer 1	+	12	12	8	4	Layer 1: 3, 3, 2, 1
		16	15	10	5	Layer 2: 3, 2, 2, 1
$3 \times$ layer 2	+	9	6	6	3	Layer 3: 2, 2, 1, 1
		25	21	16	8	Layer 4: 1, 1, 1, 0
$2 \times$ layer 3	+	4	4	2	2	
		29	25	18	10	
$1 \times$ layer 4	+	1	1	1	0	
		30	26	19	10	Ideal level

Fig. 15.8 Layer distribution.

Layer dispersion interlevel

Tape	Layers	Applied	Excess on T_i	Remove T_i
1.	2, 2, 1	14	40	54
2.	2, 1, 1	6	34	40
3.	1, 1, 0	0	20	20

Dispersion

T_0	T_1	T_2	T_3	
I	14	11	6	Last ideal
I	+ 40	+ 34	+ 20	Excess (new)
—	54	45	26	Depletion

Fig. 15.9 Dispersion of 125 strings.

six applications of layer 2. The 34 excess strings on T_2 develop in the same way as do the 20 excess strings on T_3. The strings to be removed (Remove T_i in 15.9) are the sums of excess strings and applied layers.

Figure 15.10 shows the partial passes. For partial pass 1, an initial three-way merge removes all excess strings from T_3. Only excess strings are removed from T_3 because layer three was never applied. Then T_3 is reversed to receive strings from a two-way merge that removes all necessary strings from T_2. (See Fig. 15.9. Forty strings have been taken from T_2 now.) T_2 is reversed to receive remaining strings from T_1. These are 14 remaining nonexcess strings. In general, for partial-pass one, a $(k-1)$ merge to T_0 depletes the "remove" counter at $T(k-1)$. This tape is then reversed to receive strings from a $(k-2)$ merge until the "remove" counter of $T(k-2)$ is depleted, etc. The "remove" counter is the sum of the excess and nonexcess

Cascade and Compromise Tape Merges

Partial pass	T_0	T_1	T_2	T_3	
0	54A	45A	26A		
1	20D (3L)	34A	25A	6A	Strings in 3-way merge to T_0. Remove 20 excess strings from T_3, T_2, T_1.
	20D (3L)	14A	5A	{ 6A / 20D (2L) }	Strings in 2-way merge to T_3. Remove 14 excess strings and 6 non-excess from T_2. Remove 20 excess from T_1.
	20D (3L)	0	{ 5A / 14D }	{ 6A / 20D (2L) }	14 strings copied to T_2. Removing 14 non-excess from T_1.
2	6D (3L)	14A (6L)	5A	{ 6A / 6D (2L) }	3-way merge of 14 descending strings to T_1.
	0	14A (6L)	{ 5A / 6A (5L) }	6A	2-way merge of descending strings to T_2.

Fig. 15.10 Partial passes. Removing strings and returning strings.

strings to be removed. The strings written during the partial pass are descending strings. On tapes other than the depleted input tapes there will be a mixture of ascending and descending strings.

Partial pass 2 must achieve the previous level. It does so by redistributing the descending strings. A $k - 1$ merge to the tape with no descending strings clears one tape of descending strings and establishes one tape at an ideal level. In Fig. 15.10, tape T_1 receives strings from a three-way merge to achieve 14 ascending strings. The tape depleted of descending strings is T_2. This tape will now receive strings from a two-way merge. When the two-way merge is completed, an ideal level is achieved, and all strings are in ascending order. The number of descending string redistribution phases depends on the initial dispersion.

Each partial pass moves 114 string lengths in this example. Four additional normal Cascade Merges will bring the total of string lengths moved to 728. Six passes of the dummy method would have moved 750 string lengths.

When there are very many excess strings relative to nonexcess strings, the effectiveness of the partial pass is reduced. The best candidates for the partial-pass method are those in which ideal levels are exceeded by relatively small numbers of strings. Figure 15.11 shows a situation in which a significant reduction in string movement is achieved.

15.5.3 Pass Pairs

A copy pass is avoided by forcing an even number of passes. When the pass-pair approach is associated with partial passes, the savings in string movement may be significant. The partial-pass, pass-pair technique treats

No. of strings	T_0	T_1	T_2	T_3	T_4	Status
85	I	30	26	19	10	Last ideal
103		36	32	23	12	Depletion

(a) String population

1. Layer 1: 3, 3, 2, 1
2. Applications of layer 1: 2
3. Excess strings moved (103–85): 18
4. Non-excess strings moved (number of layer applications): 2
5. Total partial pass movement: $20 \times 2 = 40$
6. Strings moved for four complete passes: 412
7. Total moved: 452
8. Total moved if dummy method used: 515 (five passes)

(b) Partial pass summary

Fig. 15.11 Partial-pass success.

all ideal levels that would result in an odd number of passes as interlevel nonideal dispersions. When the layer dispersion is used, odd-pass levels do not even occur. In place of the full data copy, correct direction is ensured by partial-pass adjustment.

If merges and copy require the same time, then the pass-pair technique is obviously superior. For any particular merge, if merge passes take more time than the copy, then the reduction in string movement may not be profitable. If merge phases are seriously compute-bound, for example, the copy may be preferred, even though more strings are moved. With Cascade it is also possible to predict the number of passes and, if appropriate, to reverse the direction of initial strings.

15.6 STRING LENGTH

Unlike Polyphase, the backward Cascade has no negative effect on the expected string length of a Replacement Selection sort. With Replacement Selection it is always worthwhile to compare the number of strings that will be developed when strings are of length 2, with the number when they are of length 1.5. It may be that the additional strings of Polyphase will force more passes and consequently more data movement even on a tape configuration which is hypothetically more efficient for Polyphase.

15.7 INTERNAL EFFECT OF CASCADE

The internal effect of Cascade merging is similar to that of Polyphase. Provision must be made for $(k - 1)$ input streams, with the potential negative effect that they may have on block size, buffering, and consequent overlap.

Since the number of comparisons is reduced as merge order decreases, I/O–CPU ratios will not remain stable throughout the merge. A $(k-1)$-way phase may be compute-limited but I/O limitation may be achieved at $(k-2)$ or some lower order of merge.

15.8 I/O SUBSYSTEM

As an "unbalanced" merge, the Cascade is dependent on the ability of the I/O subsystem to sustain concurrency of read/write in an unbalanced pattern. Two-channel balanced systems will not run a Cascade Merge efficiently.

15.9 ADDITIONAL REMARKS

A table showing the string-passing power of Cascade Merge is presented as Fig. 15.12. For indicated numbers of tapes, the indicated numbers of total strings can be reduced in the number of passes shown. If dummy/copy is used, an additional pass should be charged when the number of passes shown in Fig. 15.12 is odd. All strings within the ranges of the column are merged in the number of passes required to merge the next higher number. This is not true for the partial pass-pair technique.

No. of tapes	3	4	5	6	7	8	9	10
1 pass	2	3	4	5	6	7	8	9
2 pass	3	6	10	15	21	28	36	45
3 pass	5	14	30	55	91	140	204	290
4 pass	8	31	85	190	371	658	1,086	1,865
5 pass	13	70	246	671	1,547	3,164	5,916	11,372
6 pass	21	157	707	2,353	6,405	15,106	31,938	68,764
7 pass	34	343	2,032	8,272	26,585	72,302	173,322	416,834
8 pass	55	773	5,859	29,056	110,254	345,775	>1M	>1M
9 pass	89	1732	16,866	92,091	457,379	>1M		
10 pass	144	3894	48,570	318,671	>1M			

Fig. 15.12 Cascade merge, string-passing power.

15.9.1 Piggyback and Early Release

It is possible to disperse on fewer tapes than are desired for the merge. Additional strings are piggybacked on the back of one tape and copied over when dispersion is complete, and the extra tape(s) become available for the merge. The reader may wish to refer to Chapter 14. It is also possible to release a tape prior to the last pass by not placing strings on it until the merge is committed to a second pair of passes.

15.9.2   Rerun

The end of a backward-read phase is a natural rerun point for the Cascade Merge. At this point the tape just depleted is at load point and, if dismounted, presents a record of just-moved strings. A malfunction in the next phase may be repaired by remounting and repositioning other tape units. Dismount at the end of each pass gives a pass-to-pass rerun capability.

15.10   COMPROMISE MERGES

First mention in the general literature of the existence of a family of merges of which Polyphase and Cascade represent extremes is commonly credited to D. E. Knuth. In a letter to the editor of the *Communications of ACM* in October, 1963, Knuth [30] described and gave an example of a "compromise" merge. The name is derived from the position the merge occupies between the extremes represented by Polyphase and Cascade. This "betweenness" derives from the pattern of merge phases undertaken during a merge pass. A $(k - 1)$ Polyphase Merge always has one phase per partial pass. A $(k - 1)$ Cascade Merge has $(k - 1)$ distinct phases per pass.

A Compromise Merge is a merge which has more than one phase per pass but less than $(k - 1)$. For any value of $(k - 1)$ with $k > 4$, there will be more than one compromise merge between Cascade and Polyphase. There is a continuum of merges. At one extreme there is the merge with only one phase, for example, a four-way phase; there is the merge with four-way and three-way phases; the merge with four-way, three-way, and two-way phases; the merge with four-way, three-way, two-way, and one-way phases.

Each of the passes of the Compromise Merge is a partial pass, a feature in common with Polyphase; however, more strings are moved during a Compromise Merge, a condition leading to greater efficiency than with Polyphase for some values of k and N. Similarly, for some values of k and N, more strings participate in higher-order merges, leading to greater efficiency than with Cascade.

Figure 15.13 shows a Compromise Merge with $k = 6$ and with three phases. The definition of ideal levels will be described in the next section. The merge process is identical to Cascade except that a pass is terminated before all data has been moved.

15.10.1   String Dispersion

The realization of ideal levels for Compromise Merges reflects their intermediate position between Polyphase and Cascade. The computation of the number of strings required on each tape is performed for some number of

	T_0	T_1	T_2	T_3	T_4	T_5	String lengths moved	Phase
	–	44	40	36	25	13		–
I	13 (5L)*	31	27	23	12	0	65	1 5-way
	13 (5L)*	19	15	11	0	12 (4L)*	48	2 4-way
	13 (5L)*	8	4	0	11 (3L)*	12 (4L)*	33	3 3-way
II	9 (5L)	4	0	4 (14L)*	7 (3L)	8 (4L)	56	1 5-way
	5 (5L)	0	4 (13L)*	4 (14L)*	3 (3L)	4 (4L)	52	2 4-way
	2 (5L)	3 (12L)*	4 (13L)*	4 (14L)*	0	1 (4L)	36	3 3-way
III	1 (5L)	2 (12L)	3 (13L)	3 (14L)	1 (48L)*	0	48	1 5-way
	0	1 (12L)	2 (13L)	2 (14L)	1 (48L)*	1 (44)*	44	2 4-way
	1 (39L)*	0	1 (13L)	1 (14L)	1 (48L)*	1 (44L)*	39	3 3-way
IV	(Last)	1 (158L)					158	
							579	

*Tapes not participating in merge.

Effective data passes $= \dfrac{579}{158} = 3.7$

Fig. 15.13 Compromise merge with three phases.

tapes as though dispersion were being made for Cascade, and for other tapes as though dispersion were being made for Polyphase.

Consider Fig. 15.14. For tape populations (k) of 4 through 9, values for $(k - 1) = M$, minimum order (W), and number of phases (P) are shown. A number of simple relationships present themselves. The number of phases to a pass is the difference between the number of tapes on the system and the desired minimum way of merge ($P = k - W$). Conversely, the minimum way of merge for a pass is the difference between the number of tapes on the system and the number of phases desired ($W = k - P$).

During any compromise pass, some tapes will deplete and others will not. In Cascade, all tapes deplete. In Polyphase, only one tape depletes. The tapes which are depleted during a pass receive strings as if a Cascade Merge were to be performed. During a pass, P tapes will be depleted and P tapes will therefore receive strings, as for a Cascade Merge. $M - P$ tapes will not be depleted and will receive strings as if a Polyphase Merge were to follow.

Figure 15.15 shows (on line 1) the initial dispersion of ($k - 1$) strings common to all ($k - 1$) merges. There are seven tapes, and the minimum order of merge is to be 4. There are therefore to be three phases per merge pass, and three tapes are to receive strings in a Cascade-like fashion.

In a Cascade dispersion, tape 1 at a level holds a number of strings equal to the sum of all strings at the previous level; T_2 at a level holds a number of strings equal to the sum of all strings at the previous level excluding the tape with fewest strings; T_{k-1} (the tape with fewest strings)

Total tapes k	Dispersion tapes M	Minimum order W	Phases P	
4	3	3	1	Polyphase
4	3	2	2	
4	3	1	3	Cascade
5	4	4	1	Polyphase
5	4	3	2 ⎫	Compromise
5	4	2	3 ⎭	
5	4	1	4	Cascade
6	5	5	1	Polyphase
6	5	4	2 ⎫	
6	5	3	3 ⎬	Compromise
6	5	2	4 ⎭	
6	5	1	5	Cascade
7	6	6	1	Polyphase
7	6	5	2 ⎫	
7	6	4	3	
7	6	3	4 ⎬	Compromise
7	6	2	5 ⎭	
7	6	1	6	Cascade
8	7	7	1	Polyphase
8	7	6	2 ⎫	
8	7	5	3	
8	7	4	4 ⎬	Compromise
8	7	3	5	
8	7	2	6 ⎭	
8	7	1	7	Cascade
9	8	8	1	Polyphase
9	8	7	2 ⎫	
9	8	6	3	
9	8	5	4	
9	8	4	5 ⎬	Compromise
9	8	3	6	
9	8	2	7 ⎭	
9	8	1	8	Cascade

Fig. 15.14 Compromise merges.

	T_0	T_1	T_2	T_3	$T_4(T_w)$	T_5	T_6	
1.	1	1	1	1	1	1	1	Level 1
2.		1	1	1	3	2	1	Cascade for T_4-T_6
3.		6	5	4	3	2	1	Next cascade
4.		2	2	2	2	2	1	Next polyphase
5.		4	4	4	3	2	1	Level 2 Compromise
6.		4	4	4	3	2	4	
7.		4	4	4	3	8	4	
8.		4	4	4	12	8	4	
9.		4	4	13	12	8	4	
10.		4	14	13	12	8	4	
11.		15	14	13	12	8	4	
		Polyphase			Cascade			

$$\left.\begin{array}{l} T_6 L_3 = T_1 L_2 \\ T_5 L_3 = T_1 L_2 + T_2 L_2 \\ T_4 L_3 = T_1 L_2 + T_2 L_2 + T_3 L_2 \end{array}\right\} \text{Cascade}$$

$$\left.\begin{array}{l} T_3 L_3 = T_4 L_3 + T_6 L_2 \\ T_2 L_3 = T_4 L_3 + T_5 L_2 \\ T_1 L_3 = T_4 L_3 + T_4 L_2 \end{array}\right\} \begin{array}{l}\text{Polyphase}\\ \text{Level 3}\\ \text{Compromise}\end{array}$$

Fig. 15.15 Dispersion for Compromise.

at a level holds a number of strings equal to the number of strings on the tape with most strings at the previous level.

On line 2 of Fig. 15.15, the three tapes which will be depleted on the first pass of the merge are T_4, T_5, and T_6. These have received strings as if for a Cascade merge. Line 3 shows the ideal level for Cascade, which would be achieved if Cascade dispersion were to continue. Line 4 shows the ideal level which would be achieved if dispersion were being made for Polyphase.

Tapes 1 through 3 have not received strings for the new level. These are tapes which will not deplete and which will receive strings in a manner similar to that for a Polyphase dispersion. The number of strings on the last tape to receive strings in a Cascade manner represents a "base," from which the number of strings for remaining tapes will be computed. The base tape in Fig. 15.15 is T_4. Line 5 of Fig. 15.15 shows, moving from right to left, that the number of strings on T_3, T_2, and T_1 is the sum of the base number and T_6, T_5, and T_4 at the previous level. In general:

1. The last tape to receive strings from a Cascade-like dispersion is T_w.

2. The number of strings on T_w is a base number, which will be used in computing string population for all tapes to the left of T_w.

3. The number of strings on tapes to the right of T_w at the previous level are added to T_w to determine the number of strings on tapes to the left. In Fig. 15.15, T_w on line 5 is T_4 with 3 strings. Therefore, the second level for T_3 is

$$T_w + (T_6 \text{ at level 1}) = 3 + 1 = 4;$$

the second level for T_2 is

$$T_w + (T_5 \text{ at level 1});$$

and for T_1 it is

$$T_w + (T_4 \text{ at level 1}).$$

Similarly for level 3, T_w is T_4 with 12 strings. To compute strings for T_1, T_2, T_3, at level 3, add 12 to the T_4, T_5, T_6 populations for level 2.

The new level is shown on line 11 in Fig. 15.15. The analogy to Polyphase dispersion is the addition of a single value across a population of tapes. The effect of the computations is to reduce the number of strings on those tapes which will not be depleted during the initial merge pass. Lines 6 through 11 of Fig. 15.15 show the tape-by-tape development of level 3. Tapes 6 through 4 are the tapes receiving strings, as for Cascade. Tapes

T_0	T_1	T_2	T_3	T_4	T_5	T_6		
—	15	14	13	12	8	4		Start
4*	11	10	9	8	4	0	6-way	
4*	7	6	5	4	0	4*	5-way	Pass 1
4*	3	2	1	0	4*	4	4-way	
3	2	1	0	1*	3	3	6-way	
2	1	0	1*	1*	2	2	5-way	Pass 2
1	0	1*	1*	1*	1	1	4-way	
0	1	0	0	0	0	0	6-way	Last

*Not in merge or output tape.

Fig. 15.16 Compromise merge of three phases.

3 through 1 receive the Polyphase-like increment. Figure 15.16 shows the merge which would follow if string formation ceased at this point. There are three phases per pass, a minimum way of merge of 4.

In Appendix C we present an algorithm in PL/I for the computation of ideal levels for all $(k - 1)$-way merges which deplete input before merging. This algorithm differs from the discussion in this section only insofar as the value of P represented by the variable PHASES is 1 less than the P in Fig. 15.14. PHASES = 0 is a Polyphase dispersion; PHASES = $(M - 1)$ is Cascade. An interesting detail, which is apparent in Fig. 15.14 as well as in the algorithm, is that numbers of phases equal to $(k - 1)$ and $(k - 2)$ both cause Cascade dispersions since the phase eliminated is the copy phase for the $(k - 2)$ value.

15.10.2   Backward Compromise

In order to perform backward Compromise, it is necessary either to copy strings from all tapes not depleted during a pass or to rewind all tapes not depleted. The effect on the efficiency of the process depends on the characteristics of the system. The proper placement of dummies may relieve this requirement.

16
Oscillating and Crisscross Merges

16.1 INTRODUCTION

The tape merges discussed so far do no external merging until all input data has been formed into strings by a dispersion pass. This chapter will describe two merge techniques in which string formation and string merging alternate with each other throughout the ordering process until input data is depleted. The two techniques are known as Oscillating and Crisscross. Both techniques are "unbalanced" and achieve orders of merge greater than $k/2$. Both are applicable to disk as well as to tape environments.

16.2 OSCILLATING MERGE

The Oscillating Merge was initially conceived to be absolutely dependent on read-backward capability, and an introductory description is simplified by that assumption [41].

The initial operation of the dispersion produces $(k-2)$ ascending strings. The input tape contains the remaining strings; one tape is empty. When $(k-2)$ ascending strings have been produced, a $(k-2)$-way merge is undertaken backward to form one string of $(k-2)$ string lengths in a descending direction. Next, $(k-2)$ more strings are dispersed from the input tape. One of these ascending strings is placed on the tape which received the results of the prior merge. This is done to ensure that a following $(k-2)$-way merge will place a descending string of length $(k-2)$ on a different tape. When the second merge is complete, a further dispersion of $(k-2)$ strings is made and a third merge undertaken. At this time there are strings of length $(k-2)$ dispersed across $(k-2)$ tapes. Now a second-order merge is undertaken. This merge forms a string of length $(k-2)^2$.

The Oscillating Merge may be summarized as follows:

1. Disperse $(k-2)$ strings on $(k-2)$ tapes.
2. Perform a $(k-2)$ merge, forming strings of length $(k-2)$.
3. Repeat (1) and (2), $(k-2)$ times.
4. Perform a $(k-2)$ merge, forming strings of length $(k-2)^2$.
5. Repeat (1) through (4), $(k-2)$ times.
6. Perform a $(k-2)$ merge, forming strings of length $(k-2)^3$.
7. Repeat (1) through (6), $(k-2)$ times.
8. Continue dispersing and merging in the same fashion until input is depleted.

The read-backward merge may produce a descending final string, which may have to be copied to produce an ascending string if the following application requires it.

Figure 16.1 shows the progress of an Oscillating Merge which has reached the point where $k-2$ second-level merges have been performed. At this time, the first third-level merge is undertaken. A string of length $(k-2)^3$ is formed. The process then begins the dispersion of strings once more.

16.2.1   Characteristics and Power of Merge

The order of the merge is $(k-2)$ throughout all merge phases. The increase in string size follows a $(k-2)$ power series. In the five-tape example of Fig. 16.1, string sizes, after various level merges, are 3, 9, 27.

The number of data passes is $\lceil \log_{k-2} S \rceil$. Twenty-seven strings dispersed to three tapes, as in the example, will pass the data three times ($\log_3 27$). (A two-way balanced merge using T_1 through T_4 would take five passes for the same data.) If ideal levels are not reached, then higher-level merges will be of lower order and produce shorter strings.

The selection of tapes for dispersion of strings is arbitrary, except that the tape which has received strings from the last merge must receive a string from the new dispersion.

A striking characteristic of the method is the number of tape-direction reversals that must be undertaken throughout the process. Each merge phase moves tape in the opposite direction, requiring the tape unit to switch its mode, reposition the read head, and prepare to read. After the dispersion

T_I	T_1	T_2	T_3	T_4	Comment
—	1A	1A	1A	0	Disperse $k-2$ strings
	0	0	0	1D (3L)	$k-2$ merge (1)
	1A	1A	0	1D (3L), 1A*	Disperse $k-2$ strings
	0	0	1D (3L)	1D (3L)	$k-2$ merge (2)
	1A	0	1D (3L), 1A*	1D (3L), 1A*	Disperse $k-2$ strings
	0	1D (3L)	1D (3L)	1D (3L)	$k-2$ merge $(k-2)$
	1A (9L)	0	0	0	$k-2$ second level (1)
	1A (9L)	1A	1A	1A	Disperse $k-2$ strings
	1A (9L), 1D (3L)*	0	0	0	$k-2$ merge (1)
	1A (9L), 1D (3L), 1A*	1A	1A	0	Disperse $k-2$ strings
	1A (9L), 1D (3L)*	0	0	1D (3L)	$k-2$ merge (2)
	1A (9L), 1D (3L), 1A*	1A	0	1D (3L), 1A*	Disperse $k-2$ strings
	1A (9L), 1D (3L)*	0	1D (3L)	1D (3L)	$k-2$ merge $(k-2)$
	1A (9L)	1A (9L)	0	0	$k-2$ second level (2)
	1A (9L)	1A (9L), 1A*	1A	1A	Disperse $k-2$ strings
	1A (9L), 1D (3L)*	1A (9L)	0	0	$k-2$ merge (1)
	1A (9L), 1D (3L), 1A*	1A (9L)	1A	1A	Disperse $k-2$ strings
	1A (9L), 1D (3L)*	1A (9L), 1D (3L)*	0	0	$k-2$ merge (2)
	1A (9L), 1D (3L), 1A*	1A (9L), 1D (3L), 1A*	0	1A	Disperse $k-2$ strings
	1A (9L), 1D (3L)*	1A (9L), 1D (3L)*	1D (3L)	0	$k-2$ merge $(k-2)$
	1A (9L)	1A (9L)	0	1A (9L)	$k-2$ second level $(k-2)$
	0	0	1D (27)	0	$k-2$ third level (1)

*Rightmost string is accessible.

String lengths moved (not counting distribution pass) = (3)(3)(3) + (9)(3) + (27)(1) = 81.

Data passes = 81/27 = 3.

Fig. 16.1 Oscillating merge.

of $(k-2)$ strings, all tapes must be reversed to perform the $(k-2)$-way first-order merge. They must be reversed again after the merge in order to receive new strings. There are two reversals per tape for every sequence of dispersion and first-order merge. The number of times this sequence will occur is of course determined by the amount of data. This method and Crisscross are the only methods in which the number of reversals is so large, running to large multiples of the number required for other methods (of which Cascade is closest). In a machine which has no ability to overlap reversals and in which reversal time is significant, the large number of reversals can be a serious penalty to performance.

16.2.2   Read-Forward Oscillating

On systems with no backward-read or with very bad reversal characteristics and good rewind capability (overlapped rewind at high multiples of read times), a read-forward version of the Oscillating Merge is reasonable [25]. The tape system must have a capability for intermixing writing and reading on the same tape. On many tape systems such intermixing carries the danger that a new write will obliterate an existing record.

Figure 16.2 shows a read-forward merge. The reader should follow the discussion on that figure. The initial dispersion is $(k-1)$, but subsequent dispersions are $(k-2)$, excluding the tape which holds a long string formed by a merge. Each dispersed string is usually followed by a rewind, so that the rewind of tapes may be characteristically overlapped by the write of a successor string. The $(k-2)$ merge of Comment 2 forms a long string behind an established short string on T_4. The merge tape is now rewound. After a $(k-2)$ dispersion, the short string on T_4 participates in a $(k-2)$ merge with new strings, leaving its space vacant. The method depends on using this vacant space. When Replacement Selection is used with forward Oscillation, it is necessary to limit string size so as to guarantee that all strings will fit in vacated string space. Each first-level $(k-2)$ merge leaves one dispersed string untouched, placing a long string behind it. As dispersion and merging continue, a point is reached where a $(k-2)$ second level is undertaken. See Comment 9 on Fig. 16.2. This happens after $(k-1)$, not $(k-2)$, dispersion cycles.

The rewind delays in the merge are limited to one string length after every $(k-2)$ first-level merge and $(k-2)$ string lengths for a second-level merge. The merge design of Fig. 16.2, described here as presented by Goetz [25], minimizes rewind time and the delays due to rewinding. It is not generally possible to implement Oscillating as a read-forward technique in a more straightforward manner.

T_0	T_1	T_2	T_3	T_4		Comments
I	1*	1*	1*	1	1.	Disperse $k-1$ strings on T_1–T_4. Rewind T_1, T_2, T_3. Asterisks indicate load point at tape at end of dispersion or merge.
	0*	0*	0*	1* 1(3)	2.	Perform $k-2$ merge to T_4. Rewind all tapes. Parentheses indicate string lengths
	1*	1*	1	1* 1(3)	3.	Disperse $k-2$ strings on T_1 through T_3. Rewind T_1 and T_2. T_4 is not referenced.
	0*	0*	1* 1(3)	0* 1(3)	4.	Perform $(k-2)$-way merge from T_1, T_2, T_4 to T_3. Rewind all tapes; vacant position develops on T_4.
	1*	1	1* 1(3)	1* 1(3)	5.	Disperse $k-2$ strings on T_1, T_2, T_4. Rewind T_1 and T_4. Note that new string occupies vacant position on T_4.
	0*	1* 1(3)	0* 1(3)	0* 1(3)	6.	Perform $(k-2)$-way merge to T_2. Rewind all tapes. Positions on T_3 and T_4 become vacant.
	1	1* 1(3)	1* 1(3)	1* 1(3)	7.	Disperse $k-2$ strings on T_1, T_3, T_4. Rewind T_3 and T_4.
	1 1(3)	0 1(3)	0 1(3)	0 1(3)	8.	Perform $(k-2)$-way merge to T_1. Recognize presence of $k-1$ strings of proper length and suppress rewind. Tape read heads on T_2–T_4 are in front of long string at this point.
	1* 1(3) 1(9)	0*	0*	0*	9.	Perform a $(k-2)$-way second-level merge onto T_1. Rewind all tapes.
	1* 1(3) 1(9)	1	1	1	10.	Disperse $k-2$ strings, etc.

Fig. 16.2 Forward oscillating merge.

16.2.3 String Length and Internal Impact

The Oscillating Merge design impacts the dispersion of strings in a number of ways. Since only $(k-2)$ strings are merged, there is a slight relaxation of the space requirements for blocking and buffering. However, this benefit is offset by the nature of the interaction between the merge and the dispersion. The size of strings will be restricted by the attempt to hold in core all required sort space plus the program code for the merge and the dispersion. The alternative is to overlay sections of merge and dispersion code. The decision to restrict sort area size or to overlay from phase to phase is a trade-off which must be investigated for each implementation on each configuration. The total delay to the sort due to overlay overhead must be estimated and compared with the increase in data-passing due to the development of shorter and consequently more numerous strings.

When Replacement-Selection sorting is used, the impact of the method becomes quite severe. In order to maintain an expected string size approaching twice the elements in the sort area, the tournament must be maintained and replenished. When the tournament is reestablished after a merge, its first string is a string roughly of size (1.7 × sort area). It will be necessary to initiate the tournament after each first-level merge so that one string in $(k - 2)$ will be of length (1.7 × sort area). Worse, the last string of a dispersion will be only of length (1 × sort area), since replacement cannot be applied while the tournament area is used for merge space. For every dispersion of $(k - 2)$ strings, there will be one of length (1.7 × sort area) and one string of length (1 × sort area). The method will therefore produce more strings than an equivalent Cascade or forward-read Polyphase. Backward Polyphase, of course, has a similar problem. The use of storage space to avoid "flushing" the tournament does not constitute a solution, since any tournament in storage would necessarily be smaller than a tournament used for Cascade, Polyphase, or Balanced.

16.2.4   Rerun

A rerun point is established whenever all data so far dispersed is collected on one tape. This occurs at the time of the first instance of a merge at any level. The sort may be "backtracked" to the last instance of a single sequence, and input may be backspaced to redisperse from that point. More discrete rerun points may be defined by repositioning to the condition prior to the last first-level merge, or last $(k - 2)$-way merge of any level. Repetition of this merge and dispersion from that point can offset a restart.

16.3   CRISSCROSS MERGE

The Crisscross Merge is a method suitable for either disk/drum devices or for backward-read tape. To the author's knowledge, its only implementation in a sorting package developed by a computer manufacturer is as a disk technique for an OS/360 sort. It is an efficient tape technique as well and can be described among the tape merges. Like the Oscillating Merge, the method depends on an alternation of string formation and string merging. The methods differ, however, in the pattern of the merges. Little comparative or analytic material exists on the relative efficiency of the Crisscross and Oscillating Merges.

16.3.1   The Crisscross Merge on Tape

The method initially disperses $(k - 2)$ strings from left to right on $(k - 2)$ tape units. When $(k - 2)$ strings have been dispersed, a $(k - 2)$ merge is

	T_0	T_1	T_2	T_3	T_4		Strings
1.	I	0	1D (3L)	1D (3L)	1D (3L)	Base level (Level 1)	9
2.	1	1	0		1	Disperse k−2	
3.	0		1D (3L)	1D (3L) 1D (3L)	1D (3L)	Merge k−2. First level 2 string	12
4.	1		1D (3L) 1	1D (3L) 1D (3L) 1	1D (3L)	Disperse k−2	
5.	0		1D (3L)	1D (3L) 1D (3L)	1D (3L) 1D (3L)	Merge k−2 Second level 2 string (k−3 level 2 strings now available)	15
6.		1A (9L)	0	1D (3L)	1D (3L)	k−3 strings at level 2 merge with 1 string at level 1	

Fig. 16.3 Crisscross merge.

undertaken to form a string of length $(k - 2)$. This process of alternate dispersion and merging is done $(k - 2)$ times, until there are $(k - 2)$ strings of length $(k - 2)$ dispersed across work tapes. This status, called a "base level," is achieved in exactly the manner of an Oscillating Merge. Line 1 of Fig. 16.3 shows the status of the merge at the formation of the initial base level.

When reaching a base level, instead of performing a $(k - 2)$-way second-level merge, as in Oscillating, the Crisscross method reinitiates the formation of strings. The essential goal of the technique is to postpone as long as possible the formation of long strings and the involvement of long strings in merges. Essentially each merge is a partial pass which involves only part of the strings on the input tape.

After the first base level is achieved, $(k - 2)$ strings are dispersed (line 2, Fig. 16.3) and merged (line 3) to form a string the same size as the strings of the base level. This string is placed behind a base-level string. Dispersion of strings is resumed (line 4), and when $(k - 2)$ strings are dispersed, another merge of initial strings is performed (line 5). At this point the dispersion of strings on tape is 0, 1, 2, 2. The condition for another merge has been reached. This condition is that $(k - 3)$ strings of the length of the base level are formed and available for merge with a string of the base level. The partial merge of line 6 forms the first string of length $(k - 2)^2$. When there are $(k - 2)$ strings of length $(k - 2)^2$, a new base level will have been formed.

Figure 16.3 suggests that there are three merge processes. One, an Oscillating Merge involving $(k - 2)$ dispersions of $(k - 2)$ strings and $(k - 2)$ merges of order $(k - 2)$, establishes the first base level. Another merge process, the dispersion and merge of initial strings, occurs $(k - 3)$ times to produce the conditions of line 5. This dispersion-and-merge process occurs whenever a base level is achieved and continues until $(k - 3)$ strings of a new level have been formed. When the condition of line 5 exists, the partial level-building merge of line 6 is undertaken. This merge is performed whenever a base level has not been reached and there are $(k - 3)$ strings of the length of last base strings formed by the lower-level dispersion and merges.

The selection of tape units to receive strings is fundamental to the method and not at all obvious. This method is unlike other methods in that the order and permutation of string levels across specific tapes of the tape population are quite critical.

The process that establishes the first base level disperses strings "from left to right" and merges on the tape to the right of the $(k - 2)$nd dispersed string. First strings are dispersed on tape 1 through tape $(k - 2)$, and the first long string is formed on tape $(k - 1)$. The next dispersed string is placed

	Merge pointer (MP)	Dispersion merge pointer (DM)	Direction switch	T_0	T_1	T_2	T_3	T_4	
1.	2	–		I	0	1D (3L)	1D (3L)	1D (3L)	Base level
2.	2	2(2,1,4,∪3)*	I	I	1	1D (3L) 1	1D (3L) 1	1D (3L)	Disperse $k-2$
3.	2	3		I	0	1D (3L)	1D (3L) 1D (3L)	1D (3L)	Merge $k-2$
4.	2	3(3,2,1,∪4)		I	1	1D (3L) 1	1D (3L) 1D (3L) 1	1D (3L)	Disperse $k-2$
5.	2	4		I	0	1D (3L) 1D (3L)	1D (3L) 1D (3L)	1D (3L)	Merge $k-2$
6.	2	4	←	I	1A (9L)	0	1D (3L)	1D (3L)	$k-3$ strings at level 2 merge with level 1
7.	3	3(3,2,1,∪4)		I	1A (9L)	0	1D (3L)	1D (3L) 1D (3L)	
8.	3	4(4,3,2,∪1)		I	1A (9L) 1D (3L)	0	1D (3L)	1D (3L) 1D (3L)	
9.	3	4	←	I	1A (9L)	1A (9L)	0	1D (3L)	
10.	4	4(4,3,2,∪1)		I	1A (9L) 1D (3L)	1A (9L)	0	1D (3L)	
11.	4	1(1,4,3,∪2)		I	1A (9L) 1D (3L)	1A (9L) 1D (3L)	0	1D (3L)	
12.	4	1	←	I	1A (9L)	1A (9L)	1A (9L)	0	

*∪ = union, "merge to"

Fig. 16.4 Tape use in crisscross $k - 2 = 3$; $k - 3 = 2$).

in tape $(k-1)$, and then additional strings are placed on T_1, T_2, ..., etc. Tape $(k-1)+1$ wraps around to become tape 1.

At the time the first base level is established, an internal pointer, here called the *merge pointer*, points to the last tape to receive strings during the $(k-2)$-way merge which established the base level. Figure 16.4, duplicating Fig. 16.3 in part, shows the dispersion of the base level and the value of the merge pointer (MP). The last tape to receive strings was T_2.

The first instance of the second merge process now occurs. The dispersion of strings and the merge are controlled by a pointer, which is here called the DM pointer, set initially equal to MP. For every series of $(k-3)$ dispersions of $(k-2)$ strings, DM is set to MP at the time of the first dispersion. The first string in Fig. 16.4 is placed on T_2. DM is then decremented so that strings are placed right to left. Line 2 shows the dispersion of strings on T_2, T_1, and T_4. There will be a union of strings on T_3. Line 3 shows this merge. Lines 4 and 5 show the dispersion and merge of another set of initial strings. DM is initially 3, so strings are placed on T_3, T_2, T_1 and merged to T_4.

On line 6 the basic cycle has been completed; there are $(k-3)$ new strings of proper length eligible for merge with a string at the previous level. A partial merge pass is undertaken.

The partial merge has two forms. In one, strings are merged on the tape to the left of MP; in the other, strings are merged on the tape to the right of MP. There is a direction switch, initially set (line 6), for a merge to left. For this reason, strings of length 9 go to T_1. When there are multiple consecutive partial merges, as can happen in the later stages of the merge, direction is alternated. The initial partial merge after a series of dispersions and merges is always to the left. After a partial merge a determination is made as to whether another should be immediately undertaken. If no base level has been achieved and there are $(k-2)$ tapes with strings of the length of the last level, a partial merge is undertaken. Otherwise MP is advanced by one and the dispersion/merge sequence is undertaken. DM is set to the new value of MP on line 7, and dispersion gets under way.

Lines 7 and 8 show the $(k-3)$ merges of initial strings. The conditions for a partial merge have been achieved; there is no base level and there are strings of equal length to merge. The partial merge is conducted to the left, onto T_2. At the end of the merge the conditions for an immediate successor are not met, MP and DM are adjusted, and the dispersion/merge is again entered. Lines 10 and 11 show its performance. Line 12 shows the partial merge that follows and forms a new base level.

As the process ages, a variety of string sizes develops on the tape units. When data is depleted, a $(k-2)$-way merge forms a final string. There may be requirements for partial passes, depending on string dispersion and direction at the time of the merge.

16.3.2   Comparison With Oscillating

For string populations of significant size, the method will outperform Oscillating because it will pass long strings fewer times.

For very small numbers of strings, the reduction in string movement will not occur since there will be insufficient levels for the partial merges to be effective. Twenty-seven strings on a five-tape system will be the same for both methods. For 100 strings on a six-tape system, Crisscross will move roughly 10 percent fewer strings. The method will never be worse than Oscillating, and where reasonably large string populations are expected, it is to be preferred. The impact on string size and number of strings is the same for both methods.

17
Tape
Merge
Overview

17.1 INTRODUCTION

The caution with which an overview of tape merging must be approached cannot be overemphasized. There is no "best" tape merge, although the literature abounds in claims and counterclaims. Much of the serious investigation into the properties of merges and the conditions which affect performance is disconcertingly formal and, consequently, inaccessible to many workers. In this section we will attempt to describe the operational characteristics of the major tape merges and the characteristics of data and hardware to which tape merges are sensitive. Some general statements about performance are also included.

The fundamentals of merge performance may be cast into two questions: (1) How much data is moved by the merge? (2) How long does the merge take to move it? Often elapsed time may be reduced by increasing data movement because the hardware characteristics of a host system allows for very efficient movement of data. For example, we have seen that a four-way merge may outperform a six-way merge because of blocking, buffering, and internal comparison considerations.

For the purpose of designing a merge, two sets of information are required. One set is information about the data the merge will process and the hardware on which it will run. The other is the set of measurements by which to compare one merge technique with another, one order of merge with another for the same technique, one partial pass or adjustment approach with another. This discussion, which begins with a description of the measures of a merge, develops guides for the investigation of merges rather than conclusions about performance.

17.2   MEASURES OF A MERGE

The usual measures of a merge are (1) the number of phases, (2) the effective order of the merge, and (3) the amount of data moved. The number of phases of a merge is significant because there are time-consuming functions associated with phase ends. Rewinds and reversal must be counted and timed. The analyst must be careful to determine that interphase merge costs do not offset the efficiency of a merge that seemed to be preferable to another. The "effective order of a merge" does not always provide an intuitive appreciation of the number of phases a merge will require, and often rests on theoretical generalizations which do not reflect the actual number of strings to be moved for particular string populations or particular systems.

The author's preference is for the measure of *data passes* as the most computable and useful first statement of the benefits of one merge or merge design over another. The number of data passes is the number of records moved by the merge divided by number of records in the file. The measure exists in various forms [12,13]. The concept of "number of string movements" is closely related. When the size of each string is fixed and identical to that of all other strings, a count of the total number of string movements gives a perfectly accurate statement of the data passes of a merge. When strings are not of the same size, a count of string movements will provide an acceptable *approximation* of the data moved. The replacement methods of internal sorting which lead to variable string sizes do not normally produce strings of wildly varying sizes. The measure of data passes is often referred to as the "number of passes per string"—the number of times a string is moved during the merge.

17.3   FACTORS INFLUENCING PERFORMANCE OF A MERGE

A number of considerations relating to the data and the hardware affect the data-passing characteristics of a merge. The data considerations are the following:

1. Volume of input data

2. Number of strings produced

3. Knowledge of number of strings

4. Relations between dispersion and merge

5. Natural merge and/or string extension techniques

6. Data structure.

Each of these factors will influence the amount of data moved by a merge. Hardware factors that influence the amount of data moved (as opposed to the amount of time required to move the data) are the number of tape units (determining the way of the merge) and memory sizes.

17.3.1   Volume of Input Data

A single pass over 160,000 fifty-word records will move more data than ten passes over 10,000 twenty-word records. In addition, the volume of data affects merge design in other ways. The use of a $(k - 1)$ technique may be infeasible when the amount of data involved will cause a tape "overflow." Since the dispersion of data on tape during the course of the merge (or even during dispersion) for $(k - 1)$ tapes tends to the unbalanced, there is an increased likelihood of attempts to write too much on one tape. In vendor-provided sorting packages offering a choice between Polyphase and Balanced, one of the parameters of that choice is the volume of data. There is a danger of overflow even with Balanced merges. In certain situations, premature aggregation of records on a tape in later passes may occur. For example, 100 strings on a six-tape system with no partial pass will place 89 percent of the data on one tape on the next-to-last pass. In general, however, the exposure to overflow is less in Balanced Merges.

   If the volume of data is such that a single input tape and a single output tape are not sufficient, then data passing will be affected by design decisions. If the analyst determines that only initial and final data will require more than one tape, he may decide to let the merge proceed at the maximum way of merge, spending the time to mount a new input tape and a new final output tape. However, he may decide to restrict the order of the merge so that he can perform tape swaps on input and final output. The reduction in order of merge may or may not result in more data passes, depending on the number of strings and on the particular orders of merge involved [18].

   In some situations the volume of data is too large for a single sort/merge. Then the strategy is to partition the file, sort/merge each partition, and merge the final output tapes. If various final string lengths are produced by the several sort/merges, there is a merging strategy to minimize the work of the final merge. The optimum merges described in Chapter 9 apply.

17.3.2   Number of Strings Produced

The specific number of strings produced affects the efficiency of tape merging. Certain merges that are "usually" less efficient than other merges for a given number of tapes may become more efficient for a very small number or a very large number of strings.

The merges in this text are sensitive to the number of strings and, beyond that, to the occurrence of particular numbers: the so-called perfect distributions. Sensitivity to perfect numbers may be reduced by partial-pass techniques, but a merge that is "usually" less efficient than other merges for a given number of tapes may become more efficient for its perfect levels across that number of tapes. Merges vary in their responses to bad levels. Balanced merging will cost an additional full pass over data if a perfect power of merge order is not achieved and no partial pass is used. Cascade may also exact a heavy penalty for imperfect levels. Polyphase merging imposes the least penalty for imperfect levels. The relative efficiency of Polyphase over other methods for imperfect levels is greater when the number of strings is small.

Consider a 20-tape system with 145 initial strings. A Balanced Merge will require $\lceil \log_{10} 145 \rceil = 3$ data passes to sequence the file. A Polyphase Merge will order this data in 2.54 data passes. Despite the greater theoretical effective merge power of Balanced with $k = 20$, the Polyphase is more efficient because it has a perfect level while 45 is a bad level for Balanced. The importance of partial passes for Balanced can be illustrated by the fact that, with partial passes, Balanced can order the 145 strings in 2.36 data passes. With 100 strings, a perfect level for Balanced, Balanced outperforms Polyphase by 2.00 data passes to 2.37.

A similar example comparing Polyphase and Cascade may be given. With eight tapes, Cascade is superior to Polyphase for 140 strings, Polyphase to Cascade for 45 strings. In general, Polyphase tends to be more effective for fewer strings and Cascade for more strings because of the string-passing power of Cascade. Many more strings must be put out to get to a new Cascade level, and this tendency becomes even more pronounced at higher levels. Even for smaller string populations, however, there are instances of Cascade superiority when the levels are very nearly perfect for Cascade. Partial-pass methods may further reduce the number of strings moved for a bad Cascade level. For a large population of tapes, the number of strings at which Cascade becomes superior is much smaller. For a small population of tapes (for example, $k = 4$), Polyphase is superior throughout.

17.3.3   Knowing the Number of Strings

Knowing the number of strings to be merged will enable one to select a method more intelligently. In addition to the impact of the number of strings on merge efficiency, there are other benefits from accurate prediction. The direction of initial strings may be selected so as to avoid copy passes for Cascade, Balanced, or Oscillating if the number of passes is known or can be estimated from knowledge of the expected number of strings. For the

Compromise methods, knowledge of the number of passes is required for any algorithm which determines the pattern of ascending and descending strings so as to avoid rewinds or copies of undepleted tapes.

Knowing the number of strings depends on knowing the volume of elements to be sorted, the method used for the internal sort, and the size of the work area given over to the internal sort. The price may be high, involving the avoidance of replacement or string-extension techniques. It is within the technology to design the merge after the strings have been formed, when a precise count is known. The difficulties in doing so stem from the fact that the blocking decisions have already been made by the dispersion algorithm and the pattern of ascending and descending strings has been set. However, the blocking can be changed during the first merge pass or whatever partial pass is necessary to prepare for the selected merge.

17.3.4 Number of Strings Produced by Replacement

An investigation of the relative merits of one merge technique must include the detail that Backward Polyphase, Oscillating, and Crisscross cause more strings to be produced by replacement selection.

The superiority of Oscillating over Cascade lies in its ability to pass more strings between levels. If the generation of additional strings forces an extra merge pass, then Oscillating may not outperform Cascade with an adjustment pass for tape populations in the range of $k = 8$ to $k = 11$. For really large tape populations, such as $k = 20$, the effect of extra strings is nil, and Oscillating is the best technique in terms of data passing unless the population of strings is large enough to justify Crisscross. A generalization for the number of additional strings produced by an Oscillating merge, using Replacement Selection, has been estimated by Goetz as

$$[2(k - 2)/(2k - 5.3)] - 1 \text{ percent additional strings [25].}$$

The Backward Polyphase is very vulnerable to additional string creation. The expected size of a string with Replacement Selection after a reversal of direction has been shown by Knuth to be (1.5 × tournament size). A horizontal dispersion algorithm will minimize the increase in strings by producing as many consecutive strings as possible in the same direction. For the small tape populations, however, in which Polyphase is the apparently preferred method, the increase in strings may be enough to make it reasonable for the analyst to consider Forward Polyphase or another method.

The impetus for Backward Polyphase is the elimination of rewind times. An alternative is the implementation of a Polyphase of less than maximum merge order, which can overlap rewinding with merging and absorb the rewind times. The merge order is $(k - 3)$, and tapes rewound during a phase (the depleted tape and the one to receive strings) are used as output and input in the next phase.

17.3.5   Natural Merge

In all statements based on the concept of data movement, straight merges are assumed. To say that a three-way Balanced Merge with 27 initial strings requires three passes assumes that there will be, in sequence, nine strings of length 3, three strings of length 9, and one string of length 27. It is entirely possible that the 27 initial strings may be reduced to one string on one pass if there is sufficient order in the data and natural merging is used. Statements about data passing represent the worst case if advantage is taken of natural merging and string extension.

17.3.6   Data Structure

The amount of data passed by a merge is affected by various structural factors. Are the records fixed or variable in length? Did the dispersion pass extend the size of the record by the addition of control fields? Did it pad out short records or fill blocks with dummy records or strings with dummy elements [22]. The basic size of a record affects such things as blocking, buffering, and size of sort area, and it consequently contributes to the fundamental parameters of the sort/merge process.

17.4   HARDWARE AND THE MERGE

The hardware factors listed by Goetz (see Chapter 12) are essential considerations of the merge designer. The fundamental issues he must resolve are the following.

1. What is the impact of memory size on the I/O- or CPU-boundedness of different orders of merge? Is blocking or buffering restricted by $(k - 1)$-way merging or by the selection of a particular merge order so that merge cycles will not be I/O-bound or will run so slowly that additional faster passes are desirable?

2. What is the map of the I/O subsystem? What kind of read/write overlap is available? Will $(k - 1)$ merging cause interferences and delays?

3. What are the rewind and reversal characteristics? Will simultaneous rewinds at high speed be possible? What is the true cost of reversing a tape unit for read?

Balanced $\lceil \log_{k/2} S \rceil$

Oscillating $\lceil \log_{k-2} S \rceil$

Cascade

Number of strings for given number of tapes, k

Passes \ k	3	4	5	6	7	8	9	10
1	2	3	4	5	6	7	8	9
2	3	6	10	15	21	28	36	45
3	5	14	30	55	91	140	204	290
4	8	31	85	190	371	658	1086	1865
5	13	70	246	671	1547	3164	5916	11372
6	21	157	707	2353	6405	15106	31938	68764
7	34	343	2032	8272	26585	72302	173322	416834
8	55	773	5859	29056	110254	345775	1M	1M
9	89	1732	16866	92091	457379	1M		
10	144	3894	48570	318671	1M			

Polyphase

Number of strings (number of passes) for given number of tapes, k

Passes \ k	3	4	5	6	7	8	9	10
1	2(1.0)	3(1.0)	4(1.0)	5(1.0)	6(1.0)	7(1.0)	8(1.0)	9(1.0)
2	3(1.6)	5(1.6)	7(1.6)	9(1.6)	11(1.5)	13(1.6)	15(1.5)	17(1.5)
3	5(2.4)	9(2.2)	13(2.1)	17(2.1)	21(2.1)	25(2.1)	29(2.0)	33(2.0)
4	8(3.1)	17(2.8)	25(2.7)	33(2.7)	41(2.6)	49(2.6)	57(2.6)	65(2.6)
5	13(3.9)	31(3.5)	49(3.2)	65(3.2)	81(3.1)	97(3.1)	113(3.1)	129(3.1)
6	21(4.6)	57(4.0)	94(3.7)	129(3.7)	161(3.6)	193(3.7)	225(3.5)	257(3.6)
7	34(5.6)	105(4.7)	181(4.3)	253(4.3)	321(4.1)	385(4.1)	449(4.1)	513(4.1)
8	55(6.0)	193(5.3)	349(4.8)	497(4.8)	636(4.8)	769(4.6)	897(4.6)	1025(4.6)
9	89(6.7)	355(5.9)	673(5.3)	977(5.3)	1261(5.2)	1531(5.1)	1793(5.1)	2049(5.1)
10	144(7.5)	653(6.5)	1297(5.8)	1921(5.8)	2501(6.0)	3049(5.6)	3578(5.6)	4097(5.6)

Fig. 17.1 Merge data passes.

Answers to these questions will influence decisions to use Cascade, Oscillating, Balanced, or Polyphase in either forward or backward directions. They also affect the choice of a proper order of merge within each method. More subtle determinations will influence the choice of a dispersion algorithm; for example, synchronizer or channel-release delays will encourage or discourage vertical or horizontal dispersions.

17.5 COMPARISON OF MERGES

Figure 17.1 provides a summary of data passes for Polyphase, Cascade, Oscillating, and Balanced merges for various values of k and various string populations [5,28]. No partial passes are assumed. Values for Balanced and Oscillating are given as formulas, for Polyphase and Cascade as tables. Values have been selected from Fig. 17.1 to illustrate certain behavior patterns in Fig. 17.2.

k	Strings	Passes			
		Balanced	Oscillating	Cascade	Polyphase
6	37	4	3	3	2.7–3.2
	86	5	4	4	3.2–3.7
	144	5	4	4	3.7–4.3
	250	6	4	5	3.7–4.3
	425	6	5	5	4.3–4.8
	863	7	5	6	4.8–5.3
	1296	7	6	6	5.3–5.8
10	37	3	2	2	2.0–2.6
	86	3	3	3	2.6–3.1
	144	4	3	3	3.1–3.6
	250	4	3	3	3.1–3.6
	425	4	3	4	3.6–4.1
	863	5	4	4	4.1–4.6
	1296	5	4	4	4.6–5.1

Note: A file of 100,000 records, 100 characters each, will form no more than 250 strings when sorted in a 40,000-character work area (a medium to small machine) prior to tape merging.

Fig. 17.2 Some sample comparisons of data passes.

18
Random
Access
Sorting

18.1 NATURE OF DEVICES

Random-access storage devices, despite their name, are characterized by access delay times which vary in duration. The delay times include seek time and rotational time. *Seek time* is the duration of the physical movement of a read/write mechanism. *Rotational delay* time is the time required to position the rotating surface under the waiting read/write mechanism. The reader familiar with disk and drum may skip to Section 18.2.

18.1.1 Seek Time

Seek time occurs when there are fewer read/write assemblies than there are addressable partitions on recording surfaces, and it is therefore necessary to move the read/write assemblies to proper positions. The amount of time required for the movement depends on how far the mechanism must be moved across the surface to an addressed track. Seek time is commonly measured in milliseconds.

The designer of a merge must seriously consider arm movement as a factor in performance. Each device type has its own access time curve, its own minimum, maximum, and average times. It is necessary to refer to manuals about the device to determine exact seek times for various distances.

Since seek times are so significant, much attention has been paid to increasing the amount of data available to a read/write head in any particular position. One means of doing so is to increase the recording density on the surface of the disk. Another important organizational feature of contemporary disk systems that increases data availability without arm movement is the *cylinder* concept, which requires multiple read/write assemblies that move together across the surfaces of a stack of disks. These multiple read/write heads are attached to one arm. The organization effectively extends the size of a track by making available on one arm movement a number of tracks—a "cylinder"—each on a separate surface. Although there may be electronic head-switching time associated with moving from one track in a cylinder to another, it is small in comparison with seek time. The amount of data available to the arm without a physical seek is increased by a multiple equal to the size of the stack. Despite the existence of multiple heads, the arm is still treated as a single unit, and only one head reads at a time. Parallel reading and writing from the different surfaces of the same cylinder do not occur.

There are a number of variations in disk storage design, some of which are now of only historical interest. Very early disk devices had a single arm to service an entire stack, and therefore vertical movement from surface to surface added to seek time. In some devices there are attempts to improve performance by having multiple arms. Under programmer control, the arm nearest a desired location would move to that location for an access.

Current multiarm devices provide access to greater storage capacities by associating multiple stacks with a single unit. Each stack has its own arm, and data on one drive is available only to the arm associated with that drive. Addressing involves a reference to drive, surface, and track. Such a device is capable of performing simultaneous seeks. How fully this capability is realized in a system depends on the capability of control units and channels and the architecture of I/O on the system. More remarks on this subject are included in Section 18.1.4.

Seek time may occur in very large drum systems, such as the Fastrand family of drums for UNIVAC equipment. The Fastrand drum units provide massive storage capacity, and their use, uncommon for drums, is to support large data bases. Each Fastrand II drum unit contains 192 tracks serviced by 64 read/write heads. In order to reference the drum it may be necessary to position the heads over the desired tracks. All heads are "gang-mounted" to move together; there is no independent head positioning. The heads of this unit are exactly analogous to the read/write heads of a single arm moving across a disk surface.

An option of the Fastrand system is a family of 24 tracks called Fastband. Each Fastband track is provided with a read/write head, and reference to these tracks does not involve seek time.

Configurations of multiple independent drums brought into the system through a single control unit are possible with Fastrand. Movement may

be performed concurrently across up to eight independent units. Readers
with a specific interest in sorting on Fastrand should refer to more detailed
descriptions of this unit made available by the manufacturer. The unit is
cited here as an example of a drum device involving seek time.

18.1.2 Rotational Delay Time

Both disks and drums have a rotational delay, the period of time during
which the read/write head is positioned over the correct track but must wait
for the proper read/write location to move beneath it. The rotational delay
depends on the speed of rotation of the drum or disk and on data organization. Speed of rotation varies considerably from device to device. A primary storage drum of 5000 words available with the UNIVAC Solid-State
Computer rotates at 17,500 rpm. Drums of intermediate size, such as the
IBM 2303 or UNIVAC FH–432 or FH–1782, tend to rotate considerably
more slowly. The FH–1782 (with a capacity for two million 36-bit words)
rotates a. 1800 rpm. Very large drums rotate even more slowly; e.g., Fastrand II rotates at 870 rpm.

Delays vary from zero to a full rotation. Rotational delay time is often
quoted in terms of minimum, maximum, and average. The average is commonly taken as the simple mean of the minimum and maximum.

Some drums implement the Fastrand concept by providing additional
heads for "fast" tracks. Access time for these tracks is reduced in proportion
to the number of heads. A typical number is four. On a drum with a maximum access delay of 20 milliseconds (a full rotation) a track with four heads
would have a 5 millisecond maximum access and a 2.5 millisecond average
access.

Rotational delay considerations for disks are exactly the same as for
drums. Maximum, minimum, and average times vary from unit to unit.
Total access time is of course the sum of seek and rotational delay times
for all moving-head devices.

18.1.3 Other Devices

There are other cyclic storage devices, but they are not often involved directly in sorting in contemporary systems. When the data held by tape strip
and other archival devices is to be rearranged, it is commonly "percolated"
forward to faster devices before the sort/merge starts.

18.1.4 Systems

In addition to being subject to rotational delay time and seek time, random-
access storage devices, like other I/O devices, are subject to control unit,
channel, and CPU delays. The seriousness of these delays depends on the

architecture and configuration of the host system. The architecture determines how much can be done in parallel, and the configuration determines how much of the parallel potential in the system can be achieved in the particular machine room. The considerations involved here are substantially similar to the hardware considerations discussed in connection with tape merging, but there are some new details worth examining.

18.1.4.1   One unit

The simplest configuration in which a drum or disk may be used as a sorting device has a single bulk storage unit. All reading and writing occur on that one unit. The configuration does not allow for overlap of seeks with other seeks or with any I/O function on the device. The minimum merge time will be the sum of the times for reading, writing, seeking, and rotational delays. The design problem for this configuration is to reduce the seek time to acceptable levels.

18.1.4.2   Multiple units

A system may have a set of multiple, independent drives. It is possible to conduct a k-way merge across k drives similar to that in a tape operation. Seek time involved in moving a read/write mechanism from one input string to another is eliminated. The output mechanism moves in an orderly fashion across the surface during a phase.

The manner of attachment of multiple units to the system will have a great impact on the performance of the sort. The architecture and configuration of control units and channels will determine whether there may be overlap of seek times and rotational times, and whether there is parallel reading of data. It is possible to attach each device to an independent control unit and each control unit to an independent channel, thus ensuring read/write overlap. More commonly, several drives are associated with one control unit, and the control unit may or may not have exclusive use of a channel. In some systems, a control unit may have access to more than one channel. So long as a capability exists for overlap of reading and writing, additional channel capacity is not required. As with tape, only marginal improvements are expected with more than one read and one write path.

Equally as important as the number of channels and controls units are the details of where functions are performed—the rules for "hand-shaking" across the I/O subsystem. A single channel that can transfer a seek address to a control unit and then service another device until the seek is completed may operate with the efficiency of multiple channels that "lock" to a device for the entire seek time. Similarly, a control unit that can initiate a seek on one device and then release itself to initiate a seek on another may be as effective as multiple control units. A designer must investigate the interface characteristics of channel, control unit, and device very carefully to determine what overlap potentials exist.

18.2 BASIC MERGE CONCEPTS

A merge on a disk can be characterized in the following ways.

1. It may be of an order as large as the number of strings: S strings may participate in a one-pass S-way merge.

2. It may select particular strings to participate in any particular merge phase and may arbitrarily redefine merge order from phase to phase.

3. It may be dominated by seek time. This fact may render theoretically optimum merge patterns inefficient. Therefore minimizing seek time is an important major goal of design. In the seek environment, a merge that moves considerably more data than another may be superior in performance because elapsed seek time is significantly lower. The data transfer rates of disks are very fast. An IBM 2314 can transfer 312,000 characters per second. Reading a block of 1000 characters from a 2314 requires about 3 milliseconds. Average seek time plus rotational delay is 87.5 milliseconds. There is clearly a big difference between this situation and that with a tape device, where seek time (i.e., start time) is on the order of 5 to 10 milliseconds, and reading 1000 characters takes about 11 milliseconds.

18.2.1 Balanced Merge

The various elements of the problem of merge design will be introduced with Balanced merging, a technique derived from the Balanced tape merge. The demonstration device will be a single-disk device.

The technique of Balanced merging on disk involves the definition of a number of work areas on the disk. If three areas are defined, strings are dispersed into two of them in some manner, and therefore when input is depleted, two areas hold all the strings and there is an empty area. Next, the strings of one area are combined with one another in an m-way merge into the empty area. When this merge is complete, the strings of the other area are merged together into the area vacated by the first merge. At any time there is an area free to receive strings. A pass of the merge is completed when the strings from every area that initially received strings have been merged into another area.

Figure 18.1 shows an initial dispersion and merge pass of a Balanced Merge using three areas. The order of merge is arbitrarily selected as 4 for this example. A series of four four-way merges of 16 strings from area 1

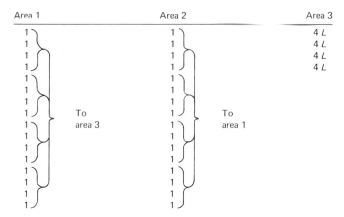

Fig. 18.1 Balanced disk merge. First move, from Area 1 to Area 3. Pass 1 ends when Area 2 strings are merged to Area 1.

develops four strings of length 4 in area 3 and leaves area 1 empty. Strings from area 2 may then be merged into area 1. At the end of the pass, areas 1 and 3 contain strings of length 4, and area 2 is empty. The first pass of the merge is complete since the full data file has been moved.

At the end of a pass, conditions then existing determine whether an additional pass of this type is required. If the total number of strings is smaller than the order of merge, a final interarea merge forms one string. If the number of strings is larger than the merge order, an additional intermediate pass is undertaken. Figure 18.2 shows the remainder of pass 1, an additional intermediate pass 2, and the final pass. The final pass differs from intermediate passes in that strings from different areas are combined to form an ordered string. The order of the final merge is the number of remaining strings.

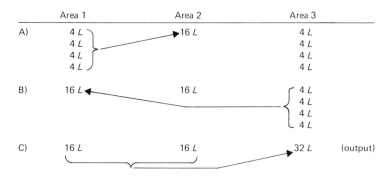

Fig. 18.2 Pass 2 and last pass of balanced disk merge.

18.2.2   Seeking in a Balanced Merge: Intra-area Merging

The merge of strings within an area rather than between areas minimizes arm movement. It is an attempt to reduce both the number of times and the distance that an arm must move during the merge.

If space for each area is allocated on a cylinder basis, then all strings in an area are available to a read/write arm without a seek. If more than one cylinder is required, then arm movement is still constrained by the number of cylinders (e.g., limited to a two-cylinder area). If all strings fit in one cylinder, all reading of strings for the merge is accomplished without seek time, regardless of the pattern of contribution of particular input strings to an output string. In a one-arm merge it is necessary to move the arm for writes to a nearby cylinder, and much of the effect is thus lost, but arm movement is still less than for merging across areas.

Intra-area merging makes the merge order independent of the number of areas defined for strings. Any order of merge may be undertaken on the strings of an area. The precise order of merge is selected with an eye to balancing data movement against access time by controlling the number of seeks performed, and reflects the storage available for blocks from each input string.

18.2.3   Order of Merge

The minimum amount of data movement is achieved by the highest order of merge. Running on a theoretical system with absolute random access to any data and with no limit on memory size, the best merge order is always the largest, and the best strategy is to merge all strings on one pass [9,27].

In the discussion of tape merging, we were very much aware of the limitations on merge order imposed by machines. In disk merging the same constraints apply with even greater force. The factors that affect the selection of an order of merge are:

1. Size of primary storage;
2. Average seek time; and
3. Data transfer time.

The size of primary storage is critical to selecting the order of merge. With any given amount of core, the space that can be occupied by any string is a function of the order of merge. As Fig. 18.3 shows, with a 16-way merge the maximum number of records from any string that can be held in primary storage is 6, whereas with a two-way merge 50 records can be held. The number of records that can be held per string is the block size that can be associated with the merge.

Order (m)	Block size[1]	Blocks per string[2]	Accesses per intra-area merge[3]	Intra area merges per pass	Block access per pass[3]	Block access total[4]	Data passes[5]
16	6	84	1344	1	1344	1344	1
8	12	42	336	2	672	1344	2
4	25	20	80	4	320	640	2
2	50	10	20	8	160	640	4

Notes:
[1]Number of records fitting in core at one time. For this example, with 10,000 characters of core available for input and 100-character records, 100 records can be in core at any time. With 16 input strings, six records from each string can be in core at any time. This is computed by dividing merge order into the number of records in core.
[2]Number of blocks to form a string 500 records long. There are 16 initial strings, or 8000 total records; therefore, each string contains 500 records.
[3]An intra-area merge is one m-way merge of m strings to an output area. A 16-way merge pass has one 16-way merge and accesses 84 X 16 blocks. A two-way merge pass has eight two-way merges to move the file. Accesses per intra-area merge = m X number of blocks per string. Accesses per pass = accesses per intra-area merge X number of intra-area merges per pass.
[4]A 16-way merge has one pass, a two-way merge four passes, etc. Total accesses = number of passes X accesses per pass.
[5]File is moved each pass.

Fig. 18.3 Block sizes for various merge orders with associated accesses and data passes.

The block size determines the number of blocks required to form a string. The number of blocks per string determines the access delay time, since accessing is accomplished on a block basis. Set against the reduction in data passing associated with higher-order merges is the likelihood of more seek times whose magnitude dominates the merge.

Figure 18.3 shows the total number of block accesses required by various merge orders in reducing 16 strings to one string. To determine the best order of merge for 16 strings, it is necessary to find the order of merge for which the sum of data transfer time and seek time is minimum.

To time various merges, one must know how many passes there will be and how long each pass will take. The four merges of Fig. 18.3 are "timed" in Fig. 18.4, where total times include only time to read and to seek blocks. On a single-arm device, write time would be additional, and seek time would occur between every read and write. The calculations of Fig. 18.4, though a simplification, show the tendency for data passes and access times to work against each other.

m	Total access	Access time, ms[1]	Records read[2]	Read time, ms	Rank	Total time (seek + read)
16	1344	16,800	8,000	5,280	2	22,080
8	1344	16,800	16,000	10,560	3	27,360
4	640	8,000	16,000	10,560	1	18,500
2	640	8,000	32,000	21,120	4	29,120

Notes: 1. Accesses X 12.5 ms per access in this example.
2. Read time = 0.66 ms per 100 character record.

Fig. 18.4 Times for various merges; fast access.

For this example, the four-way merge is the best by a considerable margin. Next best is the 16-way merge, and the poorest is the two-way merge. The 16-way merge is dominated by access time. The two-way merge is dominated by reading time.

The parameters which affect this result are the access time and read time, the ratio between them, and the size of core. The assumption of an average access time of 12.5 milliseconds is critical to the result.

When the ratio of transfer time to seek time is different, the results will be different. Figure 18.5 shows the same conditions with one exception: an access time of 87.5 ms is used. The two-way merge is now a close second. By comparing Figs. 18.4 and 18.5, one can see the relation between the merges and the influence of parameters.

m	Total access	Access time, ms[1]	Records read	Read time, ms	Rank total	Time
16	1344	117,600	8,000	5,280	3	122,800
8	1344	117,600	16,000	10,560	4	128,160
4	640	56,000	16,000	10,560	1	66,560
2	640	56,000	32,000	21,120	2	77,120

Note: [1] Accesses × 87.5 ms per access in this example.

Fig. 18.5 Times for various merges; slow access.

It is not always clear what should be used as an estimate of access time. Published average access times tend to be very conservative. If some information is at hand about size of file and available space on disk, a better estimate of average access time may be developed. If only 10 tracks of a 200-track surface are used, published average access times are very gross estimates. In a multiprogramming environment, or any environment where the sort loses control of the arm, the use of very conservative numbers is best.

18.2.4 Selecting Merge Order

The text has described a merge design in which all phases move all data in fixed full passes. The selection of merge order for such a design is most elegantly discussed by N. A. Black [3], from whom the presentation in the succeeding sections has been taken, in substance. While this discussion does not exhaust all aspects of optimization, it provides insight into the general approach to the problem.

The nominal merge time is the sum of the times for all passes. The time of a merge pass depends on the order of merge and the effect of that order on machine performance. The machine parameters of access time, transfer

Core size, bytes	Read time, ms	ALPHA (seek + delay)/read
1,000	6.6	13.25
10,000	66	1.33
20,000	132	0.66
30,000	198	0.44
40,000	264	0.33
50,000	330	0.26
100,000	660	0.13
250,000	1,650	0.05
500,000	3,300	0.03
1,000,000	6,600	0.01

IBM 2311 data rate = 156 kilo bytes; IBM 2311 mean seek + rotational delay = 87.5 ms.
Data rate for 1 byte = 6.6 μs (approximate).

Fig. 18.6 Computation of ALPHA for IBM 2311 disk file.

time, and core size can be related by a single value unique to each set of system configurations. This single value provides a rough but usable guide to merge optimization. Figure 18.6 shows the ratio of seek time to read time, ALPHA, for various available core sizes of an IBM S/360, for the data rates and access time of an IBM 2311 disk. Read time is developed by multiplying the character read rate by the number of characters required to fill available storage. As core goes up, the proportion of seek time to read time decreases, and more data may be read per seek.

The values of the ratio ALPHA depend on the data transfer rate of the I/O subsystem, the value used for average seek time, and the size of core to be filled. In the development of ratios, average seek time may be used. If the analyst has some reason for preferring another number, he may of course use it.

The ALPHA ratio is used to develop a nominal pass figure of merit when we assume read/write/compute overlap and I/O bounding. The figure of merit for any pass is (ALPHA $\times$ m) + 1. The number of passes depends on m in the usual way. Larger m's (merge orders) will result in fewer passes. The total time of the merge is

$$\sum_{1}^{k} (\text{ALPHA} \times m_k) + 1,$$

where k is the number of passes. This number is merely the sum of all pass times.

The relationship is developed by a number of simple steps. The first determines what possible range of values for k (number of passes) may be applied. If the maximum order of merge is used, only one pass will occur. If a minimum order ($m = 2$) is used, the number of passes is equal to $\lceil \log_2 \rceil$ of the number of strings. The number of values of k is relatively small. For a population of 1000 strings, k can vary from 1 to 10.

Knowing the range of k, one can calculate for each k the merge order:

$$m^k = S.$$

Figure 18.7 shows the calculation of merge orders for 128 strings. To complete the merge in seven passes, a merge order of 2 will suffice ($2^7 = 128$); to complete it in six passes, a merge order of 3 is necessary ($3^6 > 128$); etc.

No. of passes, k	Merge order, m
1	128
2	12
3	6
4	4
5	3
6	3
7	2

Fig. 18.7 Possible merge orders for 128 strings.

The exact proper order of merge is not an integer. However, it is impossible to plan merges of nonintegral order (although the effect of merging, in terms of string reduction, may not be an integer). Therefore, for most merge designs there must be some number of merges at *less than the maximum* order of merge. The next step in merge design is to determine how many of these lower-order merges are necessary. Figure 18.8 summarizes, for each value of k, the number of m-way and $(m - 1)$-way merges required to reduce 128 strings. To maintain stability, only m and $(m - 1)$ merges are used; therefore, total merge time is the sum of times for m-way and $(m - 1)$-way passes.

No. of passes, k	No. of m-way passes	Merge order, m	No. of $(m - 1)$-way passes	Merge order, $m - 1$
1	1	128	—	—
2	1	12	1	11
3	1	6	2	5
4	2	4	2	3
5	4	3	1	2
6	3	3	3	2
7	7	2	—	—

Fig. 18.8 Reducing 128 strings.

The final step in determining merge order is to apply the information in Fig. 18.8 for file-passing time. The minimum value for this equation will demonstrate the proper merge order. Figure 18.9 reveals that one 12-way merge forming 11 strings (10 of length 12 and one of length 8), followed by an 11-way merge, is the optimum design for 128 strings with

Rank	k	m	No. of passes of order m	m − 1	No. of passes of order m − 1	(ALPHA × m) + 1	(ALPHA × (m − 1)) + 1	Sum of order m passes	Sum of order m − 1 passes	Total passes
7	1	128	1	127	0	17.64[1]	—	17.64	—	17.64
1	2	12	1	11	1	2.56	2.43	2.56	2.43	4.99
2	3	6	1	5	2	1.78	1.65	1.78	3.30	5.08
3	4	4	2	3	2	1.52	1.39	3.04	2.78	5.82
4	5	3	4	2	1	1.39	1.39	5.56	1.26	6.82
5	6	3	3	2	3	1.39	1.26	4.17	3.78	7.95
6	7	2	7	1	0	1.26	—	8.82	—	8.82

Note: 1. (ALPHA × m) + 1 = (0.13 × 128) + 1 = 17.64

Fig. 18.9 Final calculations for determining merge order (ALPHA = 0.13).

ALPHA = 0.13. The reader will observe how closely this result depends on
ALPHA. Values of ALPHA determine the desirability of various merge
designs based upon relationships among core, memory size, and I/O data
speeds.

Now let us apply this technique to the earlier example. If we use 10,000
characters (as in Fig. 18.3) and the IBM 2311 timings of Fig. 18.6, the value
of ALPHA is 1.33. This high value of ALPHA will lead us to expect that
lower-order merges will be more profitable. In fact, from Figs. 18.4 and 18.5,
we expect merges of the order of 4 to be optimum. Figures 18.10 and 18.11
confirm that the four-way merge is optimum, as well as confirming the
relative order of the two-, four-, and 16-way merges. Since the merge design
precludes sequences such as an eight-way merge followed by a two-way
merge, the eight-way is not included in the comparison.

k	m	$m - 1$	Passes	
1	16	15	1	16-way
2	4	3	2	4-way
3	3	2	3	3-way, 2-way
4	2	1	4	2-way

Fig. 18.10 Black's method for examples of Figs. 18.3, 18.4, and 18.5 (ALPHA
= 1.33; number of strings = 16).

18.2.5 Merge Design: Variable Merge Order

If the constraint of a stable merge order is removed from merge design,
a much larger population of designs becomes admissible. For 16 strings,
the design combining an eight-way with a two-way merge (which was ad-
mitted in Figs. 18.3 and 18.4 but excluded in the last section) becomes
admissible.

The designs in which merge order is constant, however, will be optimum
in all cases where the MINB design is optimum. The MINB design is the
merge in which a minimum number of seeks is performed throughout the
life of the merge. MINR design is the merge design in which a minimum
number of records is passed throughout the life of the merge. The size
of the ALPHA constant will determine the selection between MINB and
MINR designs, based on core size, seek time, and transfer time. When
MINR designs are optimum, the highest-order merge within the selected
maximum merge is always desirable (otherwise the MINR characteristic is
lost). When MINB designs are optimum, the desirability of maintaining
nearly identical merge orders from pass to pass means that designs of the
type described in Section 18.2.4 will be preferable to all others. Selecting
a merge order by the computation of ALPHA and the application of the
formula (ALPHA $\times m$) + 1 should provide an excellent design in those
cases where full merge passes are to be performed.

k	m	No. of passes of order m	m − 1	No. of passes of order m − 1	(ALPHA × m) + 1	(ALPHA × (m − 1)) + 1	Sum of order m passes	Sum of order m − 1 passes	Total passes
1	16	1	15	0	22.28	–	22.28	–	22.28
2	4	2	3	0	6.32	–	12.64	–	12.64
3	3	2	2	1	4.99	3.66	9.98	3.66	13.64
4	2	4	1	0	3.66	–	14.64	–	14.64

Fig. 18.11 Black's method, continued (ALPHA = 1.33).

18.2.6   Partial Passes

Given a population of strings that are theoretically all equally available, how should a merge be designed to reduce them to one string in the shortest time? Figure 18.12 shows a population of strings of various sizes and an optimum merge of the strings of the population. Each three-way merge (order of merge is selected for the reasons given in the preceding sections) is labeled *A* through *E*, and the columns of string population show the string list after given merges. Parenthesized strings are created by the given merge; starred strings were created by a preceding merge. The two critical features of the merge are the following.

1. Shortest strings are merged first.
2. The order of merge is always maximum for later merges.

List of strings in size order	(A) First merge	(B) Second merge	(C) Third merge	(D) Fourth merge	(E) Fifth merge
			List after		
19	26	29	38	81	(337)
19	27	34	52	99	
26	28	36	67	(157)	
27	29	38	81		
28	34	52	(99)		
29	36	67			
34	(38)	(81)			
36	52				
52	67				
67					

Note: New strings in parentheses.

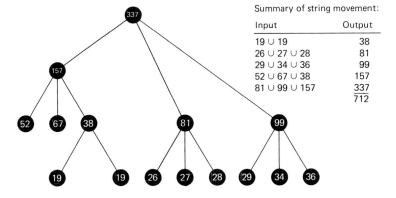

Summary of string movement:

Input	Output
19 ∪ 19	38
26 ∪ 27 ∪ 28	81
29 ∪ 34 ∪ 36	99
52 ∪ 67 ∪ 38	157
81 ∪ 99 ∪ 157	337
	712

Fig. 18.12 Optimum merge (new strings are in parentheses).

The reader who has read this text in sequence will recognize here the application of the tree concept, as discussed in Section 9.4.5. In connection with random storage devices, the concept of the optimum merge has great potential. The effectiveness of the merge of Fig. 18.12 can be seen in Figs. 18.13 and 18.14. The former shows a balanced three-way merge on the population of strings of Fig. 18.12 without any ordering to string size in (a) and the best balanced merge in (b). The tree in (b) shows the strings processed in order from smallest to largest. The difference between the two merges is not nearly so significant as the difference between the first bal-

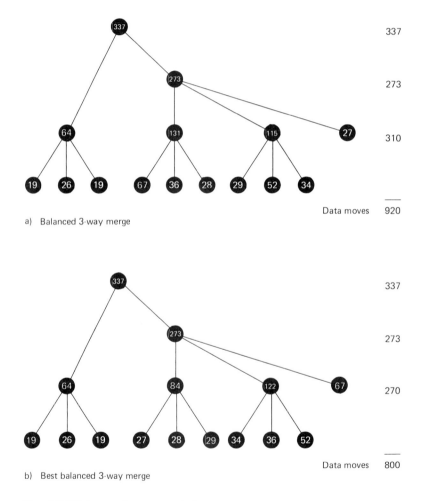

a) Balanced 3-way merge

b) Best balanced 3-way merge

Fig. 18.13 Improving merge balance.

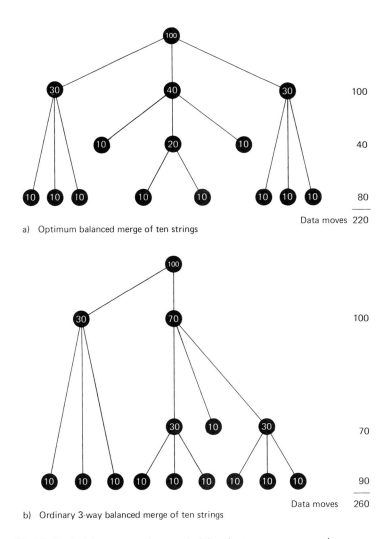

a) Optimum balanced merge of ten strings

b) Ordinary 3-way balanced merge of ten strings

Fig. 18.14 Optimizing merge by maximizing last-pass merge order.

anced merge and the optimum tree. The power of the optimum merge is
further demonstrated in Fig. 18.14. With 10 strings of identical size there
is more than a 15-percent decrease in data moving between the balanced
and the optimum tree. Thus the need to maintain the last pass at maximum
order is the critical factor in taking advantage of string size. Note, however,
that when string sizes are identical and the number of strings is a power
of the merge order, the optimum merge and the balanced merge are identical.

Figure 18.15 shows an interesting effect of applying partial adjustment passes to a balanced merge. Part (a), identical to Fig. 18.14(a), shows that the balanced merge can be transformed into the optimum merge by a partial pass that reduces the number of strings to a power of its merge order. The tree of this merge is optimum. The partial adjustment pass transforms the balanced tree into the "best" tree by ensuring that the order of merge for the last pass will be maximum. In general, a partial adjustment pass will transform a balanced merge of order *m* into the best string movement tree (the optimum tree) when the number of strings overshoots a perfect power by a small amount. In Fig. 18.15(b), a partial adjustment pass has been

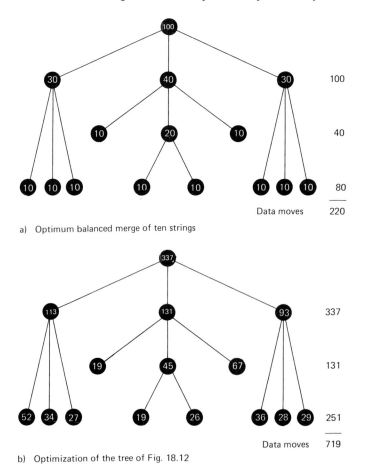

a) Optimum balanced merge of ten strings

b) Optimization of the tree of Fig. 18.12

Fig. 18.15 Use of partial pass to optimize balanced merge.

performed on the first two strings. The shape of the tree is the same as that of the optimum tree of Fig. 18.12. The number of records moved is different because of string sizes. In the worst case for this set of strings, the balanced merge would move 793 strings (the partial pass would merge strings of lengths 52 and 67). In the best case, of course, 712 strings are moved.

The number of strings to include in the partial pass, or the order of merge for the first merge of the optimum tree, is given by Burge (and others) as:

1. Order of partial merge = optimum order if (number of strings − 1) mod (optimum order − 1) = 0. (*A* mod *B* is the remainder obtained when *A* is divided by *B*.)

2. Order of partial merge = ((number of strings − 1) mod (optimum order − 1)) + 1, otherwise. That is, the order of partial merge is *remainder plus* 1.

For example, with 11 initial strings and 3 as the order of merge, there is no remainder in the division of 10 (number of strings − 1) by 2 (order of merge − 1), and therefore the partial merge is also of order 3. With 10 strings, however, the partial merge would be of order 2 since ((9 modulo 2) + 1) is 2.

The conclusion which can be drawn from this section is that, with strings of reasonably stable size, a Balanced Merge with a partial pass adjustment is an excellent approximation of the optimum merge from a record-moving point of view.

18.2.7   Seek Optimizing: Interleaving

An approach to the control of seek time is interleaving, which arranges components of strings in such a way that seek time is minimized.

In sequential dispersion, the blocks of strings are written on contiguous space as they are developed. In interleaved dispersion, space is left between blocks during the writing of a string. The space is used for blocks of other strings; thus a block of one string is physically followed by a block of another, or a track full of blocks of one string is followed by a track full of blocks of another. Block interleave and track interleave are identical when block size equals track size. The "width" of the interleave may vary. A merge order of 10, for example, will not necessarily interleave blocks of 10 strings; it may interleave by lower orders.

The rationale of the interleave is based on the nature of the merge process. Blocks of strings are accessed one at a time as they are depleted. To the extent that there is a tendency for a block of string 1 to be depleted, followed by a depletion of block of string 2, etc., it is convenient to be able to acquire blocks from different strings with minimum arm movement.

The "randomness" of data has an impact on interleaving. Since a given amount of space must be left within writes of a string, it is necessary to know string length in advance. The need to know string length will lead, on dispersion, to the use of an internal sort that produces shorter fixed-size strings, or to the truncation of all strings at a given size, perhaps $2F$, where F is number of records in the work area. Shorter strings would be artificially dummied to that length. The impact of truncating Replacement Selection strings is not too serious, but the need to maintain known string sizes continues throughout the merge.

When block size is less than track size, a problem occurs in interleaving. During the course of merging it may be necessary to start writing on a track at a position other than the first physical space of the track. Without the support of formating techniques, which make a record-level access possible, a considerable amount of space may be lost on a track. One technique for finding a starting location is to write an entire track of the first block written in the track. This in effect defines a track format, which can then be used to locate specific positions. A considerable amount of extra writing may occur. On devices with format tracks, which may be prespecified, the problem is less severe.

The interleaving of strings is useful whenever orders of merge higher than the number of read/write arms are desirable. Interleaving may be associated with balanced merging, as we have described it in this section, or with other merge designs that do not relate number of work areas or arms to order of merge.

One further problem of interleaving is the difficulty in undertaking initial partial passes. A solution to this problem and to that of restricted initial string sizes may be in delaying the interleaving until after the development of all strings. Strings can be dispersed sequentially and a directory can be built to represent location and length. Knowing the number and length of strings, one can determine the optimum interleave format.

18.2.8 Write in Place

One problem not solved by interleaving in the one-drive situation is the interference between reads and writes during merge passes. An arm must be moved for a write and returned for a read. A technique for reducing arm-movement time is to write elements of output strings in the space just read from. Using this method, one is said to *write in place*. The method may be associated with interleaving by forcing a specific pattern of reads and writes, such that every I/O operation is a read/write sequence. Since newly formed strings are written according to the pattern of reads performed across the population of input strings, they are not necessarily contiguous. A method of block chaining seems to be a hardware requirement for this technique.

Because of the discontinuity of strings that develops with writing in place, it is natural to associate the technique with interleaving in an attempt to reduce read/seek time while eliminating write/seek time. Spaces vacated by a series of reads, when interleaving is used, will tend to be continuous and allow longer in-place writes. The details of forcing alternate reads and writes lie in buffering techniques or in scheduling techniques applied in such a way as to guarantee that there is always a block to be written when reading occurs.

A difficulty with writing in place is the necessity of providing a mechanism of some sort for checkpointing.

18.2.9 Multiple Arms

There are a number of approaches to multiple-arm environments, which differ from one another in the way areas are defined and used across devices. The usefulness of a technique is partially a function of the number of devices available. When many independent devices are available, it may be desirable to use the drives as sequential devices in a "tapelike" fashion. An implementation of the Crisscross technique is appropriate in situations where the optimum order of merge is low. If the optimum order of merge is not more than the number of arms, a tapelike merge can be performed with one area per device. Input comes from $(k - 1)$ devices onto a kth device. After the beginnings of the strings are located, there is no seek time. Writes can be overlapped with reads, and the pattern of input reads is like that of Crisscross Merge of tape.

If the optimum order of merge is considerably higher than the number of devices, the Crisscross approach becomes less attractive. Placing multiple areas on a device to sustain higher-order merges may increase seek time. A 15-way merge on eight drives, for example, involves the access of 15 data areas by seven arms. Each arm moves back and forth between two string streams in an inefficient way. There may be sort situations in which a Crisscross Merge of order equal to the number of arms will outperform a Crisscross Merge of much higher order because of the reduction in seek time.

With multiple arms, higher-order Balanced merges may also profit from reduced seek time, although most known contemporary designs other than Crisscross do not profit significantly from the presence of more than two drives. The dramatic change in performance comes from having separate arms for reading and writing. This separation not only avoids arm movement but allows reads and seeks to be undertaken concurrently. A Balanced Merge from one area to another obviously profits from having the input and output areas on separate disks.

One use of more than two disks that is available to a Balanced Merge (with or without interleaving) involves extending the string-holding area by concatenating cylinders of separate devices. The "logical cylinder" so created would then have n arms servicing the merge of the strings within the area, and there would be n times as much space available to an area before an arm movement was required.

The logical-cylinder concept may also be combined with writing in place. Since there is no need to separate writing and reading to reduce arm movement when writing in place, the advantage of multiple arms comes from overlap and the possibility of scheduling reads and writes for minimum arm movement.

Obviously, the advantages of multiple-arm design depend on having all arms continuously available. In some multiprogrammed or teleprocessing installations, the arms may be preempted temporarily by the operating system supervisor. If preemption is to be expected, the sort designer cannot count on achieving the theoretical benefits of multiple-arm control.

18.3 ADDITIONAL CONSIDERATIONS
IN RANDOM-ACCESS MERGING

The previous sections have described the nature of merging in a random-access environment. Determining the order of merge, merge design, controlling seek time, and some approaches to multiple devices were discussed. There remain several topics which cluster around random-access merging, not applying to any particular technique but constituting a general set of considerations and possibilities that have to do with the nature of merging, programs, and machinery. They are discussed here in order to complete a consideration of random-access merges within the context of this chapter.

18.3.1 String Size: Directories

The size of strings may be variable for random-access merging unless interleaving is used in the dispersion. When interleaving is used, string sizes may vary if the interleave is undertaken after dispersion [1]. A directory of all strings and string sizes allows the "on-the-fly" development of interleave distances. In general, when string sizes are allowed to vary, a directory may be necessary if merging is performed (as in Balanced) on multiple strings in the same area (as opposed to separate merging areas). The starting location of strings must be made known to the system in order to enable it to conduct the merge.

For any method based on approximation of an optimum tree, the size of all strings must be known. It has recently been suggested by Frazer and Wong that strings be extended during the dispersion phase by the use of disk storage and an extension of Replacement Selection. An implication of such an extension of strings is that the program can intentionally produce a population of strings of various sizes. Control of the variation in string size can consequently lead to an optimization of merge time by making the optimum tree more effective. If strings of particular size are produced, it may be possible to disperse them in an optimum way so that short strings are particularly accessible.

18.3.2   Buffering

The overlap of read/write/compute in the random access environment is achieved with buffering, and buffering decisions are therefore as critical here as they are in the tape environment.

Buffering is based on merge order (m) in the random-access environment. Double buffering ($2m + 2$) is closely approximated by the preselection of floating buffers, using ($m + 2 + X$) buffers, where X is the number of floating buffers and may vary from 1 to ($m - 1$). The value of X determines how good an approximation of double buffering the preselection will be [43]. The proper buffering level will change with the speed of the CPU, the ratio of CPU to I/O-system speeds, and the size of core. In general, the designer should only very grudgingly grant buffers beyond $X = 2$ if the higher values will cost merge power. An investigation must be made, as for tape, of the relative performance of higher-order merges and lower-order merges in terms of overlap time. Block size and optimum merge must be honed and modified in light of buffer allocation.

18.3.3   Blocking

One possibility that must not be overlooked is the use of dynamic blocking: changing block sizes throughout the merge from pass to pass on the basis of the predicted order of merge of the next merge pass. One technique is chaining. Short blocks may be written by the dispersion phase. During the progress of the merge, the short blocks are combined and read in chained multiples to form large blocks appropriate for the merge order.

18.3.4   Look-Ahead Seek: Arm Scheduling

Two I/O techniques that are not unique to a sort/merge environment are the look-ahead seek and arm scheduling. With preselection, it is possible in most systems to start the seek of the next block to be read before it is required, thus overlapping seeking with processing. It is possible to start the seek before the buffer is available. Arm scheduling involves the accumulation of read and write requests and their execution in order of distance. A queue of requests is ordered in such a way that arm movement is smoothed, and seeks between I/O operations are minimized. When the concept is applied to multiple arms, the arm nearest a desired record is sent to the position. However, devices with multiple arms that can access the same record are now uncommon.

18.3.5   Tag versus Record Sorting

A great deal of attention has been given in the literature to the question of tag versus record sorting in a disk environment. Tag sorting involves the construction of an auxiliary file of tags. Instead of merging records, the procedure merges only the tag file. When the merge is complete, the records are fetched in sequence from storage and placed on the final output device.

Certain advantages can be gained with tag sorting. It can reduce data movement (in terms of bits moved), reduce the number of blocks (and, consequently, the number of seeks) for any merge order, or increase merge order because of the reduction in blocks required to represent the file. The disadvantage lies in the seek times required to access the records to create an ordered file [39].

18.3.5   String Extension and Redefinition

The probability that the data being sorted is not "random" is very high. As a result of the mechanisms used to collect data, and perhaps because of previous sorting operations, there will frequently be a good deal of usable "order" in data files [19]. The use of Replacement Selection as a dispersion pass method will take advantage of data order. Section 12.7.1 describes string-extension techniques and their limitations on tape. The disk environment provides for a considerably more elaborate application of string extension and string redefinition. The reader will find material conceptually related to this section in Chapter 9's discussion of the use of string-definition tables.

The underlying concept for String Extension in a disk environment comes from Frazer's work in "Natural Selection." By performing Replacement Selection sorting on strings already formed by Replacement Selection, one can control the number and size of strings so as to optimize subsequent merge phases.

An important realization of the idea of string redefinition is embodied in an IBM produced sort, 5740–SMI, under the name PEER technique. The sort system contains a number of sort techniques. The PEER technique provides significant improvements over other (classical) techniques when the initial file is partially ordered. The PEER technique is used only for random-access storage environments.

The essential goal of the technique (described here in a general form) is to reduce the number of strings by recognizing where strings can be appended or interspersed. The dispersion phase creates a number of indices for each string. Each index represents a string segment. A string segment, in PEER for example, is defined as that portion of a string which fits on a device track. Figure 18.16 represents a string-segment index. A single string is defined by an index number, the low and high key values in the segment, the number of records in the segment, and the address of the segment on the disk, as shown in the figure. The entries describe a string of 30 elements ranging in value from key = 2 to key = 49.

Segment no.	Low key	High key	# Records	ADDRESS
1	2	27	10	ADDRESS
2	23	36	10	ADDRESS
3	37	49	10	ADDRESS

Fig. 18.16 String-segment index in PEER technique.

When the dispersion phase is complete, a string-redefinition phase sorts the string-segment indices in ascending sequence of low-key values. The sorted indices are used to define new strings. In Fig. 18.17 a sorted list of indices contains 16 string segments. This list is used to define two new strings, one containing 12 segments, the other containing four segments. The process by which this is done is as follows:

1. Place the index entry with the lowest key in a new string list.

2. Determine whether the next index entry can extend the new list; i.e., is the low key of the next entry equal to or higher than the high key of the entry on the new list? Append the next entry if it passes the test; if it fails, place it in a second new list defining another string.

3. Test each subsequent index entry in the original list to see whether it can be appended to one of the new string lists. Start additional lists as necessary.

The reduced population of defined strings is next partitioned into work areas. The number of strings in each partition is determined by the merge order. In PEER, the first partition is assigned $(m - 1)$ strings; all other partitions will contain m strings. If $m = 4$, there will be three strings in the first partition and four strings in all other partitions. Each partition is now ordered, using the indices and Replacement Selection. The string population is reduced to correspond with the strings defined by the index lists. The actual reduction in string population depends upon order in the data.

Indices		
#	L	H
3	2	4
5	3	7
4	8	11
6	9	16
7	11	17
1	23	29
2	29	34
8	34	39
9	38	43
10	41	45
13	46	49
15	49	57
11	53	60
12	61	68
14	70	74
16	78	93

New strings		
#	L	H
3	2	4
4	8	11
7	11	17
1	23	29
2	29	34
8	34	39
10	41	45
13	46	49
15	49	57
12	61	69
14	70	74
16	78	93

#	L	H
5	3	7
6	9	10
9	38	43
11	53	60

Fig. 18.17 String redefinition in PEER technique.

When the string population is reduced by the string-definition phase, a merge phase is entered, which reduces all partitions other than the first to one string. If the number of strings in the first partition plus the total remaining strings (one each) is not greater than the merge order for last pass, then the last-pass merge is performed. If this is not true, then another string-redefinition phase is undertaken. The sequence: string redefinition/ intrapartition merge is continued until the string population is not greater than the last-pass merge order.

Part 3
Sorting Systems

19
Generalized
Sorting
Systems

19.1 GENERALIZED SORTING

Generalized sorting systems, which are provided by the manufacturer or purchased from a software supplier, are available for most computing systems. Such generalized systems aim at broad applicability, ease of use, and reasonable efficiency.

A sorting system that has broad applicability can be used to sort files of divergent characteristics. Different volumes of data, record sizes, key sizes, etc., can be accommodated. The system includes a language in which a user can describe the data to be sorted and the machine resources he wants for the sort. The system interprets the user's description and develops a running sort for that data. Some systems produce a complete sort program; others customize a prototype sort. The ease of use of the system is determined by the quality of the sort language and by the operational characteristics of the programs associated with the system. Provision for restart, checkpoint, and good operating messages enhances the usability of a sort system.

Because a generalized system must accommodate many different data and machine situations, there is some loss of efficiency. Systems attempt to overcome the usual cost of generality by providing alternative sorting methods within the same package and by providing (more or less) sophisticated algorithms to determine preferred order of merge, blocking, and buffering features for a particular sort/merge.

Over and above variations in data and configuration, the developer of a generalized sorting system must consider the nature of other components of the software package. He must decide what components of the package should be used in the sort, what aspects of the sort should be described

297

in job control language, whether the sort can be used as a subroutine within a program, whether that program can be written in any language, etc. In addition to having sort and I/O expertise, the sort developers must have an appreciation for a program-generation techniques, language structures, simulation, and a host of other software techniques. The development of a good generalized sort involves all the skills and insights in the repertoire of systems programming.

19.2 BASIC STRUCTURE

The generalized sort/merge consists of a sequence of phases.

1. Assignment
2. Sort
3. Merge
4. Last-pass merge

19.2.1 Assignment

The *assignment phase* (also called *control phase*) is the initial sort module. Its specific functions vary, depending on sort parameters presented at sort run time and on interfaces with the operating system. Among the specific functions which an assignment phase may perform are the following.

1. Read sort control cards and develop a table representing sort parameters. The sort parameter table contains information which may be used by the initialization functions of the sort and merge phases.
2. Interface with the operating systems to determine how much memory is available, and place this value in a parameter table.
3. Interface with the operating system to determine what I/O device assignments have been made, or to request assignments.
4. Generate proper comparison or key-definition functions on the basis of description of data.
5. Select internal sort technique.
6. Calculate sort area size and buffering.
7. Calculate block sizes and order of merge.
8. Optimize dispersion algorithm to maximize channel overlaps.
9. Determine merge technique to be used.
10. Request loading and linkage of user "own code" sections.
11. Provide estimate of sort run time.

How much an assignment phase actually does depends on the specific design of a package and the interface of the sort with the rest of the operating system.

19.2.2 Sort

Sort packages differ in the amount of specialization the *sort phase* undertakes for itself and the amount done for it by the assignment phase. The amount of specialization required is a function of the flexibility of the package and the extent of optimization. At minimum, a comparison routine must be generated, the I/O item advance (block and deblock) must be specified, and the address of devices must be given to I/O routines.

Some method of introducing user-provided "own code" programming must be present. The mechanism may be the assembler, a compiler, or the loader. This procedure will be discussed in more detail later.

19.2.3 Merge

All merge passes except the last are included in the merge phase, sometimes called the *intermediate merge phase.* As with the sort phase, it may specialize itself or be specialized by the assignment phase or by some combination of both. The things that must be done are to select device addresses, organize a comparison facility for an m-way comparison (for an m-way merge), and include user "own code" programs. User "own code," called EPOC (each pass own code), is not always permitted for the intermediate phases of a sorting package.

19.2.4 Last Pass

The last pass of the merge is distinguished from the others because of its interface with the user, who specifies output block size and may modify I/O unit availability. The record must also be in a format the user specifies; as a consequence of the changes in the last pass, the buffering, item advance, and storage layout must be uniquely specialized.

19.3 PARAMETERS AND THEIR USE

Systems will vary in their scope and flexibility. The parameters that must (or may) be provided reflect differences in design goals and optimization techniques. There are three major parameter-determined functions of the sort merge.

1. Comparison
2. Item advance
3. Input/output

All generalized sorts require some customization in each of these areas [16].

19.3.1 Comparison

The comparison function is the module of code which compares the magnitude of keys. In a generalized package, three kinds of data are required to prepare the comparison function for operation.

1. Position and length of each key to be compared
2. Encoding of each key
3. Sequence of the sort (ascending or descending)

Some sort systems, such as the sort/merge for the operating system EXEC 1 for UNIVAC 1107 [20], allow for a user-defined collating sequence in preference to the built-in machine sequence.

The encoding of the key describes its mode and format. On machines of mixed representation (decimal, binary, etc.), the representation must be described. For the sort/merge associated with IBM's OS/360–370 called SM–023 [7,14], the following descriptions are possible:

CH	Character	FI	Fixed point
ZD	Zoned decimal	BI	Binary
PD	Packed decimal	FL	Floating point

There are a number of choices about what may be done with a description of data encoding. The choices depend in part on the architecture of the machine. If the machine has quite a variety of format controls and comparisons, the comparison function for one format may look quite different from that for another. Many machines have Compare Binary, Compare Logical, and Compare Alphabetic instructions, which can evaluate relative magnitude directly without the necessity of conversion. If such instructions do not exist, some conversion procedure must be applied, so that the key will reach the comparison in the proper format.

How much is done with format parameters depends partially on the generation techniques used by the sort system. More specialization may be applied when the comparison function is generated as an assembly macro than when it is specialized by parameters at load time. A technique that may reduce the effort of producing highly specialized code at run time is selection of one of a number of canned comparison functions based on the description of the data.

The position of the key must be included in the encoding for both character and word machines. In word-oriented machines, keys may span several words, may be packed several to a word, or may depend or not depend on the sign of a word in which they appear. Field definition involves indicating the bit positions of the word or words which contain the key.

The length and origin of a key affect the comparison coding. Some systems may choose between register-to-register comparisons or register-to-memory comparisons on the basis of key length. Word machines will have to parametrize or generate key isolation instructions on the basis of key position.

Figure 19.1 shows the description of keys for three sort systems: the sort for IBM's OS/360–370 [14], UNIVAC III/SORT III [18,19], and the UNIVAC 1050 sort. For IBM's System/360–370, the statement appears on a sort control card which is read at the run time of the sort. For the other two systems, the key description is an assembly system macro. The OS/360 statement describes two key fields. A major key is three bytes long, beginning in byte 1 of the record. This field is zoned in format and is to be ordered in ascending sequence. A secondary field begins in byte 7 of the record. It is 34 bits long and is represented as taking 4 bytes (32 bits) and 2 extra bits. It is a binary field to be sorted in ascending order.

```
IBM OS/360 Sort
   SORT FIELDS = (1, 3, ZD, A, 7, 4, 2, BI, A)

UNIVAC III/SORT III
   COMPARE * KEYS 0, 9, 15, 2, 3

UNIVAC 1050
   SORT1, 2, D, B
   KEY1, 3, 1
   KEY2, 4, 7
```

Fig. 19.1 Key descritpion for three sort systems.

The UNIVAC III statement describes two keys for the Compare Macro. One key is held in the entire 24-bit first word of the record. The second key runs from bit 9 of word 2 (the third word) to bit 15 of word 3 (the fourth word). This information is represented by 9, 15, 2, 3.

The UNIVAC 1050 statement uses a SORT1 macro to describe the fact that there are two keys, one decimal and one binary. The KEY1 and KEY2 macros give length and position of the keys.

Other systems have still other ways of describing the fields that represent keys. Regardless of format, the intent is to make it unnecessary for a user to code a comparison function of his own for every file. If the generality of the system does not include all possible key locations and encodings, the system should provide a mechanism that allows the user to embed his own comparison algorithm in the context of the sorting package.

19.3.2   Item Advance

The function known as *item advance* introduces new elements into the sort or takes them from the sort. It is essentially an I/O interface function. It must have information about how large the record is and whether the record is of variable length (if variable-length records are accepted by the sort).

When the system allows records of either fixed or variable lengths, and/or when a record may change its size during the sort/merge, more extensive information is required. The optional RECORD statement of the OS/360 sort/merge furnishes a representative example.

The statement allows for the classification of the records as F (fixed) or V (variable) and for the provision of a set of lengths. For fixed-length records, three values may be expressed: the length of records in the input set, in the sort, and in output. The parameter provides for the possibility that "own code" processing inserted by the user may change record lengths during the progress of the sort/merge. A statement of this type, for example, RECORD TYPE = F, Length = (100, 50, 60), describes fixed-length records of 100 bytes in the input file, changed by "own code" to 50 bytes before being sorted, and changed in the last pass to 60 bytes. For variable-length records, there must be corresponding values for maximum length at the various times. In addition, values representing minimum length of a record on input and modal length are accepted.

The item advance activity provides blocking and deblocking for the sort/merge and interfaces with the I/O support packages. Since blocking and deblocking constitute a common system function, descriptive mechanisms for record length may be expected to exist in several places besides the sort control language. In the presence of a fully developed operating system, the description of record lengths is a feature of the control language. Assembly language and higher-level languages (particularly COBOL and PL/I) will also provide record description capability. The designer of a sort control language must determine whether to provide this capability in the sort language as well. The decision depends on the relationships among the sort system, the operating system, and the programming languages. The critical criterion is the availability of record description information to the sort generating function at the time of sort generation. In OS/360–370 the record statement for sort control is optional and is used only when record lengths change. Record definition is obtained by the sort system from the data description facilities of the job control language (JCL).

19.3.3 Input/Output

The area in which the relationship between sort control and general job control becomes most vague is input/output. A complete description of I/O for the sort/merge function must include the following items.

1. Location of the input file
2. Devices that will hold intermediate strings
3. Devices that are to hold output
4. Block size of input
5. Block size of output
6. Format details of input and output data files
7. Size of the file to be sorted
8. Amount of space available on any random-access storage devices used in the merge

Readers familiar with a contemporary operating system of any size will recognize that, with the exception of (7) above, all I/O requirements can be supported by the job control language. In systems with no job control language, the ability to describe the I/O environment will reside in programming language. It is therefore only in a stand-alone "black box" sort that a private I/O language must be defined and supported.

In sort system design for the allocation of I/O, one point of interest concerns how much the user *may* specify, how much he *must* specify, and what is left to the sort control elements and the operating system. These considerations are particularly critical for number of work units, amount and organization of storage space on random-access units, and distribution of devices across channels. To the extent that users may or must specify these quantities, they must receive considerable guidance from the vendor. This guidance commonly comes from vendor-provided user manuals, which explain more or less completely the implications of various decisions on sort performance. With flexible sort packages, it is pretty much a requirement that the user have some basic understanding of the sort/merge process in order to obtain the full potential of the technology in the system. This is true whether the medium is sort language or JCL.

The user designation of the number of work devices determines the maximum order of merge for tape merges. The distribution of tapes across channels may be determined by a user-specified parameter that determines the amount of read/write overlap that may be achieved. Many contemporary sort systems base a choice of tape techniques in part on the number of tapes available and on the number of channels.

User definition of space on disk is also highly critical. Each disk merge has a space assignment algorithm, but the user determines the number of usable devices (arms) and determines the distribution of devices across channels. In addition, the user may be asked (as in OS/360–370) to define work space on devices. Arm movement may be affected by the number of work areas defined and the size of each.

In contemporary operating systems, I/O specification is commonly done as part of run-time job control. Flexibility of device definition is naturally desirable, and it is best achieved by delaying I/O control parameters until run time.

19.4   ADDITIONAL PARAMETERS

Key characteristics, record characteristics, and I/O environment are the basic parameters of a generalized sorting package. Numerous other parameters may be provided, some of the most common of which are described in this section. When a parameter does not exist in a system, the related service or function may not exist, or it may be provided on the basis of closed internal systems decisions. For example, some sorts offer a variety of merge types but do not allow the user to specify which merge he wants. The choice of a merge is based on some package algorithm. Some sorts allow an optional specification of merge type by a user. The lack of this parameter in the user language, however, is not an indication of the flexibility of the package or the availability of various merge types.

19.4.1   File Size

File size is an almost universal parameter for a sort. The user may be forced to provide at least some guess about file size, or it may be optional. The sort package may make varying use of the information, depending on how ambitious and flexible the optimization features of the sort are.

The file-size parameter may be used as part of merge technique selection. The dispersion of strings across devices and storage spaces changes from merge to merge, and some techniques have inherently more capacity than others. The parameter may also be used to approximate the number of strings that will be produced and to select a merge and/or merge order based on the number of strings expected to be formed.

19.4.2   Memory Size

A common parameter for sorts that are intended to operate in potentially multiprogrammed environments is a specification of the amount of primary storage that will be made available to the sort. This quantity usually represents the space available for the entire sort function and is used in the calculation of the sort to determine work area, block size, order of merge, buffering, etc.

A variant form of the core-size parameter allows the user to specify the size of the sort work area he wants during the dispersion pass.

19.4.3 Checkpoint

The establishment of points in the sort/merge process that represent fall-back positions is an important feature of package design. A common check-point procedure built into the sort is to establish checkpoints at the end of the dispersion pass and at the end of each merge pass or partial pass. Different merges vary in the details of timing when checkpoints are taken, but all are similar in attempting to provide points of fallback so that if a malfunction occurs, the user is protected against having to lose more than a data pass. The program procedures that support a checkpoint function are not "no-cost" items. Space must be provided for the representation of data at certain stages of the merge. Time must be taken to mount devices (for tape) and to write control information. In most packages, the user may either delete the checkpoint function or call specifically for it to be included. In OS/360–370 SM–023, the CKPT keyword is a run-time parameter call-ing for checkpoint protection. The specific action of the package depends on the merge technique being used.

19.4.4 Choice of Merge

Contemporary systems allow for the use of a variety of available merge techniques. The choice of a technique may be left entirely to the system, or the user may be able to force a particular merge. The usefulness of forcing a merge depends on the care taken by the system in determining a technique, and on the knowledge of the user. It is surely an option to be used with care, provided to enable the user, when he knows something about his hardware and data, to utilize his knowledge. Before exercising this option, a user should be very careful to determine exactly how the choice is made by the system. He may find that the criteria of the system are not complete enough. They may be too arbitrary. Some detail of the package may defeat other details and make a different merge desirable. For example, (1) the package may impose some arbitrary limits on merge order, record size, and/or block size that may make a normally preferred technique less effec-tive, and the user may decide to force an alternative; or (2) the user may know he will create a perfect dispersion for a particular technique.

19.4.5 "Own Code"

The capability of incorporating "own code" programming has a twofold purpose in a sorting package. It provides a mechanism for describing sort-related functions that are clumsy to express as parameters to the sort, and it enables the user to expand what can be accomplished during his sort by providing functions that are not sort-oriented. An example of a sort-related "own code" is a routine to handle a key equality condition.

The variety of functions that can be performed by "own code" is determined by the general structure of the sorting package. In a sort system based on assembly of sort macros, the user provides a "frame" in his own code. He must call the sort macros and I/O macros and provide program start, end, and interface points. The user is responsible for delivering items to the sort, for obtaining items from it, for terminating it, etc. The sort is seen as a function within a user-written program.

When the sort is viewed as a "black box," the relations between "own code" and the sort are reversed. A particular number of exits or interface points (where the user can supply code) is provided. A loader-generated sort like OS/360–370 SM–023 provides program structure and a particular number of exits but allows reasonably unrestricted functions at each of the exit points.

Regardless of the direction of linkage between the sort and "own code," a set of linkage conventions must exist. Primarily the conventions involve the use of a set of labels, index register use, and entry and exit conditions. When "own code" provides the frame, the user must be told what labels to use in addressing the sort functions he is calling. When the sort package provides the frame, the user must be told how to return.

Some determinations beyond basic structure must be made in connection with "own code." An important decision is what will be allowed at the interface between languages other than the assembler language and the sort. It is necessary to support some interface between COBOL, PL/I, etc., and the sort package. The invocation of a sort capability from COBOL programs is an American National Standard, and the SORT verb is described in the COBOL language manuals. There is also a PL/I language sort interface. The decision that must be made about these interfaces is whether the same "own code" functions available with the assembler will be available to a higher-level language interface.

19.5 REPRESENTATION OF THE SORT

The total sort capability of a generalized sort/merge system is commonly represented as a collection of routines in a sort library. Depending on the operating system and sort system design, the sort library may be transformed into a running sort at compile or assembly time, at load time, or at interphase time during the running of the sort. The routines in the sort library include those which will be part of the running sort (running modules) and those which will specialize the running-sort modules at computer run time (assignment modules). Either assignment or running modules may be skeletal code. They may be generative macros. They may be fixed routines, which vary from sort to sort only in terms of where they are located in core. The following sections will describe two very different approaches to sort generation, which may serve as useful examples.

19.5.1   Assembler-Based Sort Generation

The sort library is organized into four major sections.

1. Generate section containing the Macro Generator for the comparison routine and all assignment modules required to specialize the sort, including:

    a) A routine that sets record-size parameters;

    b) A routine that sets parameters for input device and input blocking;

    c) A routine that sets parameters for output device and output blocking;

    d) A routine that sets parameters for device addresses for the merge;

    e) A routine that provides for the location of the sort work area; and

    f) A routine used for entry into the sort.

2. The sort code, consisting of running-sort modules. (Running-sort modules are the modules to be included in the operating sort.)

3. A standard drive program, which is provided when the sort is to be used as a "black box," containing necessary I/O support routines.

4. A set of linkage programs, for use when the sort is called by COBOL.

All sections exist as assembly-language symbolic code in the system library. All parameters to the sort are provided through assembly macro calls. Sort generation is accomplished by the macroassembler. If the sort to be generated is to run as a "black box" sort, the assembly code includes a call to the library for the support package driver. The user may provide a driver program of his own in assembly language rather than call for the driver package.

The sort is designed to run as a subroutine of a longer program, and the user may provide a driver as complex as he wishes. The output of the assembler is a sort that can be run without further parametrization. The assignment phase is partially embedded in the assembler.

In those installations which tend to run sorts repeatedly over the same data files, the one-time specialization of the sort involving an assembly represents no disadvantage and may provide a speedier sort. The ease with which the program functions associate with the sort is a potential advantage. The parametrization and specialization of the sort prior to load time may represent a disadvantage to very dynamic environments of larger machines.

19.5.2   System Loader-based Sort Generation

The sort capability is represented in a system library and in a sort library. The assignment phase of the sort is a part of the operating system library. The assignment-phase modules exist in loadable, relocatable form and are read from the system library when the sort is called.

The sort library is also relocatable. It was defined at system generation time from the total set of hundreds of modules representing various functions of a sort program. The sort library subset contains all modules that the installation might expect to use in the creation of any potential sort. (Modules that might be excluded from the sort library are, for example, disk-oriented routines when only a *tape* configuration exists.) The developed sort library is organized into sort, merge, and last-pass phases.

When the sort is called, the assignment phase is called from the system library by the system loader. The assignment-phase modules develop a sort-parameter table from control cards. The phase undertakes block, buffer, and merge calculations and communicates with the operating system to determine device assignment. The assignment phase then requests the loading by the system loader of the initialization modules of the sort phase.

The sort-phase initializing module readies the phase for operation. All modules, with the exception of user "own code," system-provided I/O, and interrupt control, come from the sort library. The initializer loads a series of sort-phase assignment modules. Each is loaded, performs its function, and is overlaid by a successor. The specializations to be performed involve the generation of a comparison function, the organization of storage for buffer and sort areas, and, if necessary, the organization of structure for a tree sort. On the basis of parameters in the parameter table, modules are generated, modules are selected and loaded, and, when the specialization is complete, transfer is made to the prepared running program.

The intermediate merge phase and last pass are developed in exactly the same manner. Each phase consists of a sort-library family of modules, some of which are to be selected and loaded. Some perform preliminary calculations, some develop storage layouts, and some generate code. Figure 19.2 is an example of the type of modules in the library, a partial list of the modules of the IBM OS/360 sort SM–023. It is a very small subset of the over 200 modules associated with the sort. Each of the paired routines

Name	Phase	Function
ROA	Sort	Replacement with single control field
ROB	Sort	Replacement with multiple control field
RDP	Oscillating sort	Deblock record, <256 bytes
RDO	Oscillating sort	Deblock record, $\geqslant 256$ bytes
AOR	Merge	Balanced tape merge algorithm
AOS	Merge	Polyphase tape merge algorithm

Fig. 19.2 Typical library modules (IBM OS/360 sort).

offers a choice: that is, either ROA or ROB will be used for the sort, either RDP or RDO, either AOR or AOS. This approach makes for larger libraries but minimizes the need for generation of code at object time. Where the range of possible situations cannot be feasibly covered by selection of alternative modules, the generation capability must still exist. This is the case with the comparison function.

There are several points about systems interface which should be made explicit. The system loader-based sort system relies on the system control stream for a definition of I/O environment. The system scheduler and allocator will have already assigned devices at the time the first module of the sort is loaded. These assignments are reflected in systems tables, which can be inspected in the assignment phase. The operation of the sort, therefore, depends on the presence of a control stream reader and resource allocator in the operating system. Further, the activation of the sort is accomplished when the operating system executes a request to the system loader for the load of the first module.

Another way in which the sort may depend on system services concerns user "own code." The modules produced as "own code" by the user reside neither in the system library nor in the sort library. They must be assembled and compiled into a private user library and taken from there. A parameter card indicates when user "own code" exists and is to be applied to the sort. An additional system function may then be necessary to prepare user modules for loading by the sort. The OS/360 sort, for example, must exercise the system linkage editor in order to prepare "own code" modules for the sort.

One distinguishing characteristic of this design for sort generation is its close association with and reliance on an operating system. Another characteristic is the use of relocatable machine-code-level modules, rather than assembly or compiler modules, in the library and the use of duplication and selection rather than generation as the basic technique of specializing the sort. A running sort is never "generated." Sort programs are "specialized" for each sort call and are not, as with the system previously described, reusable.

19.6 DECISIONS OF THE SYSTEM

In addition to, or as an alternative to, the parameters provided by the user, the sort system itself makes certain determinations, as suggested by the following questions.

1. What tape or disk merge is appropriate for a given sort situation?
2. What buffering levels and buffering technique are needed?
3. What block size should be used, particularly during the merge phases? It should be calculated to be optimum for the merge.
4. What is the order of merge?

In addition, there are details of program structure, overlay, table control, memory allocation, and linkage which must be generated and calculated by the system.

The specifics of the calculations involved will vary from system to system, as will the amount of effort devoted to finding an optimum balance of technique, merge order, blocking, and buffering. Some sort systems attempt more than others.

19.6.1 Choosing a Technique

For the most part, technique choice in current packages is limited to selecting the tape or disk merge to use. Internal algorithms are fixed, as are determinations of such details as whether to perform a record or key sort.

As we have seen in earlier discussions, the criteria for selecting a merge are the following.

1. Capacity of devices
2. Channel arrangements
3. Core size
4. Number of devices
5. File sizes; record size

The importance of capacity as a criterion is related to the tendency of certain dispersions to place disproportionate amounts of the file on one unit. Channel arrangements are important when there is a tendency, as there is with $(k-1)$-way merges on tape, to interference between reads and writes in balanced channel environments. Core size can be a limiting factor for Crisscross and Oscillating, which need either to have sort and merge code coexist in core, or to perform overlays. Core size must also be considered in relation to the increased number of strings that may be produced. The parameters of file size and record size are also relevant to the expected number of strings; and as we have seen, the expected number of strings affects the choice of one merge over another.

19.6.2 Calculating Main Memory Usage

The discussions in Chapters 12 and 13 about order of merge, buffering, and blocking are naturally relevant in the context of generalized sort/merge operations. A function of the assignment phase or merge initiation phase is to determine order of merge, blocking, and buffering for any sort to be generated.

The calculations at their simplest involve a number of determinations all reminiscent of Chapters 12 and 13.

1. *Size of program.* The size of the program depends on the size of the generated routines and of the routines selected from the library. It may be sensitive to number of I/O devices, size of user "own code," record length, key format and length, merge type, and merge order. Program size is subtracted from core size available, to determine the size of space to be used by I/O and the sort area. Calculations for sort, merge, and last pass naturally give different results for program size.

2. *Block size and buffering.* Space required depends on order of merge, read/write/compute overlap capability of hardware, and machine balance. The buffer strategy should depend on the CPU-to-I/O speed ratios of the machine, the core size, and the sorting techniques. It is usually fixed in a package. Double buffering, with each merge device having two buffers, is a common strategy. The reader is again referred to the discussion of buffering in Chapter 12.

Block size is easily calculated when a fixed-buffer policy exists. There may be reasons to adjust block size from its initially calculated size. For example, if the dispersion phase is unable to produce a sufficiently large block, the largest block that can be written by the sort is used. There may be systems reasons for rounding block size up or down, or for preferring particular block sizes. Track size or a multiple of track size is a reasonable block size for a disk system. A system-provided convenient block size of this type is often a good starting point for calculation.

3. *Merge order.* The merge order for tape systems is based on the number of available tape units. Some systems will attempt to predict the performance of a lower-order merge.

On disk systems the order of merge is based on policy and situation. If a system enforces a maximum way-of-merge policy, merge order is always equal to the number of strings; but this feature is not found in contemporary systems, which attempt to balance merge order, blocking, and buffering. Merge order may evolve from other calculations. For example, if system track size is 3600 bytes and there are 90,000 locations of core, 24 is a good first guess at merge order. If there are two output buffers, merge order is reduced to 23. Depending on the buffering strategy, merge order can be reduced until a fit is found. For example, with double input buffering, 3600-byte blocks, and 90,000 locations of core, m becomes 11.

19.7   GENERALIZED SORT DEVELOPMENT AND HARDWARE

From time to time throughout the text, we have discussed the impact of machine detail on sorting. Channel and CPU architecture and interconnection will naturally affect the performance of programmed code that embodies a sort or a sort/merge technique. The implementation of a generalized sorting capability is no less affected by the nature of a host machine than is that of a special-purpose sort. The following sections will serve as an introductory discussion of the impact of hardware on sorting from the viewpoint of system development.

19.7.1   Product Line

Current practice among some manufacturers of computing systems is to offer a family of devices which represent a continuum of points on a price/performance chart. These families are characterized by some of the following distinctions.

1. There may be a number of different central processing units of different MIPS rates (millions of instructions per second) but of effectively identical architecture. The goal of the manufacturer is portability of programs from one CPU to another, as the user's need for CPU power increases.

2. A number of different memory sizes may be associated with various CPU's in the product line. The total range of memory size may be from 16K characters to 4,000K characters. Various CPU's in the line will have different ranges of core size. For example, Model 1, the slowest CPU, may have from 16K to 64K; Model 2, from 32K to 256K; Model 3, from 64K to 512K; Model 4, from 256K to 2000K, etc.

3. A number of different memory organizations and interfaces may be associated with various CPU's. One interface may move a single character at a time from memory to CPU; another may move four at a time, another eight, etc. Various methods of interleaving or distributing addresses among

banks of memory may be used in the attempt to redress the inherent imbalance between CPU and memory speeds.

4. Various channel types and various channel speeds may be associated with the various CPU's. Two common classes of channel—the time-shared, multiplex channel for slow devices, and the burst selector channel for fast devices—have quite different behavior characteristics. Various manufacturers provide channels of different speeds. Furthermore, each CPU in the product line may permit several different types of channels to be simultaneously attached and active. The operating capacity of these machines is often a function of the actual mix of devices of varying speeds in a specific installation configuration.

5. Various methods of interfacing channels with CPU's may be used in the product line. The degree of independence of channels and the amount of logic shared with a CPU may vary from model to model of CPU, and may be different for different channels. Different programming capability may be present in the I/O subsystem, depending on the channel interface. Command and data chaining and various forms of gather-write and scatter-read may be present at different levels in the system.

6. Various I/O devices of different types and speed may be associated with various CPU's in different numbers. Larger CPU's may have very high-speed, large-capacity drums configurable into a system. These drums may not be made available with smaller CPU's because of their high speed.

7. Special features may be associated with various CPU's or I/O subsystems at various levels in the product line. Examples of such features in a CPU are expanded instruction sets, special high-performance handling of certain instructions, various degrees of instruction look-ahead, fast buffer memories for data or instructions or both. Examples of such features in an I/O subsystem are special I/O controllers that relieve the CPU of some of the burden of I/O control, such as the Channel Controller for the IBM 360/67 or the IOC of the UNIVAC 1108. The amount of effective overlap of I/O and computer can be vastly increased by the presence of additional devices of this type.

8. The mix of the device environment of the system may be enriched to accommodate history and compatibility. There may be a number of tape devices of various densities and speeds. In addition, it may be possible to read tapes of the manufacturer's previous generation of tape units that were written in different densities and perhaps in different recording modes. It may be necessary, from a manufacturer's standpoint, to offer compatible tape drives to handle tapes which are to be read or written on the drives

of another manufacturer. Lack of backward-read, unattractive reversal speeds and rewind speeds, or other constraints may make an alternative sort technique necessary for the foreign tape devices, if they are to be extensively used by the customers for holding sort files.

There is a great deal to think about before coding a sort. To some extent, operating-system strategies may make the design job simpler by defining groups of machines to be supported by the basic operating system, as well as those to be supported by the intermediate operating system and the high-level operating system. These definitions make the sort design job easier in that they limit the ranges of machine configuration over which a sort package must operate with some reasonable efficiency.

19.7.2   Techniques for the Product Line

Since a family of known sorting techniques must be applied to a population of system configurations in some meaningful way, the sort development activity must bring some insights into the systems. Roughly, these insights are:

1. A determination of the tendency to CPU-boundedness in the lower levels;
2. A determination of the expected minimum configuration that must be supported by the sort;
3. A determination of the importance of changes in CPU speed, core size, and number of channels to sort performance; and
4. An identification of new equipment or new interfaces which suggest the possibility that a new technique or a serious modification of an old one may be needed or that an abandoned technique may be profitably revived.

In addition to studying the performance implications of projected hardware, the sort developers must have some idea of the populations of the various expected configurations. If a larger percentage of users will have small core and no read/write/compute overlap, Replacement Selection may not do as the method for the dispersion pass. It may be possible to support sorting only on certain devices or on some minimum number of devices, because of the expectation that there will be no smaller configurations installed.

A great deal of study and a number of fairly sophisticated tools are required to determine what repertoire of techniques to apply to the product line. Simulation of the performance of various sort techniques on various models and configurations for a variety of files may be undertaken to determine whether it is worthwhile to provide special techniques for particular configurations or groups of configurations, and what the decision points should be in choosing a technique. A corollary consideration is the choice of a family of optimization points: core sizes, channel configurations, and CPU's at which various sort designs are most efficient.

19.7.3   Input/Output Details

The I/O subsystem must be carefully studied to determine how best to approach I/O, particularly for the merge. Dispersion techniques for disk merges are as sensitive to seek- and rotation-time optimization as to data-passing optimization, and therefore I/O techniques are as important as merge technology in determining the performance of a merge. A new device, a new channel, new I/O logic are definitive events in the sort development function. Blocking factors will be affected by track and band sizes; the ability to chain commands or data will affect dispersion techniques; the specifics of channel/control unit/device connect and release times will affect merge design and may vary from model to model in the system.

19.8   SORTS AND SOFTWARE

The sort package will be affected by the software environment required for the product line for which the sort is planned. The impact of the operating system and its facilities, utilities, and languages on sort-system structure has been discussed in various subsections earlier in this chapter. The following sections will touch once more on the nature of those relationships, particularly on those between the sort and I/O support.

19.8.1   The Sort and the Operating System

The relationship between the operating system and the sort is important because it has performance as well as design implications. It would be glib to say that no sort designed for a host operating system will be as fast as the same sort implemented for an unsupported machine; but certainly the use of operating-system support functions and the relinquishing of certain elements of control over hardware do not increase the speed of a sort. A stand-alone sort would be difficult to justify in the contemporary atmosphere of system development. For a small machine, however, it is still possible. For larger machines the possibility of running the sort as part of a multiprogramming situation is often important.

    The specific interfaces between the operating system and the sort are of two major types. One is between the sort and those preactivation procedures of the operating system which prepare the machine environment for the operation of the sort, as for any program. The other type of interface involves those "on-the-fly" services which the operating system provides as a symbiotic, coresident service module. The exact set of services rendered by an operating system in its preliminary role of scheduler–allocator–loader, and in its concurrent role as supervisor–I/O controller, depends on the design of the operating system. Roughly, the following services are performed before any element of the sort begins operation.

1. The operating system reads the job-control language, which calls for the execution of the sort and develops an internal representation.

2. Using whatever service-priority- or resource-utilization-dependent mechanisms are associated with scheduling, the operating system selects the sort for activation.

3. The operating system allocates tapes/disk/drum to the sort, posting system-control tables. The user of the sort frequently has some control over allocations, and this control is important to performance. The number and type of devices may be indicated in the job-control stream to various degrees of detail. A user may ask for 10 tapes; he may ask for 10 particular models of tape; he may ask for 10 tapes and indicate which tapes are on which channel; or he may ask for 10 tapes with specific addresses. If the tapes he requests are not available, the start of the sort may be delayed. In some cases, they may be preempted for use by the sort, depending on scheduling and allocation policies of the operating system. In all cases, a user interested in the best possible operation of the sort does well to express specific device preferences rather than rely on the allocation mechanisms of the operating system. Since such mechanisms are general, they are often quite helpless to define an optimum allocation for the sort. These remarks are especially critical in a multiprogramming environment; however, it should be noted that any optimization achieved by one user may adversely affect the performance of other multiprogrammed jobs.

4. The operating system allocates core space for the sort. This space may be all that is required for running the sort/merge or only the amount that is required for operating the first elements of the assignment phase. If the latter, additional core is requested by the sort as needed.

5. The operating system provides an interface with the operator and uses it to request tape and disk mounts as needed by the sort.

6. Prior to the running of a sort, the operating system will have offered some utility services for the construction of the sort to be run. A library management system may have brought sort modules and user modules together, or they may be gathered during execution. The system loader may be used to load and to link edit modules into a form for loading and running.

The following services can be provided for the sort while it is running.

1. Opening and closing files. A standard operating-system service that provides for label checking and interface with the operator.
2. Loading additional modules for a phase, or loading a new phase.
3. Acquiring and/or releasing core space, disk space, drum space, as required, from phase to phase or during a phase.
4. Submitting I/O commands to the control system and handling interrupts.
5. Managing buffer control and other "block level" services not included in (4). A "block level" I/O service is provided by I/O software which furnishes blocks to a user in response to READ commands.
6. Handling record advance (block and deblock) not included in (5). Record advance is provided by GET/PUT commands.
7. Creating checkpoint and restart modules.
8. Providing status record counts or other data about the progress of the sort to the operator at a console.

These ongoing services are provided by system macro instructions available to the sort coder (as to any other program). All the interfaces, ongoing and preactivated, may be presented to the sort designer as part of an already existing operating system with which the sort must run; or they may be dynamically negotiated as OS design and sort design progress together.

Obviously, the design of the sort-control language will be affected by the facilities of the job-control language of the system. The designer of the sort should have a very significant say about the job-control language. He is interested in the degree of user control over allocation. He is interested in the performance and capability of the system loader, because it is a potential operational component of his sort. Since the sort is one of the very critical application packages provided (or offered) with a system, and since sort speed is an important factor in determining the desirability of a hardware system for a user, the sort-design group should be able to influence many elements of a system that superficially seem unrelated to sorting. Using the approach that "what is good for the sort is good for the system," the sort designer should insist on approval over what he relies on.

19.8.2   The Sort and Input/Output Support

An operating system generally offers I/O services at three levels. The first is the GET/PUT level, where the problem program interfaces with the operating system to receive an item or release an item. It corresponds to the sixth of the ongoing supports listed in the previous section. There is a read/write level of support, on which the problem program interfaces with the operating system to call for the read or write of a block to or from a particular buffer. This level of support differs from GET/PUT in that the problem program is responsible for any item advance and for block formation on output. The lowest level of I/O support is that of the device handler. The problem program is responsible for all aspects of the I/O operation except direct communication with the device. It is shielded from some device details and eccentricities and from the interrupt system, but it is otherwise in control of I/O. In the IBM OS/360–370 system, this level is roughly the EXCP level. (Very small machines may allow the programmer to have direct communication with I/O devices, but these machines usually do not have generalized sorting systems.)

One of the very critical design decisions that must be made for a sort package relates to the level at which to interface with the I/O support of the system. The usual decision is to interface at the lowest possible level and to maintain, to the extent possible, private sort control of all I/O operations—except, perhaps, those for the reading of control cards and the output of system messages. The disadvantage is the additional work required to implement and test I/O coding, complicated by dependence on device details and the need for modifying the sort to supply code for devices announced later. Since there is no advantage in principle to rejecting services graciously provided for user programs, they should be rejected only out of necessity. In the early stages of many sort designs, there are whimsical statements of the intention to use system I/O with an escape clause of one type or another, such as "unless there is a decay of more than x percent in performance."

The fundamental difficulties with the use of high-level I/O support are that the code of such support is characteristically slower than private I/O, and that useful capabilities may be excluded. The slowness and exclusion come from the nature of generalized support. The functions supported by an I/O package rarely represent the total capability of a contemporary channel. If an attempt is made to provide full hardware capability, it usually involves great expense.

The designer of the sort may penalize his sort's performance by using the I/O package entirely, in that he may be unable to perform certain functions he should perform in the interests of efficiency. For example, independent look-ahead seeks may not be supported by the IOCS but may be critical to the performance of the sort.

There are a number of fairly obvious considerations governing whether high-level support can be used. If I/O support is being developed for a system while sort development is in progress, some of these points can be negotiated, but no amount of negotiation can resolve them all, and the sort will pay some penalty. The considerations may be summarized in the following questions and comments.

1. Is there any device function required by the sort that is not supported by IOCS at all?

2. Is there a loss of core due to additional buffering levels, which might reduce sort performance by leaving less space for higher m or larger blocks?

3. Is there some coding penalty in the I/O support routines? For example, I/O support may execute a large number of instructions to determine that any command it has received is properly formed. The sort could perform the I/O function without such checks since, once it is tested and debugged, its read/write patterns are known. Similarly, it may be necessary to recognize certain subfunctions in the generalized support package which the sort I/O routine need not take time to identify, because it knows they cannot occur.

19.8.3   The Sort and Multiprogramming

It is conceivable that a user, with extraordinary care and planning and with keen insight into the hardware, the sort, and operating system, can arrange to run a sort in a multiprogramming environment so that it will take the same amount of time that it would take running alone. The requirements for accomplishing this minor miracle are that neither the core, the devices, nor the device space used by the other programs in the mix could be profitably used by the sort, and that no delays are introduced into the operation of the sort because of the presence of other programs. Such delays are due to arm movement at the request of other programs, which causes additional seek time; to unavailability of channels, control units, or devices being used by other programs; to the interruption of the sort by the operating system in order to service I/O requests previously submitted by other programs. These delays may occur even if the sort has the highest possible priority relative to the other programs. Their severity is a function of both the level of contention for resources and the burden of the operating system mechanisms that support the multiprogramming capability. To the extent that the sort must share channels and arms with other programs, it is vulnerable to further delays. If the sort runs at a lower priority than that of other programs in the mix or at the same priority, the delays will become more severe. During the operation of any program, there may be periods when CPU calculations wait for the completion of I/O and cannot proceed, and periods when I/O is idle until the CPU has generated a need for a read or a write. A high-priority program will lose control of the CPU or I/O subsystem only in those intervals when it would not normally be using them. A lower-priority program will experience delays in access to the CPU because another program of higher priority will not release it. The potential delays in multiprogramming therefore fall into two categories:

1. Delays due to the deprivation of resources and sharing of resources, which increase processing time and waiting time for completion of I/O; and

2. Delays due to suspension while the CPU is serving higher-priority programs.

The sensitivity of performance to both types of delays is obvious when the goal of a program is to stay write-bound. To some extent the user can control contention delays by selecting coresident programs on the basis of minimum interference, and controlling core and device assignment and priority carefully so as to minimize the effect on the sort.

There may be sorts whose completion is not critical. These may be run as "background" jobs with mixes of characteristically compute-bound jobs. The effect will be to increase utilization of equipment as the sort soaks up channel cycles. The multiprogramming environment is ideal for a sort whose completion time is not critical.   .

Between the extremes of an urgent need to avoid any delay at all and total unconcern for time of completion, there will be a variety of tolerances for delay and larger numbers of situations in which the multiprogramming delay is acceptable. It is strictly a user-administration problem, and its resolution lies in the priorities of each center. The responsibility of the developers of sort and operating systems is to provide the user with a meaningful set of choices and the mechanisms for implementing them.

19.8.4   The Sort and Programming Languages

The programming languages available with an operating system potentially interface with the sort in a number of ways.

1. The language may be that in which the sort itself is written.
2. The language may be that in which the sort library is maintained in the system.
3. The language may provide mechanisms for calling the sort and/or providing parameters to it.
4. The language may provide "own code" functions within the structure of the sort package.
5. As part of its specification, the language may provide a sort function, which may be supported by the sort package.

The languages of a system may include an assembler and a number of higher level languages, such as PL/I, COBOL, FORTRAN, ALGOL, APL, LISP, etc. Of these languages, COBOL and PL/I have defined language-level relations with the system sort, and are therefore particularly interesting for this text.

19.8.4.1   The language of the sort

It is necessary to determine what language will be used to write the sort. A reasonable choice would be the assembly language, because it gives the programmer the most control over the quality of code. The need to generate some elements of the package also favors the use of assembly-level code. In higher-level languages, partial word or bit manipulation is frequently impossible or at best extremely clumsy, both in expression and in generated code. In the author's opinion, quality of code and range of expression make the assembler an attractive vehicle for creation of the sort.

In recent years there has been a great deal of discussion and even some development of systems programming in higher-level languages which have the necessary expressive power and which generate adequate code. If such a language exists in the environment of the sort development activity, it should certainly be investigated, but the burden of proof lies with the language. Because the sort is a very performance-oriented, generalized application, the approval must be conservative, even though many other applications have been successfully transferred to languages that are less machine-oriented.

19.8.4.2   Language of the sort library

Given a family of language processors in the system, the sort library may be maintained in any language of the family. Library collection and even generation functions can be performed, for example, by COBOL library processors or PL/I compile-time generation facilities.

In practice, it is usual to find the sort library represented at assembly level or at some sublanguage level of the system. A sublanguage level is some representation of procedure which cannot be expressed—or is not normally expressed—by a programmer. In contemporary practice, a programmer does not write procedures in actual machine-loadable, relocatable modules, and that representation would be a sublanguage form.

There may in fact be a number of sort libraries associated with a sort. When there is a system-generation procedure, with which a system and sort are defined by a user as a subset of the operating system's total capability, there may be a total sort library in assembly-language form. System generation—"generating the generator"—collects all modules to be included in the operational library and assembles them into a sublanguage level, which is the representation of the user's sort generator. System generators are associated with very large and generalized systems, such as OS/360. Generation of the generator should be an infrequently performed function.

The representation of the operational sort library necessarily has implications for the running sort-generation process. If the library is in assembly language, user coding may be added in that language; and the total sort, including sort-library elements and "own code," can be produced by a single assembly. If the library is kept at a sublanguage level, user-provided code must be assembled or compiled separately, and the services of a linkage editor or loader must be used to produce the total sort.

19.8.4.3   "Own code" language

Any language is a potential source of "own code" modules. This capability is independent of any interface between the sort package and the compiler. The interface between COBOL and the language of the sort package or between PL/I and the sort language is a feature separate from the possibility of writing code in the language, compiling the code, and then providing the compiled code as "own code." To effect this capability, the host operating system must allow programs written in different languages to be organized into a larger running program. The organization is accomplished by means of universal linkage conventions, naming conventions, libraries, and linking loader functions. The capability of providing "own code" in any language to fit into the exits of the sort is not identical with the next topic of discussion—the capability of invoking the sort package from within a COBOL or PL/I program.

19.8.4.4   COBOL and the sort

The COBOL language contains a means for invoking a sort [17]. Although a complete discussion of this language feature is best obtained from COBOL language specifications, this section will serve as an introduction to the interface and some of the considerations surrounding it. It is a reasonable requirement that a sort package support the COBOL verb SORT so that a COBOL sort need not be separately developed. The COBOL/sort interface defined by the language makes the sort function a called internal routine of a COBOL program. No restriction is placed on what a COBOL program may do before passing records to the sort or after receiving records from the sort. The language provides for a one-by-one passing of records to a sort file by an input procedure which releases a record to sort. The input procedure is a COBOL procedure named in the sort verb. In the language specification, there are no procedural restrictions on what the input procedure may do before passing a record. However, particular implementations of the sort interface can limit what it may do. A sort package organized as a subroutine need not place any restriction on an input procedure, but one organized around fixed functional exits may find it necessary to restrict an input procedure. The COBOL input procedure is often used as a given exit of the sort phase. Functions which cannot be performed by that exit may not be performed by the input procedure.

The input procedure interfaces with the sort phase of the sort. The verb RELEASE provides a pointer to a record; the sort then accepts the record and returns to the COBOL input procedure. When control passes from the input procedure, the sort assumes control and operates until the last pass. On the last pass, records are returned to an output procedure written in COBOL, which corresponds to a last-pass exit.

The specific nature of the sort algorithm will naturally affect the details of the COBOL/sort interface. Replacement Selection is an ideal method for accepting one record at a time and selecting a winner. Each time a record is released, the tournament can be run, a winner selected, and the new record used as replacement. Actually any selection or insertion sort will operate cleanly at the interface. Exchange-based sorts (e.g., Quicksort), which require a collection of records to be ranked, will operate more clumsily, since it is necessary to collect a work area of released records before actually operating the sort algorithm.

Some parameters are provided in the data division, which describes the records to be sorted, and in the control stream associated with the COBOL program.

The actual COBOL verb SORT provides the names of the input and output procedures and a definition of control fields. It specifies whether ascending or descending sequence should be associated with a control field.

The COBOL compiler is aware of the files identified with the sort and will generate parameter tables as required by the sort interface and the sort initialization module. The verb SORT may generate a link to a COBOL interface module in the sort or the system library, which then undertakes to place the input and output procedures in appropriate exit positions. An alternative approach is to support the interface with the operating system. The RELEASE verb would release to a sort interface module which would only build a sort file on some device. No sorting is undertaken until the end of the record-by-record transfers of the input procedure. Then the operating system is asked to ATTACH the sort. The sort file is sorted as though it had been called in the normal operating-system-supported way as a subtask. At face value, this is an unattractive approach, since an extra pass of the data is required. However, it may be that considerably more core can be given to the sort, that merge passes will be saved, and consequently that the sort/merge will run faster without the necessity of executing the input procedure.

19.8.4.5   PL/I and the sort

The PL/I language can invoke a sort function, but at the time of writing the capability is not general and identical in all versions of PL/I. It is very system-dependent, using specific sort-module names and exit names associated with a particular sort package. The compiler generates a parameter table from parameters written as a character string and associated with a standard call to a named module of the sort package. In the OS/360–PL/I interface the call may or may not involve PL/I coding of a named OS/360 sort exit, depending on whether the records to be sorted already exist in a data set or whether they are to be generated or modified by the PL/I program calling the sort.

19.9   SORT TIMING

The ability to estimate how long a particular sort will run is an absolute requirement both for a producer of sorts and for a user. The producer requires some insight into how attractive his new sort system will be; the user needs the information in order to plan for machine use, and he often wants to know how competitive sorts compare in performance.

Associated with each delivered sort package is a manual giving tables of timings for various situations of the sort, or a set of formulas for calculating timings—or both [5, 15, 22]. The tables of timings published by the sort producer may themselves be generated by a program that uses a set of formulas, rather than by observed runs with a live sort. All sort timings are approximations, naturally. The quality of the estimate is a function of the sophistication of the timing formulas and the amount of detail included. Even at best a static timing formula is subject to serious error. Variations in performance due to variations in data order will occur and may be significant. In addition, marginal variations in hardware performance within hardware tolerances may accumulate, to distort the performance of a sort on a particular channel and particular tapes at a particular time. Marginal pathologies, such as an unusually high, recoverable-tape error write rate (which will not cause checkpoint but which will add to the time required to perform write) will also distort sort performance.

The general expression of the time required to perform a sort is given as:

Sort time = Initialization time + Dispersion time + Merge time
+ Last-pass time.

In order to quantify these relationships, it is necessary to describe the elements that constitute each phase and to associate times with particular functions. In general, there are three components of time for a computer program: input time, output time, and process time. Depending on the hardware overlap characteristics of the systems running the sort, total times for a phase will be:

1. SUM (process time, input time, output time) when there is no overlap.
2. MAX (process time, input time, output time) when there is complete read/write/compute overlap.
3. MAX (input time + output time, process time) when there is a read/write or computer overlap capability.
4. MAX (input time + process time, output time + process time) when there is a read/compute or write/compute capability.

The development of a good timing formula takes considerable effort. It involves the calculation of a series of constants representing times for certain functions, and the application of various insights to the relationships between component functions of the sort. Work remains to be done to find the best approach to formula development and the expression of relationships. The material in this section and those that follow is meant to be illustrative and to introduce necessary considerations in developing sort-timing formulas. The reader should not by any means conclude that the relationships developed here are necessarily the best possible.

19.9.1 Dispersion-Phase Timing

The time for the dispersion phase depends on the following factors.

1. Record length
2. Block size on input
3. Block size on output
4. Characteristics of the input device
5. Characteristics of the output device(s)
6. Memory size
7. Size of control field
8. Buffering technique
9. Number of output devices
10. Space for "own code"
11. Time to execute "own code"
12. Sort method
13. Memory and CPU speeds
14. Number of records
15. Initial order of records in the file

A sort timing procedure may be more or less parametrizable. A computer program representing a sort-timing procedure may request parameters describing all the factors listed above, or some may be assumed. Constants or expressions describing hardware characteristics may be provided to a timing program, or they may be represented internally, with only the device names provided by parameter. A developer should base his decisions about how much information must be provided to a sort-timing program on the *intended use* of the program. If it is to be used to time known sorts in an equipment environment that is stable, parametrization can be held to a minimum. If there is a need to estimate sort time with new equipment or to introduce new methods for comparison, parametrization will naturally be more extensive.

Reasonable input to a sort-timing program for a developed and distributed sort package includes the following elements.

1. Record length
2. Block size on input
3. Device names
4. Memory size
5. Size of control field
6. Number of output devices
7. Number of records

This list is a reasonable approximation of what might be provided as parameters for the sort. The sort-timing program itself contains details of equipment performance and of the method used, and it calculates, as the sort might calculate, output block size, sort area size, etc.

19.9.2   Merge-Phase Timing

The process of timing the merge is similar to that of timing the dispersion pass. The elapsed time for each merge pass is calculated, that time is multiplied by the number of merge passes, and the product is merge time.

The input and output estimates are developed exactly as for the dispersion phase. The processor times are developed similarly, except that merge modules are used, rather than sort modules. Rewind and reversal times must be included, and costs for "own code" and checkpoint calculated.

Today's user of a sort is more apt to see tables than formulas, but should have some idea of how sorts are timed. The ability to estimate times is, of course, essential for the developer of a sort.

19.10   SORT SIMULATION

A program system designed to simulate sorts is becoming a standard tool in development laboratories for testing new hardware as well as new sort ideas. The usefulness of the tool is partially determined by the ease or difficulty of describing new equipment or new sort methods. The advantage of simulation over static timing is that there need be no assumptions, guesses, or averages. Each event can be clocked, its start and stop times can be known for each access, and each I/O event can be discretely timed. The time estimates developed by a simulation can be very close to actual running times. Another important feature of simulation is the possibility of introducing pseudofiles of different ordering, to observe the effect that various orders have on sort algorithms.

A reasonable approach to simulation is to run the internal sort under clocking and to simulate I/O events. The simulation of an I/O event involves the timing of the process activity needed to initiate it and a calculation of how long the I/O would require, given the address to which it is going, block size, device characteristics, etc. The effects of buffering and the dynamic interaction of I/O and CPU can be clearly seen. Delays can be identified and buffering strategies evaluated.

The parameters given to a simulator are essentially those given to a static timer. Since the simulator is generally intended for use in a development environment, rather than in a user environment, the knowledgeability of its users can be assumed to be high. For this reason and because flexibility is an underlying requirement, parametrization should be quite extensive. The merge technique, merge order, blocking, buffering, internal sort technique, device type(s), CPU models, and operating system are all common parameters.

It is important that the hardware characteristics of the system be represented in such a way that new devices can be described and introduced with reasonable effort. Unless there is an orderly and reasonable mechanism for expanding the simulator's capability, the simulator will be unable to provide the required services when it is necessary to show the impact of a new disk file, a new merge algorithm, or a new channel.

20
Special
Systems
Considerations

20.1 RELOCATION AND VIRTUAL MEMORY

The concept of relocation rests on the distinction between the names a program uses to refer to objects and the machine addresses those objects may occupy. Assemblers and compilers create "relocatable" code, which cannot be executed until it has been processed by a loader that develops actual machine addresses. Once loaded, the relocatable program operates like any other.

Alternatively, a hardware mapping device may be used to develop actual addresses dynamically. Because the current technology of dynamic relocation has direct and dramatic implications for sorting, any text dealing with sorting should include some discussion of relocation. First, we must present some rudimentary ideas about the concept of relocation.

20.1.1 Paging

In Fig. 20.1 a 13,000-word program is shown distributed in memory so that the first thousand words of the program are in memory locations 1000–1999, the next thousand words in locations 8000–8999. The remainder of the program is distributed throughout memory in 1000-word chunks. The choice of memory locations for each 1000-word portion of the program is arbitrary, based only on availability of space. The 1000-word chunks are commonly called pages, and the box interposed between program and memory is called a paging device or a relocation device.

The memory is thought of as a collection of fixed-size pages, and allocation of memory is by pages. Contemporary systems rely largely on software

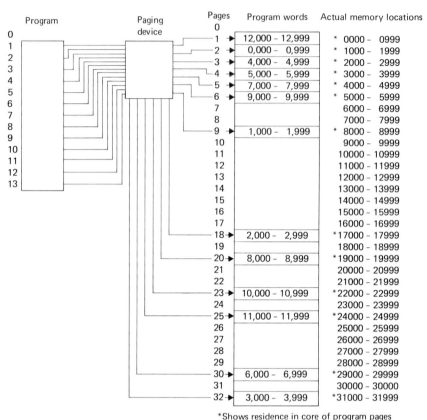

Fig. 20.1 Allocation of memory by pages.

to find empty page space distributed around memory, and to load program pages into memory pages. Hardware support is provided for the address translation when space has been found and assignments have been made.

In a paging system, the address portion of an instruction word has a page reference part and a page displacement part. With a page size of 1024, the maximum page displacement, or relative address, on a page is 1023. Since it takes 10 bits to represent the value 1023, 10 bits are required for the representation of the page displacement part of an address. The additional bits of the address constitute the page reference. A system with 16-bit addresses can have 64 pages of 1024 words each.

The page reference portion of the address is fed to the paging mechanism. The output of the paging mechanism is the core location of word 0 of the page. A hardware associative memory is a fast way of transforming page numbers to real core addresses [6]. With the decimal address 11643 used as input to the associative memory shown in Fig. 20.2, the paging location function can be observed. The page reference, 11, is a search tag, which is used to search the tag portions of the associative memory until a match is found. The contents of the matching entry point to memory page 24. The first word of page 11 of the program is at core location 24000. The output is formed and used as an addend to the page relative address, 643. The true effective address is 24643. Index register modification of the address, if any, was performed earlier. The principal advantage of the paging mechanism lies in the speed at which the mapping is accomplished; it also provides a signal when the page being searched for is not found. The circumstances under which a "fault" occurs, what needs to be done about it, and the impact on performance occupy the major attention of system designers.

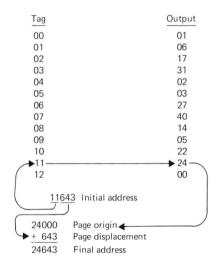

Fig. 20.2 Associative-memory paging.

Associative memories are expensive; important economies may be realized by page mechanism designs which shorten the size of an associative memory or the width of entries in the memory. Such approaches are embodied in a number of machines. Two very simple but elegant designs are represented by the Ferranti ATLAS and the IBM 360/67. The ATLAS associative memory holds only the page numbers. The real core locations are inferred from the entry where the match is made. The 360/67 holds only eight entries. The page and real addresses of the current eight most recently referenced pages are recorded in the associative memory. When a page not in the associative memory is referenced, a table in regular core is used for translation. This "paging" table in real memory holds the real memory address of a page. The page selection portion of the program address is used as an index to the table. For example, 01 would use the contents of the second location of the page table as the real core base address, 11 would use the twelfth, etc. When a page not in the associative memory is referenced, an entry is formed for it in the associative memory, and subsequent references use that entry. On every page reference, the associative memory is inspected. If the entry is not there, the page table algorithm is used.

20.1.2   Demand Paging

The paging mechanisms of a system increase the effective utilization of primary storage by preventing fragmentation of core into areas of unusable size. At any particular time, only a subset of the pages of a program may actually be in primary storage. The remaining pages reside on a backing storage device, a disk or a drum. When a reference is made to a page not in primary storage, a page fault signal is given and action is taken to bring in the page. This procedure, called *demand paging,* is intended to support programs too large for available primary storage. It is particularly attractive in multiprogramming or time-sharing systems. Demand paging is a dynamic memory-allocation technique, making it possible to bring pages into primary storage only as they are referenced. The programmer (or terminal user) has no idea how much of his program is actually in storage at any particular time or where it is located.

A number of performance questions naturally surrounds the idea of demand paging. A 64K program may very well run more slowly on a 32K machine than it would run on a 64K machine. There are potential delays when nonresident pages are brought in only as they are referenced. The degree of delay depends on the speed of the paging operation and on the number of such operations involved.

The number of page requests a program makes depends on several individual factors, all of which contribute to the quality of the dynamic mapping of pages required and pages available. Prominent among the factors affecting the paging load of a demand paging system are the following.

1. *The locality of a program.* Demand paging as a workable technique depends on the concept of locality. Over short intervals of time a program's references tend to be clustered. Instructions tend to be from consecutive locations; references to data may be to the same area of an array or to a particular record whose fields are contiguous. Different programs will display different patterns of locality. Some programs will shift "neighborhoods"

slowly; others will switch rapidly from one locality to another; still other programs have so little locality that their reference patterns are unpredictable. A program with little locality may make references to any particular address in the entire program with almost equal probability. In such a program, the collection of a preferred set of resident pages—pages "in the neighborhood"—is impossible. The greater the tendency of a program to reference within one neighborhood, the fewer pages need to be in core at any particular time. Each program has its own "parachor," the number of pages required in core to avoid unacceptable delays due to overpaging.

Consider the addressing pattern of a Lincar Selection sort versus Binary Insertion. Linear Selection moves item by item down a list, changing locality in an orderly fashion. The tree structure of Binary Insertion causes a broad address reference pattern, with jumps from midpoint to quartile point to endpoint, as inspection is made.

2. *Page size.* The size of the system page will affect the number of paging requests by increasing or decreasing the probability that a reference will not be on the same page as the last preceding reference. With small pages the proportion of a program represented by each page will be small, and the program may tend to move from page to page very quickly. If the amount of real storage is limited, page-to-page movement will increase the need for demand paging, because newly referenced pages will not be in core. On the other hand, overlarge pages may be inefficient because of dead-page space. If only half of a 4K page is used, the remaining 2K space is effectively unutilized when the page is in core. The aim of optimal utilization of memory is undermined by the fact that valuable space is "hidden" on a page.

3. *Page replacement algorithms.* When a referenced page is not found, it must be brought into primary storage. If there is space for it in primary storage, there is no problem. If there is not, a decision must be made as to the resident page that will be displaced. The various strategies for selection are called page replacement algorithms. The goal of such an algorithm is to minimize paging by replacing those pages which are out of the current neighborhoods of all active programs and which consequently have the least probability of being referenced in the immediate future. If too many pages that will soon be recalled are overlaid, the paging burden will become severe. When the same pages are repeatedly brought in, a condition known as "thrashing" develops.

The simplest and most direct strategy is first-in, first-out (FIFO). When a page is to be brought in, the page that has been in core the longest is replaced. The assumption is that the older pages have higher probability of having fallen out of the neighborhood. A more dynamic approach involves an attempt to observe which pages are still in the neighborhood. A record is kept of page use, and those pages least recently used (LRU) are made the top candidates for replacement. LRU algorithms vary in detail. There are differences of opinion about how complex a replacement strategy is needed for effective results.

A concept combining the fundamental idea of LRU with the idea of parachor is called a *working set*. At the end of a service interval, the pages referenced in the interval, or in some defined last subinterval, are noted. These most recently used pages are counted, and the size of the neighborhood is the space they require. Control will not be passed to a program until this amount of space can be guaranteed for it. The scheme is excellent for a multiprogram environment, in that it allows dynamic growth and contraction of a program's space requirements and prevents thrashing by attending to both the reasonable space needs and the reasonable page selection needs of a good replacement scheme.

There are myriads of refinements that can be applied to replacement algorithms. A preference for replacing pages that have not been changed is often part of the strategy. When an unchanged page is overlaid, no write-out is necessary; the paging operation is thus reduced to a read instead of a read/write. Relative program priority can be used in a multiprogram environment to allow high-priority programs to "cannibalize" other programs instead of overlaying its own pages.

20.1.3   Virtual Memory

A concept associated with paging is that of *virtual memory*. Since a program is mapped into real memory only as required, the programmer, in developing his program, can consider that he has a considerably extended amount of space. The IBM/370-168, using OS/VS 2.2, gives each user 16 million locations of virtual memory. The programmer can implement a program as if it is to have millions of core locations available when it runs. The effect of the virtual-memory concept is to relieve the programmer (and compiler) of the necessity for planning complex overlay structures. The paging mechanism takes over the task of overlaying pages of code with other pages as they are needed. The presence of vast amounts of virtual memory has implications for input/output as well. Data files that would ordinarily be handled by I/O commands and conventions may be represented as resident in virtual storage, and the paging mechanism becomes in effect the I/O mechanism. No explicit I/O commands are used by the programmer, who merely lays out his files in locations in virtual memory.

Thus the concept of virtual memory suggests that it would be possible to order a larger population of files using strictly internal sort techniques. As the concepts of the transparency of real resource and logical resource advance, it might be possible to conduct all sorting in this way. However, since the behavior of virtual-memory overlays varies, the sort technologist must discover what internal sorts are appropriate for the ordering of files in virtual memory [2].

20.1.4 Programming Style

While simplifying the details of programming by providing hardware management of overlays and (potentially) I/O, virtual memory introduces a new set of programmer disciplines and stylistics. Structure and programming style will have an effect on the performance of programs in a paging environment. Good programming techniques of the past may become bad techniques in a paging environment. For example, branching to subroutines may increase paging if the subroutines are outside the pages that reference them. Therefore replicated in-line coding for subroutine functions may be preferred in a paging environment. The programmer may wish to consider paging behavior in his organization of data structures, in the way he plans modules and linkages. How much a programmer needs to know to plan a good program for a paging environment is still a matter of dispute. Should he know page size, so that he may align program elements in the same page? Should he know the replacement algorithm? How careful must he be with his addressing patterns? In general, a program planned for a virtual-memory environment will perform better than one in which this aspect of the machine is disregarded. The programmer who wants to sort efficiently must certainly pay attention to the behavior of sort algorithms in a paging environment, as well as to the structure of his program [4].

20.1.5 Sorting and Paging

The sorting problem presented by paging is twofold. On the one hand, it is necessary to determine how the known sorting algorithms will behave when they have less core than is required for their code and data. On the other hand, it is necessary to explore the possibilities for replacing external ordering techniques by virtual-memory sorts.

Until now the number of comparisons and the number of data movements of an internal method were the criteria for its evaluation. To these must now be added a measure of the *amount of paging* the sort method will require at various levels of reduced core space. The amount will depend on details of encoding, the number and size of elements to be sorted, the size of keys, the size of pages, and the replacement algorithm. Each sort algorithm will have its own characteristics of locality and its own parachor, determined in part by the pattern of its addressing and by conditions of the data. Some insights have been developed for certain sorts in particular systems, but to attempt to make a comprehensive statement about sorting on paging machines would be premature.

The effect of various replacement algorithms on particular sorts is not generally known, nor has the profitability of virtual-memory sorting been extensively mapped against external merges of different designs and orders. New sort techniques aimed particularly at virtual memories have not yet emerged. It is possible to take the position that a sort should never be run except when it has the core it requires, when all of its pages will be retained in core. This position may be necessary for optimum sorting. It may be that no sorting method with good paging properties that degrades slowly in smaller real core environments can match the performance of a sort with excellent properties of comparing that causes bad paging in small core. However, there is some reason to believe that this is not true, and that the paging technique can make a positive contribution to sorting. The sections to follow will undertake a discussion of some of the considerations that have been revealed as significant.

20.1.6   Locality, Replacement, Real Space, and Time

The sort algorithm appropriate for a paging environment is one that exhibits the greatest locality and tends to reference linearly the list to be sorted. Since many of the most efficient sorts implement tree structures, with a great dispersion of addresses across the data list, there seems to be an inherent conflict between minimizing comparisons and minimizing paging. Consider the simple Exchange sort of 16 elements shown in part in Fig. 20.3. Each comparison is of an address with an immediate neighbor. Throughout the entire first pass, off-page references occur only when a page boundary is crossed, that is, when, in the linear progression of comparisons, a page is about to drop from the neighborhood. Figure 20.3 shows the progression of comparisons, the referenced pages, and the resident pages for the first-pass ranking of key 1, initially in location 13. Space for two data pages is provided, and the page-replacement scheme is FIFO. At any time, two pages of four elements each are resident, exactly 50 percent of the list to be ordered. Since space is available for initial paging of pages D and C, there is no corresponding write-out. The entire pass requires four page-ins and two page-outs. If page space for the entire list had been available, the four page-ins would still have occurred to establish the list in core. The burden of running in reduced core is the page-writing. The beneficial effect of the intensive locality of the method is evident.

Figure 20.4 summarizes the comparisons and pagings in the remaining passes of an Exchange sort. Close observation of the behavior of the sort reveals that the method displays both good and bad characteristics for paging. The bad characteristic is that the locality of the program shifts dramatically at the end of each pass. Real core at the end of pass 1 contains pages A and B, but the first page to be referenced in pass 2 is page D, and that is followed by a reference to page C. For each pass in which there is a

Location	Start	Page	End of pass 1	Page
1.	13	A	1	A
2.	6		13	
3.	4		6	
4.	9		4	
5.	11	B	9	B
6.	8		11	
7.	7		8	
8.	16		7	
9.	14		16	
10.	3	C	14	C
11.	5		3	
12.	15		5	
13.	1	D	15	D
14.	10		2	
15.	12		10	
16.	2		12	

Comparisons (Locations)	Exchanges	Page references	Pages in	Pages replaced	Pagings (R=Read, W=Write)
16 : 15	16 ↔ 15	D	D		1R
15 : 14	15 ↔ 14	D			
14 : 13		D			
13 : 12	13 ↔ 12	D, C	C, D		1R
12 : 11	12 ↔ 11	C			
11 : 10	11 ↔ 10	C			
10 : 9	10 ↔ 9	C			
9 : 8	9 ↔ 8	B, C	B, C	D	1W, 1R
8 : 7	8 ↔ 7	B	B, C		
7 : 6	7 ↔ 6	B	B, C		
6 : 5	6 ↔ 5	B	B, C		
5 : 4	5 ↔ 4	B, A	A, B	C	1W, 1R
4 : 3	4 ↔ 3	A	A, B		
3 : 2	3 ↔ 2	A	A, B		
2 : 1	2 ↔ 1	A	A, B		

Fig. 20.3 Locality in first-pass exchange sort.

reference to all pages, pages A and B must be written and read again, and pages C and D must be read and written. The good characteristic is that the method becomes increasingly local as it proceeds. At the end of the fifth pass there is no longer any reason to reference A, at the end of the ninth pass there is no reason to reference A or B, and with the establishment of C and D in core during the tenth pass, all paging stops. This characteristic of growing locality is, in general, one to be sought when sorting in a virtual-memory environment.

The switch of locality at the end of a pass suggests that a modification to the method might be in order. If even-numbered passes were to "sink" and odd-numbered passes to "lift" the elements they moved, each pass would find in core the pages it initially referenced. The elimination of the discontinuity in locality from pass to pass would make all resident pages useful. As with the single-direction sort, all paging would stop after 10

Pass	Comparisons	Pagings
2	14	4R, 4W
3	13	4R, 4W
4	12	4R, 4W
5	11	4R, 4W
6	10	3R, 3W
7	9	3R, 3W
8	8	3R, 3W
9	7	3R, 3W
10	6	2R, 2W
11	5	—
12	4	—
13	3	—
14	2	—
15	1	—
16	0	—
	106	30R, 30W
1	15	4R, 2W
	Total 121	Total 34R, 32W

Fig. 20.4 Exchange sort, remaining passes.

elements were ranked. The total paging could be reduced to 22 page-ins and 20 page-outs.

The amount of paging activity may also be affected by order [10]. If there is good order in the data and the replacement algorithm recognizes changed and unchanged pages, there will be a considerable reduction in the number of page-outs. During a pass there will be a tendency to find pages ordered and to leave them unchanged. When they are replaced in core they need not be written out, because a correct copy already exists in the backing store.

There is a further relation between the locality or tendency toward linearity in a program and the page replacement algorithm. The greater the linearity, the more relatively efficient the FIFO algorithm becomes. Because the list is being scanned in order of location, the earliest read (first-in) pages are in fact the least recently used pages, and therefore an LRU algorithm does not improve performance since it does not change the page selection. This observation applies only to data pages.

The effect of FIFO or LRU on the alternating-direction version of the sort is not trivial. Consider *A* and *B* in core at the start of the second pass, with *B* the first-in. As the pass attempts to sink a high key, it crosses the barrier between *B* and *C*. The next desired arrangement of memory is *B* and *C*. The paging operation under FIFO, however, will give *A* and *C*, because *A* was brought in after *B*. If an exchange is to take place—that is, if the last element on *B* is indeed higher than elements on *C*—then *B* must be brought back in. The apparent reduction in paging by reversing direction is achieved only with LRU. In going from *A* and *B* to *B* and *C* with LRU, the most recently referenced page is *B*, and so *C* will replace *A* immediately.

If three resident pages rather than two are allowed, overall paging is reduced, because the list to be ordered shrinks to a size which can be kept in core after the ranking of the fifth element. In this example, however, the provision of extra space is less effective than the modification of alternating directions. An important observation is that the impact of a reduction in space is felt immediately. The paging delays go from none to 23 with the loss of one page of real space and from 23 to 34 with the loss of two. Running just a little underneath total required core can be as bad as running in more severely reduced core. Conversely, the benefit obtained from adding extra pages is proportionately small. Each method will display its own sensitivities, but the fact that a sort may run in 95 percent of required core does not ensure good performance.

If the sort runs in a multiprogrammed environment, it may lose control of the CPU. During the time the sort does not have the CPU it may lose its pages in core. A pertinent consideration is the length of time the sort will have control of the CPU and the amount of change in its locality during that time. Another way of expressing this is the number of instructions the sort can execute before it loses control of the CPU. Does the sort lose control before it can finish a pass? At the end of a pass? At some point in an nth pass? The nature of the operating system's memory management and dispatching strategy becomes relevant here, to the extent that it determines whether the sort may expect to find some or all of the pages it left in core still there when it regains control. The considerations relevant to delays due to multiprogramming and those due to paging become intermingled. If, in the exchange, a sort loses both its control and its pages at the end of a pass, there is no benefit to changing direction since paging must occur when it gains control of the new pass. If the sort loses control during passes, there may be no benefit in providing additional space. For example, the presence of A, B, and C in memory rather than just A and B has no value to a sort that will lose control before it finishes referencing A. The concept of locality is associated with time, since a neighborhood changes over time. In a dynamic environment some amount of paging will be due to systems logic and some to the algorithm itself. The best a programmer can hope for is to reduce paging to that minimum imposed by the system because of its multiprogramming memory-management algorithms.

20.1.7 Tournament Sort in Paging Environment

Figure 20.5 shows the first pass of Treesort, the minimal-space Tournament sort described in Chapter 5. This sort is an interesting vehicle for an expansion of ideas about locality and the effect of order. If we consider just this first pass, the list of comparisons shows that the method is by no means

		Start	End
_A_	1.	13	13
	2.	6	16
	3.	4	15
	4.	9	14
_B_	5.	11	11
	6.	8	8
	7.	7	12
	8.	16	9
_C_	9.	14	6
	10.	3	3
	11.	5	5
	12.	15	4
_D_	13.	1	1
	14.	10	10
	15.	12	7
	16.	2	2

Comparison	Exchange	Page references	Pages in storage First	Last	Page out	FIFO pagings
16 : 8		_D, B_	_D_	_B_		2R
15 : 14		_D_	_D_	_B_		
15 : 7	15 ↔ 7	_D, B_	_D_	_B_		
13 : 12		_D, C_	_D_	_C_	_B_	1W,1R
12 : 6	12 ↔ 6	_C, B_	_C_	_B_	_D_	1W,1R
11 : 10		_C_	_C_	_B_		
11 : 5		_C_	_C_	_B_		
9 : 8		_C, B_	_C_	_B_		
8 : 4	8 ↔ 4	_B, A_	_B_	_A_	_C_	1W,1R
*16 : 8		_D, B_	_B_	_D_	_A_	1W,1R
7 : 6		_B_	_B_	_D_		
6 : 3	6 ↔ 3	_B, A_	_B_	_A_	_D_	1W,1R
*13 : 12		_C, D_	_C_	_D_	_B, A_	2W,2R
*12 : 6	12 → 6	_B, C_	_D_	_B_	_C_	1W,1R
5 : 4		_B, A_	_B_	_A_	_D_	1W,1R
4 : 2	4 → 2	_A_	_B_	_A_		
8 : 9		_B, C_	_B_	_C_	_A_	1W,1R
9 : 4	9 → 4	_C, A_	_C_	_A_	_B_	1W,1R
						11W,13R

*thrashing

Fig. 20.5 Treesort in paging (tree shown at pass end).

random in its addressing; it has locality, but a discontiguous locality sensitive to order. The method essentially progresses up a list comparing MAX (LOC, LOC − 1) with (LOC − 1)/2. As laid out by pages in Fig. 20.5, the sort requires two pages in core to complete the ordering of a triplet. One page holds the siblings and the other holds the parent. Page replacement occurs when a page boundary is crossed.

Additional paging may occur as the pass leaves the endpoint nodes, because each low element must descend the tree as far as possible, perhaps back to an endpoint. The requirement that all vertices be higher than their descendants makes for sharp reversals in the addressing pattern. Since there are only two data pages in core, one for siblings and one for parent, the sudden need to refer to siblings of siblings back down the tree causes either a sibling or a parent page to be replaced. There are 11 page-in operations after the initial load of D, the first sibling page, and B, the first parent page. The first paging-in of C (replacing B) is due to the first crossing of the page D boundary. Page B must be immediately recalled (replacing D) to compare $12:6$. Page D is now out of the normal neighborhood. Note, however, that the first starred comparison recalls D because the original contents of location 4 must be inspected to see if it must descend further. The second starred comparison must flush both current pages from memory in order to recall C and D.

This method is more sensitive to page size than a linear method and requires more resident pages to reduce paging for any set of instructions executed. The locality will shift and reshift over fewer instructions than with the linear method. Pass 2 will do three comparisons and three pagings; pass 3 will do five comparisons and six pagings. These passes are definitely thrashing, since pages are being taken out and immediately recalled, and the paging rate exceeds the comparison rate.

As the method ages, the paging will decrease because pages will become ordered and drop from the neighborhood. As the neighborhood shrinks, it will fit into available real core. Until that happens, however, the passes of this sort method will page furiously and show no locality smaller than the entire data list. Ascending order will make things worse by extending the average distance an element must travel in descending the tree.

All tree-based sorting methods that achieve tree structure by hopping around will behave in this way, and must be either modified or carefully fitted into the paging environment.

20.1.8 Impact and Magnitude of Paging

If the previous sections have given the impression that linear sorts, *per se*, are to be preferred to tree-based methods in a paging environment, that impression must be corrected. For the strictly internal sort—or a sort that appears to be internal because of virtual memory—there are tree-based techniques which do not exhibit undesirable addressing and which can be used with little penalty. Quicksort and the merge are in any event desirable methods, and they can be used effectively in a virtual-memory or paging environment with little penalty. Care can be taken with other methods to reduce paging to tolerable levels if the programmer has some idea of how much space he is likely to have.

The best a programmer may expect in a demand-paging system is that the number of paging operations is equal to the number of reference pages. This is equivalent to running with all required core, since the time to load pages once is equivalent to load time. The worst a programmer can expect is that every comparison or transfer (whichever can occur more frequently in the method) will cause a page exception (or two, if the replacement algorithm tends to overlay needed pages). Such an extreme case of paging might occur if there were room on a page for one element and room in core for one page. The burden would be a function of the number of elements. Each method has its own probability of an off-page reference, and each mapping of a method onto core has a probability that an off-page reference will not be in core. The first problem is to find a method with good addressing. The next is to organize the method so as to guarantee performance in some minimum space. The third problem is to encode in such a way that additional paging is not incurred on procedure pages.

20.1.9   Quicksort and Merge

Quicksort and the merge are both internal sorts with highly linear address patterns and low comparison requirements, and both are consequently excellent for the paging environment.

Quicksort follows a list linearly, looking for a value larger than a selected pivot. When one is found, an exchange is made with a value known to be lower. The search for a lower and the search for a higher move toward each other, and a pass ends when the two searches meet. If one page is kept for the bottom-up search and one for the top-down search, paging will occur during a pass only when page boundaries are crossed. The severity of interpass paging depends on the size of real storage, because the locality shifts dramatically after a pass. As the method progresses, partitions become smaller and at some point will fit into the space provided. The larger that space is, the sooner partitions may be entirely contained in real core and the lower the paging burden will be.

The addressing pattern of the merge becomes less local as it ages, but it remains linear. The crucial design factor is to maintain a balance between order of merge and number of pages, so that all substrings are always in core. A minimum of three pages is required for a two-way merge, that is, a page for each string and a page for the merged string as it develops. The addition of more space does not improve the performance of the merge.

Figure 20.6 shows the behavior of a first pass of Quicksort in the environment of the earlier examples. Paging occurs only when a boundary is crossed (A to B, then B to C). At the end of a pass there are two partitions.

Begin pass   End pass

1.	13	A		2	
2.	6			6	
3.	4			4	
4.	9			9	
5.	11	B		11	
6.	8			8	Partition
7.	7			7	one
8.	16			12	
9.	14	C		10	
10.	3			3	
11.	5			5	
12.	15			1	
13.	1	D		13	
14.	10			15	Partition
15.	12			14	two
16.	2			16	

Comparisons by position	Transfers by position	Reference	Incore	Pagings
	1 ⟶ TEMP			
16 : TEMP	16 ⟶ 1	A, D	A, D	2R
2 : TEMP		A	A, D	
3 : TEMP		A	A, D	
4 : TEMP		A	A, D	
5 : TEMP		B	D, B	1W, 1R
6 : TEMP		B	D, B	
7 : TEMP		B	D, B	
8 : TEMP	8 ⟶ 16	B, D	D, B	
15 : TEMP	15 ⟶ 8	B, D	D, B	
9 : TEMP	9 ⟶ 15	C, D	C, D	1W, 1R
14 : TEMP	14 ⟶ 9	C, D	C, D	
10 : TEMP		C	C, D	
11 : TEMP		C	C, D	
12 : TEMP	12 ⟶ 14	C, D	C, D	
13 : TEMP	13 ⟶ 12			
	TEMP ⟶ 13			

Fig. 20.6 Quicksort with PIVOT = 13.

This configuration constitutes a special case—one partition resident in core and fitting a page perfectly—of which advantage should be taken. As partitions are reduced to page size or below, any technique for ordering the partition is acceptable, since there are no paging considerations.

20.1.10 Replacement Selection in Paging

The overcommitment of core for an internal phase can cause severe problems for the popular technique of Replacement Selection. If the sort is organized so that the tree of the tournament refers to the records themselves (nondetached table), the probability of a reference to one page of records in record space, given random data, is as great as the probability of a reference to another. Each comparison will cause the record area to be accessed for the key. The location of the record in record space is completely unpredictable, and severe paging may occur even with small reductions in real core. For example, if three pages are available for a four-page list of records, the probability of reference to a record page not in core is 0.25. Given 16 elements on the full list, there will be $\log_2 16 = 4$ comparisons to select a winner. There will be 64 comparisons (not counting initialization) to select winners for the entire list. If there is a 0.25 probability of referencing a nonresident record space page, 16 page faults are probable. If there is room for only two pages, then $\frac{2}{4}$ of the pages will not be in core, and there is a 0.50 probability of referencing a nonresident record page. There will "probably" be 32 page exceptions. The number of page operations is the probability of referencing record pages not in core times the number of comparisons. To prepare strings for 10,000 input items, with 50 items per page, a tournament size of 10 pages, and room for eight pages in core, will lead to $\frac{2}{10} \times 140,000$ page exceptions. That is, 140,000 is 10,000 times $\lceil \log_2 10,000 \rceil$ and $\frac{2}{10}$ is the probability of referencing a nonresident element. With a 20-percent reduction in core, 28,000 page exceptions, or more than 2.8 exceptions per element sorted, will occur. The degree of decay of the sorting algorithm with only one nonresident page is $\frac{1}{10} + 140,000$ or 14,000 paging operations.

What is the real meaning of 14,000 paging operations? For 14,000 comparisons the sequence load/compare/branch occurs 14,000 times. With attendant control, key preparation, and other auxiliary instructions, perhaps as many as 15 instructions are directly involved in a comparison operation. The magnitude of 14,000 paging operations is completely different. Hundreds of instruction times for execution of paging-support instructions are referenced, and CPU time is forfeited while pages are being transferred. The paging will involve at least one read, which is attended by the usual amount of seeking, rotational delay, and data transfer time. Paging is a *milli*second-level operation, not the microsecond-level operation involved in comparisons for even small machines.

There are a number of strategies for making Replacement Selection tenable. Furthermore, they are of general interest because they involve techniques that may be applied to the paging environment in general.

One strategy is to reduce the size of the number of records in the tournament to a level that will fit in core. The impact will be to produce more strings for the merge, and the result may or may not be more merge passes. If more merge passes do result, the additional merge passes may or may not increase elapsed sort/merge time by more than the savings in paging.

The ultimate effect on time depends on the relative speed of the paging device versus the merge devices, the paging properties of the merge, the order of merge, etc.

Another strategy is to organize the tree as a detached table. All comparisons are conducted on the table (a tree tag sort), and records are not referenced until the tag table is ordered. The tree size may be constrained to guarantee that the pages containing the tag table are in core at all times. When the tag table provides a winner, the records are retrieved and passed to output. The number of paging exceptions is the probability of referencing a nonresident page times N. This value is considerably below the levels of the nondetached Tournament sort. With references to the record area occurring only to move a winner, a string of 16 winners will cause only 16 paging operations.

20.1.11   Key versus Record Sorts

If key size is small relative to record size and tag tables can be completely contained in core, detached tag sorting can reduce the number of paging exceptions to the level shown for Replacement Selection with detached table. The nondetached table sort is of almost no value for paging; since access to the record is required for each key, the paging of records does not change. Space required for the table may even result in *increased* paging.

20.1.12   External Merging

The great problem with the external merge lies in the relationship between blocking and buffering schemes and the paging schemes. Large block sizes and deep buffering schemes may produce situations in which not every string to be merged has an element available in real core. A balance must be struck between optimization of the merge and minimization of paging.

In conditions of limited core, the backup blocks will tend to be paged out and the "back-ends" of very large primary blocks may also be subject to paging. The intent of the buffering scheme may be defeated by the paging mechanisms. If average access time to a block from the merge device (disk) is considerably larger than from the paging device (perhaps a drum), it may still pay to backup-buffer and allow the paging. Recalling the block from the paging device may take less time than waiting for the block to be read from a slower device. The paging storage acts as a kind of staging device for blocks. When the paging device and the merging device are the same, the use of backup buffering becomes somewhat suspect. The buffers require extra space and do not guarantee the presence of a record in storage when the CPU requires it. But if backup buffering becomes suspect the whole concept of CPU–I/O balancing must be subjected to difficult reanalysis. This problem of preserving I/O–CPU overlap will become particularly severe in "one-level" storage systems where all I/O is "folded" into virtual memory space and data allocation is "transparent" to the sort.

The size of the block will also affect paging behavior. Blocks larger than a page are naturally more vulnerable to paging than blocks equal to or smaller than a page. An ideal situation is achieved with small blocks which fit integrally into pages. Given a block size of 2K, a page size of 4K, and a merge order of 8, then four pages can provide backup buffering, and paging will not occur if four input pages can be guaranteed. If a 4K block is preferable, some floating buffer techniques may be used. But floating buffers represent potential trouble since they introduce randomness into the buffer reference pattern.

In general, merge order should not exceed the space required to guarantee a block from every string. The major trade-off in merging in a paging environment is the balancing of normal I/O against paging I/O. Good design requires that the programmer or user know what space he is likely to have, since the calculation of merge order, block size, and buffering scheme are all dependent on being able to calculate paging rate versus I/O degradation for various amounts of space.

20.1.13   Cache

A feature similar to paging (and part of many machine designs) is the cache, a memory with an access time considerably faster than the access times of primary storage. Because of an automatic contents-management system, the existence of the cache is not apparent to a programmer. The IBM 360/85 was the first standard machine to be delivered with a cache, and the feature is available with many IBM S/370's. Memory speeds tend to be a limiting factor; i.e., a fast CPU will perform much faster when memory responses are faster. The average instruction rate is a function of the percentage of instructions that find their data in the cache.

Data is brought into the cache in blocks that contain the referenced data, and a block resides in the cache until it is overlaid with another block. The movement of data in and out of the cache is commonly supported by a hardware implementation of an LRU algorithm. The core–cache interface and the problems of residence are a perfect analogue of the core-backing storage interface supported by paging.

The cache, like paging, is based on the concept of locality in a program. Cache size is a small percentage of total core size, perhaps 1 to 5 percent. In order for the feature to be effective, it is certainly necessary that more than 5 percent of memory references find data in the cache. Upwards of 75 or 80 percent of all memory references for a significant population of programs would be considered a minimum operational requirement.

Some attention paid to addressing patterns as they relate to the effectiveness of the cache will indeed make a performance difference. A programmer in a user environment writing an internal sort risks very little with a sort, like Quicksort, whose addressing facilitates use of the cache. The use of a detached-key table that may squeeze entirely into the cache is a good approach. The records themselves will eventually be accessed for ordering at memory speeds, but all comparisons and exchanges are performed with the speed of the cache.

20.2 MULTIPROCESSORS

There are a number of machine environments with various forms and degrees of *parallelism*. Those machines which can execute more than one instruction at the same time, or execute a single instruction in more than one place at a time, present a challenge to the designer of sorting algorithms.

20.2.1 Generalized Multiprocessors

A multiprocessor is a computer system capable of keeping several instruction streams active. Various system designs support the rather general concept of the multiprocessor. Figure 20.7 shows a system of four memory banks associated with three processors through a central memory control unit (MCU). Each memory address is equally accessible to all the processors through the MCU, which also resolves contention for memory access between the processors. Each processor interfaces with two I/O subprocessors, which provide paths to the population of control units and I/O devices. In such a system there is no single "ownership" by a processor of any subset of devices; rather, all processors have access to a common pool.

The multiprocessor is characterized by macroparallelism; i.e., separate or related programs may be running concurrently without the use of any processor time-sharing (multiprogramming) technique. The effectiveness of these systems depends in part on the number of programs which can be run independently at any time and the amount of contention for the resources of the system because of the operation of multiple CPU's. In order to utilize a multiprocessor, there must be several independent jobs to do and sufficient memory and I/O to support the CPU's running the jobs, without untenable delays due to contention for memory, channels, and device space.

Many of the problems associated with multiprocessors are those associated with multiprogramming. There is a set of problems connected with running $(P + X)$ processes in P processors. They include queue disciplines, preventing pathological changes to shared system information, scheduling policies, allocation strategies, etc.

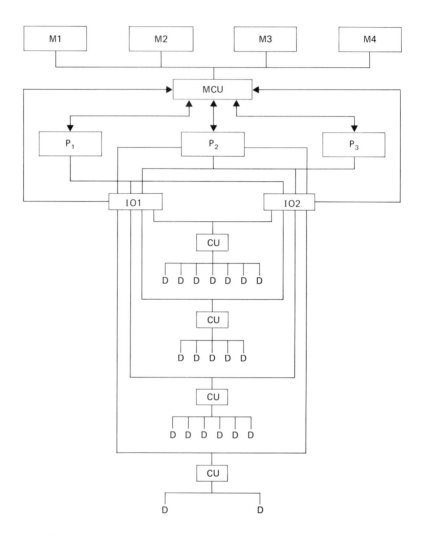

Fig. 20.7 Generalized multiprocessor.

In general, the idealized goal of a multiprocessor with N processors is to run a problem in $1/N$ of the time, but that goal is difficult to achieve for a number of reasons.

1. Contention for resources between processors will slow the performance of each processor.

2. The time cost of initiating and synchronizing parallel processors will add nontrivial overhead.

3. Problems are often not partitionable into work units of equal size.

20.2.2   Quicksort on a Multiprocessor

Quicksort, or some modification of Quicksort, seems to be a natural for purely internal sorting on a multiprocessor with an efficient processor-activation mechanism. A discussion of the application of Quicksort to multiprocessors of various sizes will illustrate a number of considerations and problems connected with effectively using multiprocessors as sorting devices.

Figure 20.8 shows the tree of a perfect Quicksort. Each application of the algorithm results in partitions of equal size. The initial partition has 64 elements, and therefore two partitions of 32, four partitions of 16, eight partitions of 8, etc., are successively generated. This idealized operation of Quicksort occurs (see Chapter 7) whenever the pivot is the list median. In an encoding of Quicksort that takes k comparisons each partition, where k is partition size, the number of comparisons will be roughly $N \log_2 N$ (in this example, 384). In the following discussion, the reduction in the number of comparisons will be used as the only measure of the effectiveness of a multiprocessor. To simplify the discussion we will make a number of dangerous assumptions.

1. The mechanism for requesting a processor has no meaningful cost.

2. When processors are requested, they are spontaneously available, and therefore no time is lost waiting for a processor to become free. If all the processors to be used for the sort are initially preallocated, this condition can be guaranteed. Such preallocation is not uncommon in multiprocessor use (real and imagined). A negative effect of preallocating will become apparent shortly, however.

3. The most dangerous assumption of all is that a processor will process a partition at a rate that remains the same, unaffected by whether other processors are active, quiescent, or working on other problems. The list to be ordered may be allocated across memory banks so that processors interfere with each other when accessing data. Instruction access interference will reach a high level when multiple processors are accessing instructions in the same bank. The existence of private caches in the system will reduce memory-contention delay. Alternatively, data may be moved to banks of memory, and thus each processor's data references will be localized to private banks. Replication of code, a private copy for each processor, and judicious allocation of copies will reduce access contention. The use of core for additional copies of code will restrict the size of lists which may be sorted, of course, by using additional space for procedure.

All the considerations above must be taken into account when an algorithm is mapped onto a multiprocessor. We will set these considerations aside for the moment, to investigate the performance of Quicksort as an algorithm on an idealized multiprocessor.

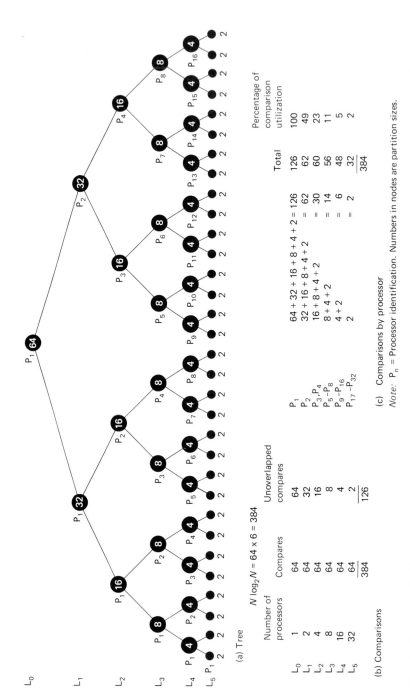

(a) Tree

$N \log_2 N = 64 \times 6 = 384$

Number of processors	Compares	Unoverlapped compares
L_0	1	64	64
L_1	2	64	32
L_2	4	64	16
L_3	8	64	8
L_4	16	64	4
L_5	32	64	2
		384	126

(b) Comparisons

Comparisons by processor

			Total	Percentage of comparison utilization
P_1	$64 + 32 + 16 + 8 + 4 + 2 = 126$		126	100
P_2, P_4	$32 + 16 + 8 + 4 + 2$	$= 62$	62	49
P_3, P_4	$16 + 8 + 4 + 2$	$= 30$	60	23
$P_5 - P_8$	$8 + 4 + 2$	$= 14$	56	11
$P_9 - P_{16}$	$4 + 2$	$= 6$	48	5
$P_{17} - P_{32}$	2	$= 2$	32	2
			384	

(c) Comparisons by processor

Note: P_n = Processor identification. Numbers in nodes are partition sizes.

Fig. 20.8 Tree of Quicksort.

Part (b) of Fig. 20.8 summarizes some performance statistics of a 32-processor system applied to the optimum Quicksort tree. At L_0 of the tree, a single processor runs the algorithm to produce two partitions. At L_1 an additional processor is acquired. There are 32 comparisons required to process partitions of size 32. Each processor processes a partition and completes the pass in 32 elapsed comparison times. Each level requires 64 comparisons, but at each level more processors are acquired, and so the number of comparisons is effectively the number required to process one partition. The sum of unoverlapped comparisons in ordering the entire 64-element list is 126. The elapsed time (the time for sequential comparisons) has been reduced from 384 to 126. Note in this example, that, for simplification, the ranked element is considered, each pass, to be part of a partition. A more precise example would show the resulting partition of pass 1 to be 32 and 31.

The $\frac{2}{3}$ reduction in elapsed comparison time must be a disappointment, since the theoretical goal of the application of 32 processors to a problem is a reduction to $\frac{1}{32}$ of the elapsed time of the single processor. This ideal goal can be achieved only if each processor does $\frac{1}{32}$ of the work and is utilized 100 percent of the elapsed time for the entire process. With 384 comparisons to be performed, an ideal split of work for each processor would be 32 partitions of 12 elements each. But Quicksort does not lend itself to the development of such a complex of partitons. Part (c) of Fig. 20.8 shows why the proportionate reduction in elapsed comparison time is so modest. Of the 384 comparisons to be performed, processor 1 performs 126, processor 2 performs 62, processor 31 performs 2. The utilization of processors, measured by the number of comparisons performed in an elapsed time for 126 comparisons, confirms that most of the processors are making only trivial contributions to the sort. This pattern is inherent in the method because of the way partitions are generated.

In a dynamic system where processors can be acquired as needed, the low utilization of processors need not be a serious throughput problem. It may be assumed that the processors are doing other productive work when they are not contributing to the sort. The dynamic acquisition of processors, of course, must imply delays—intervals of time during which partitions are not being processed because processors are busy with other jobs.

The one-processor-per-partition map of Quicksort is as unacceptable for the interests of sorting as for the interests of system throughput. Consider that two 32-processor systems can perform 4032 comparisons in the time it takes one processor to perform 126. The elapsed time of a sort requiring 4000 comparisons (125 per processor) may be less than the Quicksort requiring 384 total comparisons.

There are two very reasonable approaches to a solution for what we have just observed. One is to investigate the performance of the algorithm for reduced numbers of processors, and the other, of course, is to modify or replace the algorithm.

Figure 20.9 shows the performance of exactly the same 64-element list on four processors (the tree is somewhat distorted and is incomplete). The

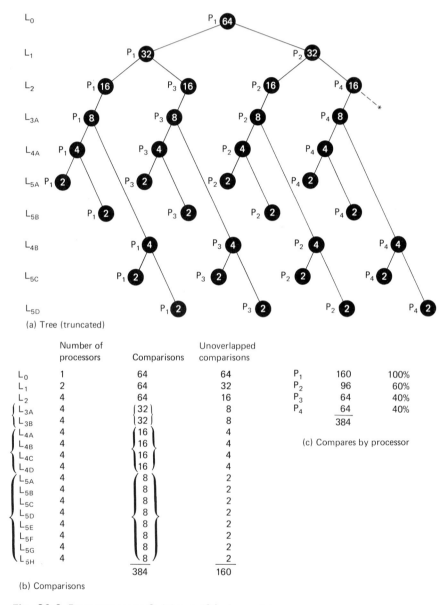

(a) Tree (truncated)

(b) Comparisons

	Number of processors	Comparisons	Unoverlapped comparisons
L_0	1	64	64
L_1	2	64	32
L_2	4	64	16
L_{3A}	4	32	8
L_{3B}	4	32	8
L_{4A}	4	16	4
L_{4B}	4	16	4
L_{4C}	4	16	4
L_{4D}	4	16	4
L_{5A}	4	8	2
L_{5B}	4	8	2
L_{5C}	4	8	2
L_{5D}	4	8	2
L_{5E}	4	8	2
L_{5F}	4	8	2
L_{5G}	4	8	2
L_{5H}	4	8	2
		384	160

(c) Compares by processor

P_1	160	100%
P_2	96	60%
P_3	64	40%
P_4	64	40%
	384	

Fig. 20.9 Four-processor Quicksort (64 elements).

processor at each node is indicated, and comparisons at each node are shown in the circles. The tree is truncated at L_2. At the end of L_2 there are eight partitions of length 8 on the partition stack. The processing of only four of them is shown on the tree. (Level 3B is not shown.) The right-hand branches of L_{4A} nodes are shown as occurring below the lefthand branches. This intentional distortion of the tree shows that the processing of left branches will occur later in time. Level suffixes, A, B, C, are now used to show the repeated application of the same processors to partitions of the same size. At L_{4A}, for example, we note the first instance of processing partitions of size 4, and at L_{4B} the second. Although not shown, L_{4C} and L_{4D} also occur. A single processor processes partitions of size 32 and size 16 only once, but it processes partitions of size 8 twice, of size 4 four times, and of size 2 eight times. In this way, four processors are mapped against 32 partitions.

The increase in the number of elapsed comparison times is a direct result of a shortage of processors. With a four-processor system, the maximum effective parallelism is four wide. To perform the 64 comparisons required for processing the 16 partitions of length 4 requires four iterations of four processors, each iteration effectively performing four comparisons. Sixteen sequential comparison times must be added for processing L_4 partitions whereas with 32 processors only four elapsed sequential comparison times are required. The sort is delayed but by disproportionately little; a reduction to $\frac{1}{8}$ the number of processors ($\frac{4}{32}$) causes a decay of less than 30 percent in performance.

Given the reduction in contention when a system is operating with four rather than with 32 processors, it may turn out that, for specific instances of a sort, four processors will outperform 32. There is, of course, an optimum number of processors for this sort, somewhere between 4 and 32. Eight processors would require 136 elapsed comparison times (Fig. 20.10), and there would be little reason for requesting additional processing.

Level	No. of processors	No. of comparisons	Elapsed times
0	1	64	64
1	2	64	32
2	4	64	16
3	8	64	8
4A	8	32	4
4B	8	32	4
5A	8	16	2
5B	8	16	2
5C	8	16	2
5D	8	16	2
		Total 384	Total 136

Fig. 20.10 Elapsed comparison time with eight processors.

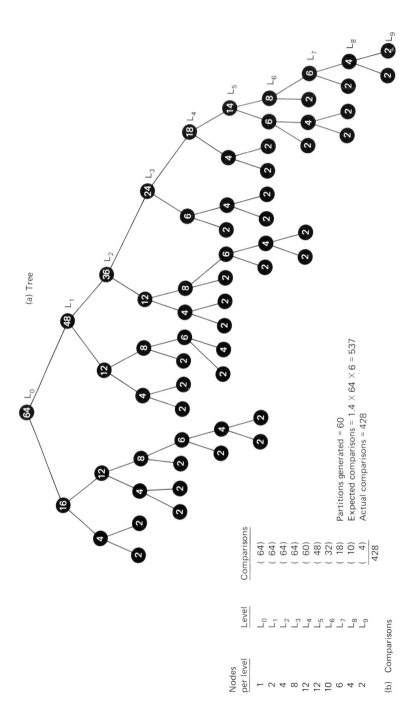

(a) Tree

Nodes per level	Level	Comparisons
1	L_0	(64)
2	L_1	(64)
4	L_2	(64)
8	L_3	(64)
12	L_4	(60)
12	L_5	(48)
10	L_6	(32)
6	L_7	(18)
4	L_8	(10)
2	L_9	(4)
		428

(b) Comparisons

Partitions generated = 60
Expected comparisons = $1.4 \times 64 \times 6 = 537$
Actual comparisons = 428

Fig. 20.11 A realistic Quicksort example.

The minimum application of multiprocessing power, the two-processor configuration, results in 224 elapsed comparison times. The second processor performs 160 comparisons and is utilized 71 percent of the elapsed 224 comparison times.

Figure 20.11 is a much more realistic picture of the performance of Quicksort. It is not an especially bad case, since the requirement of 428 comparisons is lower than the expectation of 537 comparisons. The generation of the tree is based on the assumption that the partitions generated from a partition will have a size ratio of 3 : 1 for some, 2 : 1 for others, and 1 : 1 only for intitial partitions of size 4. The tree is as arbitrary as the partitions of a real sort. In total, 60 partitions are processed.

A programmer faced with deciding how many processors to apply has absolutely no way of knowing how many partitions are going to be generated. The hapless programmer who decides to apply 32 processors for the 32 partitions expected from the perfect tree will be very surprised. The number of partitions can no more be precisely determined than can the exact number of comparisons. It is entirely data-dependent.

The tree of Fig. 20.11 reveals that there will be no opportunity at any level of the tree for more than 12 processors to be effectively operative, since no level contains more than 12 nodes. But this statistic is as hidden as any other when the decision about the number of processors to use must be made. The decision can, of course, be avoided if processors can be dynamically acquired. Another significant revelation of Fig. 20.11 is that some processors are taking on a partition of size 8 when others are taking on partitions of various other sizes and that, over the course of the sort, the release of a processor to pick up other partitions becomes unpredictable and chaotic. There is no real way of predicting *which* processor will be processing *what* partition *when*. If the shape of the tree could be known in advance, there might be a way of defining an optimum mapping of processors to partitions of various sizes, but since the tree is dynamically generated, no such strategy may be applied. Figure 20.12 shows the operation of a two-processor system and a system which guarantees that there will be no partition stacking. A processor is available whenever a partition is generated.

The mechanics of partition/processor association will have some effect on performance. The mechanism used for developing the system in Fig. 20.12 is as follows: A processor generates two partitions, always keeps the smaller, and places the larger on a common stack. A processor is committed to a series of partitions whenever it begins to work. For example, the initial processor, P_1, will process partitions of sizes 64, 16, 4, and 2 (the leftmost branch of the tree) before it becomes "available." Similarly, when processor 2 becomes active, it processes partitions of sizes 48, 12, 4, 2. The processing of a partition of size 2 completes a partition sequence, and the processor

	Comparisons		Comparisons
P_1	256	P_1	170
P_2	174	P_2	116
	430	P_3	68
Unoverlapped: 262		P_4	72
		P_5	6
			432
		Unoverlapped: 224	

(a)   (b)

Fig. 20.12 "Random" tree performance with (a) a two-processor system and (b) a system that guarantees no waiting.

then goes to the common queue (stack) for more work. When there is none, the processor may idle or be released. If processor preallocation is used, the processor will idle or loop until work appears on the stack. If dynamic allocation is used, the processor will go to some other work queue.

The difference between the two systems of Fig. 20.12 is simple. In (a), the two processors are preallocated, and a generated partition is stacked until a processor becomes free. In the guaranteed system, a processor is acquired, and consequently partitions are never queued in the stack. Figure 20.12(b) is therefore an optimum multiprocessor mapping for the particular tree of Fig. 20.11. Whenever there is work, it is done, and there is no way to further reduce elapsed time.

An interesting aspect of the tree of Fig. 20.11 is *indeterminacy*. At various points it is possible for a processor to pick up one partition or another. If P_1 is released and goes to the stack for work, and at the same time P_2 is ready to stack a partition, there is no way to predict whether P_1 or P_2 will get to the stack first. It is a "race" condition resolved by microseconds. If P_1's request for work is recognized by the system a microsecond before P_2's stacking operation, P_1 will pick up what was at the top of the stack while P_1 and P_2 were busy on their just-completed partitions. If P_2 gets to the stack first, it will put a partition on the top of the stack. When P_1 takes a partition, it will be the partition just placed by P_2. While a processor is adding to or taking from the stack, the other (or others) must be "locked out," prevented from accessing the shared resource, the stack. Current multiprocessor systems depend on the execution of a lock instruction (e.g., Compare and Swap on the IBM 370/168) to protect shared resources from pathological simultaneous reference by two processors.

A suggested modification of Quicksort for a parallel-processor environment involves the use of two pivots per pass, so that a second processor may participate in the initial partitioning of the list. A list of elements less than one bound is clustered toward the top; a list of elements greater than the other is clustered toward the bottom. Published results do not conclude that the approach is faster than Quicksort.

20.2.3   Pair Exchange on a Multiprocessor

Choosing a method for a parallel processor involves determining how much effective parallelism exists in the method, and mapping that onto an expected number of machines. The critical measure is elapsed time. A method in which there is very high utilization of every processor associated with the sort provides for most effective parallelism and is to be preferred. Consider a system in which it is possible to determine how many processors are available. The sort might then develop as many partitions as there are available processors and perform a P-way merge with one processor, releasing all others. A series of lower-order merges using as many processors as possible might be performed. The sort for each partition might well be Quicksort, but no processor would sort other than the initial partitions assigned to it.

In Chapter 2 a method of sorting called Pair Exchange was described. It had the characteristics that all comparisons are disjoint; that is, comparisons are of positions K and $(K + 1)$, then of $(K + 2)$ and $(K + 3)$. Because there is no common element between consecutive comparisons, they can be done at the same time. Given N elements, each pass of the method requires $N/2$ comparisons. With $N/2$ processors, each comparison can be executed by a different processor. Processor 1 can execute the position 1 : position 2 comparison; processor 2, the 3 : 4 comparison; ... processor 32; the 63 : 64 comparison. Each pass requires time for only one comparison and for the one exchange executed by each processor that recognizes an inversion. The method requires a maximum of $(N + 1)$ passes. A list of 64 elements will require a maximum of 65 sequential comparison times, using 32 processors. There will be over 2000 total comparisons. The number of total comparisons is four times that expected of Quicksort, but the number of effectively elapsed comparison times is half that of Quicksort at its best. Because of the separability of the work, the bad single-processor method can use multiprocessor systems effectively, giving roughly proportional increases in performance as processors are added. With 16 processors, each can perform two comparisons per pass. The total effective elapsed time per pass is two comparison times. The time for the sort doubles.

There are various programming techniques which might be useful on a parallel processor. Associated with indeterminacy is a concept called *redundancy*. It may be possible to allow processors to duplicate work and still achieve good elapsed times if the method in which the duplication occurs has maximum parallel effect. Distributive sorting may be efficient in a multiprocessor environment. The increase in speed can be very close to being proportional to the number of processors. Processors distribute into bins independently at a fast rate, reducing distribution time each pass by some number close to P.

20.2.4 Parallel Merge

One method for sorting on a multiprocessor derives from a modification of Shellsort, which guarantees that no comparisons of a pass will involve the same comparand. The method described here is due to K. E. Batcher [1]. It is closely related to work done in the area of sorting networks.

A sorting network is a hardware unit consisting of a number of comparison units many of which can operate in parallel. A network is designed to sort a specific number of inputs, and the measure of a network is the number of comparisons required to sort the given number of elements. A great deal of theoretical attention has been given to the minimum number of comparisons required to sort specific N (always very small). The first article in the general computer literature was published in the *Journal of ACM*, Volume 9, pp. 282 (1962), by Bose and Nelson [3]. The article is entitled "A Sorting Problem" and discusses the "p-operator" method for generating proper compare sequences. Hibbard, in 1963 in the *Journal*, published an ALGOL algorithm which generates proper comparison sequences for various N [12]. Since then work has been done by Floyd and Knuth [8,9], Van Voorhis [24], and others to find optimal sorting networks.

The pattern of comparisons in a sorting network can be adopted by a general purpose multiprocessor. The heart of such a sort is the development of comparison sequences which are disjoint, so that some number of processors can operate as independent comparison units. The overhead of the method is the development of proper sequences.

Batcher's "parallel merge" operates in a number of multiphase passes. Preliminary passes arrange elements so that, during the last phase of last pass, no element will be farther than one position away from its proper rank. Any element during the last phase of the last pass may be properly ranked by comparing it (and perhaps exchanging it) with its neighbor. All comparisons of the last phase are disjoint and may be done in parallel.

In a spirit similar to Shellsort, phases run with various distances between comparands. The effect of all phases prior to the last is to form an ordered list of odd-numbered elements and an ordered list of even-numbered elements. Elements of the odd list and even list are neighbors at last phase and the lists are "merged" by the last phase comparisons.

The control structure is considerably more complex than that of Shell-sort. A fundamental difference is that the number of passes and the distance between elements are controlled by separate variables. A third variable is required to define comparison sequences for a phase. A fourth variable is used to control phase-to-phase or pass-to-pass transition.

The initial value of P (pass counter) is 2^{t-1} where t is $\lceil \log_2 N \rceil$. Readers may remember that this is similar to the first value of D in Hibbard's version of Shellsort. For the list of Fig. 20.13, with 16 elements P will initially be 8. In this case $t = 4$ and $2^{t-1} = 2^3$ or 8. The initial value of D (distance) is also 8, the sequence generation variable (S) is initially 0, and the phase control variable (C) is initially 8. The algorithm for setting initial values is $P = 2^{t-1}$, D P, $S = 0$, and $C = 2^{t-1}$. P will never be reset but will decrease in value by successive integer division by two. D, S, and C will be reset to their initial values at the beginning of each pass. During a pass, at the end of each phase, they will assume different values.

	Initial list	List after phase 1	Position	$i + 1$: $i + 1 + D$	Exchanges
1.	13	13	$i = 0$	1 : 9	
2.	11	3	1	2 : 10	2 ← 10
3.	7	7	2	3 : 11	
4.	6	6	3	4 : 12	
5.	4	4	4	5 : 13	
6.	5	5	5	6 : 14	
7.	9	2	6	7 : 15	7 ←→ 15
8.	8	1	7	8 : 16	8 ←→ 16
9.	14	14			
10.	3	11			
11.	12	12	(b) Comparison sequence		
12.	16	16			
13.	15	15			
14.	10	10	$P = 8$ (2^3)	i	$i \wedge P$ (binary)
15.	2	9	$D = 8$	0	$0000 \wedge 1000 = 0 = S$
16.	1	8	$S = 0$	1	$0001 \wedge 1000 = 0 = S$
			$C = 8$	2	$0010 \wedge 1000 = 0 = S$
(a) List before and after phase 1			lim $i = N - (D+1) = 7$	3	$0011 \wedge 1000 = 0 = S$
				4	$0100 \wedge 1000 = 0 = S$
				5	$0101 \wedge 1000 = 0 = S$
				6	$0110 \wedge 1000 = 0 = S$
				7	$0111 \wedge 1000 = 0 = S$

New pass ($P = C$)
$P = P/2$
$D = P$
$S = 0$
$C = 8$

(c) Phase control

Fig. 20.13 Phase one of pass 1.

The generation of a proper comparison depends upon a pointer (i) which varies during a phase from 0 to $N - (D + 1)$. For every binary value of i which when ANDed to P has a result equal to S, there is a proper phase comparison. In order to perform comparisons in parallel, it is necessary to generate the set of comparisons to be performed at the beginning of each phase. The reader should not be disturbed if the reasons for particular manipulations of S, C, and D are not obvious.

In Figure 20.13 the comparison and exchange sequence for $i = 0$ to 7 (b) is shown next to the transformed list (a). The pass control for this phase accords with the discussion above for values of P, D, S, and C. The pointer i is limited to 7 because

$$N - D + 1 = 16 - 9.$$

On this phase no value of i has a bit in the 2^3 position, so all comparisons $i : P$ result in 0, equal to the value of S. Any time the result of the AND of i and P is not equal to S, the comparison generated by that value of i is suppressed. Comparison positions are $(i + 1)$ and $(i + 1 + D)$. The effect of the $(N - (D + 1))$ limit on i is to exclude overlapping (nondisjoint comparisons). Note that the next comparison would be $i = 8$, $9 : 17$. Even with a 17th member of the list, this comparison would not be performed because position 9 has already been involved in a comparison in this phase. The AND of i and P serves the same purpose. The AND of two quantitites produces a 0 result in a bit position whenever both quantities do not have a 1 in that position.

If there are eight processors available, phase one can complete in one comparison/exchange time plus address generation time. There are many possible ways to implement the method. A single processor can generate all comparisons and then call for eight comparing/exchanging processors. It is probably more efficient to let an initial processor take a value of $i = 0$ and then call for a processor passing a value of $i = 1$. The $i = 0$ processor calculates its comparison sequence and performs the comparison/exchange. The $i = 1$ processor calls for an $i = 2$ processor, etc. Depending on the number of available processors, processor calls may spend some time waiting on a queue. If there are eight processors for this phase, each processor (except not the 8th) will call for another processor, calculate the address of one comparison, perform, and then release. The last processor to release can attend to phase-end procedures. If there are less than eight processors, some processors may go to the "queue of waiting i's" more than once. Note that it does not matter in what order comparisons are performed.

When a phase completes, that is, when

$$i = N - (D + 1),$$

a determination is made as to whether an additional phase is required or a new pass should be undertaken. If P is equal to C, a new pass is undertaken. If P is unequal to C, an additional phase of the same pass is undertaken. A new pass is distinguished by a new value for P equal to $\lfloor C \rfloor$.

At the end of the phase of Fig. 20.13, P and C are equal and a new pass begins. The initial values for this pass are $P = 4$, $D = 4$, $S = 0$, $C = 8$. The first phase is shown as Fig. 20.14. There are 8 comparisons, with 4 comparisons excluded by the i and P rule. With $P = 4$, values of $i = 4, 5, 6, 7$ will produce a result unequal to $S = 0$. The effect of this exclusion is to suppress the comparisons $5 : 9$, $6 : 10$, $7 : 11$, and $8 : 12$. Since 9, 10, 11, and 12 are already involved in comparisons during this phase, it is necessary to exclude them. Processors which generate the comparison to be suppressed can release or go to the queue for a new value of i.

	Pass begin list	List after phase 1, pass 2
1.	13	4
2.	3	3
3.	7	2
4.	6	1
5.	4	13
6.	5	5
7.	2	7
8.	1	6
9.	14	14
10.	11	10
11.	12	9
12.	16	8
13.	15	15
14.	10	11
15.	9	12
16.	8	16

(a) List

Position	$i+1: i+1+D$	Exchanges
$i = 0$	$1 : 5$	$1 \longleftrightarrow 5$
1	$2 : 6$	
2	$3 : 7$	$3 \longleftrightarrow 7$
3	$4 : 8$	$4 \longleftrightarrow 8$
4	—	
5	—	
6	—	
7	—	
8	$9 : 13$	
9	$10 : 14$	$10 \longleftrightarrow 14$
10	$11 : 15$	$11 \longleftrightarrow 15$
11	$12 : 16$	$12 \longleftrightarrow 16$

(b) Comparison sequence

$P = 4$
$D = 4$
$S = 0$
$C = 8$
$\lim i = N-(D+1) = 11$
New phase $(P \neq C)$
$D = C-P = 4$
$C = C/2 = 4$
$S = P = 4$

(c) Phase control

	i	Excluded $i \wedge P$	S
	4	$0100 \wedge 0100 \neq 0$	
	5	$0101 \wedge 0100 \neq 0$	
	6	$0110 \triangle 0100 \neq 0$	
	7	$0111 \triangle 0100 \neq 0$	

Fig. 20.14 Phase one of pass 2.

At the end of the phase, the new phase condition $P \neq C$ exists. The proper values of D, P, S, C are:

$$D = C - P(8-4) = 4, \qquad S = P = 4, \qquad C = C/2 = 4.$$

The phase-control variables for the second phase of pass 2 are 4, 4, 4, 4, as shown in Fig. 20.15. With $S = 4$ the only comparisons undertaken by the method are those where the AND of i and P are equal to 4. With $P = 4$, i values of 4, 5, 6, 7 will give such a result. These comparisons are those excluded from the first phase. (See Fig. 20.14.)

Pass begin list and phase end list		Position	$i+1: i+1+D$	Exchanges
1.	4	$i = 0$	—	
2.	3	1	—	
3.	2	2	—	None
4.	1	3	—	
5.	13	4	5 : 9	
6.	5	5	6 : 10	
7.	7	6	7 : 11	
8.	6	7	8 : 12	
9.	14	8	—	
10.	10	9	—	
11.	9	10	—	
12.	8	11	—	
13.	15			
14.	11	b) Comparison sequence		
15.	12			
16.	16			

a) List

		Excluded		
		i	$i \wedge p$	S
$P = 4$		0	$0000 \wedge 0100 \neq 4$	
$D = 4$		3	$0011 \wedge 0100 \neq 4$	
$S = 4$		8	$1000 \wedge 0100 \neq 4$	
$C = 4$		11	$1011 \wedge 0100 \neq 4$	

lim $i = N-(D+1) = 11$
New pass $(P = C)$
$P = 4$
$D = 4$
$S = 0$
$C = 8$

c) Phase control

Fig. 20.15 Phase two of pass 2.

When this phase ends, a new pass condition exists $(P = C)$. P is halved to 2, D is set equal to P, and S and C are set to initial values of 0 and 8. Figure 20.16 shows the third pass with $P = 2$. The pass has 3 phases. Eight processors can cooperate in phase one, 4 processors in phase two, 6 processors in phase three. That the number of usable processors varies from phase to phase is a difficulty of the method. Since phase-to-phase compari-

	Pass begin list	End phase 1	End phase 2	End phase 3
1.	4	2	2	2
2.	3	1	1	1
3.	2	4	4	4
4.	1	3	3	3
5.	13	7	7	7
6.	5	5	5	5
7.	7	13	12	9
8.	6	6	6	6
9.	14	9	9	12
10.	10	8	8	8
11.	9	14	14	13
12.	8	10	10	10
13.	15	12	13	14
14.	11	11	11	11
15.	12	15	15	15
16.	16	16	16	16

a) List

	Phase 1	Phase 2	Phase 3
	$P,D,S,C=2,2,0,8$	$P,D,S,C=2,6,2,4$	$P,D,S,C=2,2,2,2$
	lim $i = N-(D+1) = 13$	lim $i = N-(D+1) = 9$	lim $i = N-(D+1) = 13$
$i = 0$	1 : 3*	—	—
1	2 : 4*	—	—
2	—	3 : 9	3 : 5
3	—	4 : 10	4 : 6
4	5 : 7*	—	—
5	6 : 8	—	—
6	—	7 : 13*	7 : 9*
7	—	8 : 14	8 : 10
8	9 : 11*	—	—
9	10 : 12*	—	—
10	—	—	11 : 13*
11	—	—	12 : 14
12	13 : 15*	—	—
13	14 : 16	—	—
	$P \neq C$	$P \neq C$	$P = C$ End of pass

*Asterisks show exchanges

b) Comparisons and pass control

Fig. 20.16 Pass 3.

sons, in general, can overlap (for example, 7 : 13 in phase two, 7 : 9 in phase three), it is not possible, in general, to operate in more than one phase at the same time. However, there are instances, as in Fig. 20.16, where it would be possible to run phases in parallel. If N is known, the complete layout of work can be preplanned, since the sequence of comparisons does not change, regardless of the initial order of the list. Figure 20.17 draws the final pass. Notice that at the beginning of the last phase no element is more than one position away from final rank.

	Pass begin list	End phase 1	End phase 2	End phase 3	End phase 4
1.	2	1	1	1	1
2.	1	2	2	2	2
3.	4	3	3	3	3
4.	3	4	4	4	4
5.	7	5	5	5	5
6.	5	7	7	7	6
7.	9	6	6	6	7
8.	6	9	9	9	8
9.	12	8	8	8	9
10.	8	12	12	11	10
11.	13	10	10	10	11
12.	10	13	13	13	12
13.	14	11	11	12	13
14.	11	14	14	14	14
15.	15	15	15	15	15
16.	16	16	16	16	16

a) List

	Phase 1	Phase 2	Phase 3	Phase 4
	$P,D,S,C=1,1,0,8$	$P,D,S,C=1,7,1,4$	$P,D,S,C=1,3,1,2$	$P,D,S,C=1,1,1,1$
	lim $i = N-(D+1) = 14$	lim $i = N-(D+1) = 8$	lim $i = N-(D+1) = 12$	lim $i = N-(D+1) = 14$
$i = 0$	1 : 2*	—	—	—
1	—	2 : 9	2 : 5	2 : 3
2	3 : 4*	—	—	—
3	—	4 : 1	4 : 7	4 : 5
4	5 : 6*	—	—	—
5	—	6 : 13	6 : 9	6 : 7*
6	7 : 8*	—	—	—
7	—	8 : 15	8 : 11	8 : 9*
8	9 : 10*	—	—	—
9	—	—	10 : 13*	10 : 11*
10	11 : 12*	—	—	—
11	—	—	12 : 15	12 : 13*
12	13 : 14*	—	—	—
13	—	—	—	14 : 15
14	15 : 16	—	—	—
15	—	—	—	—

*Asterisks show exchanges

b) Comparisons and pass control

Fig. 20.17 Last pass.

The feasibility of implementing this technique on a general-purpose multiprocessor depends in part upon the relative efficiency of processor acquisition and release. Given a first mechanism for this function, it is an attractive and efficient technique. Figure 20.18 is a summary of total comparisons versus elapsed comparison time. The method comes close to achieving the time reduction expected of a well-running, well-planned multiprocessor.

	Comparisons (Total)	Elapsed comparisons (Best)	Elapsed comparisons (2 processors)	Elapsed comparisons (4 processors)
Pass 1, phase 1	8	1	4	2
Pass 2, phase 1	8	1	4	2
Pass 2, phase 2	4	1	2	1
Pass 3, phase 1	8	1	4	2
Pass 3, phase 2	4	1	2	1
Pass 3, phase 3	6	1	3	2
Pass 4, phase 1	8	1	4	2
Pass 4, phase 2	4	1	2	1
Pass 4, phase 3	6	1	3	2
Pass 4, phase 4	7	1	4	2
	63	10	32	19

Fig. 20.18 Pass summary. "Best" assumes spontaneously available processor.

20.2.5   External Merging in a Multiprocessor

An external sort/merge on a multiprocessor can be designed so that multiple processors are involved both in the creation of strings and in the merge or only in the merge. A file can be divided into N partitions so that it is, in effect, separately sorted and merged by N processors with a final full-file collation. Each processor can produce its own strings and run its own merge. A difficulty with this approach is that memory requirements for the simultaneous residence of the sort, the merge, and separate sort areas may impact string size. A balance must be struck between the increase in number of strings and the time to write the strings. A great advantage of producing strings independently is the ability to overlap their writing.

Independent block sorts on independent processors seem to provide a feasible approach to using multiprocessor power. Short strings produced by these sorts might be profitably merged by one or more processors on their way out to intermediate devices.

Choice of merge design will depend on those considerations of data accessing and data passing which were discussed in earlier chapters. Contention between processors for channels and access arms must be considered if processors share pooled I/O. The general preferability of a one-processor, ten-way merge or a two-processor, five-way merge, etc., can be subjected to the kinds of analysis we have discussed with the merge.

20.3   OTHER HARDWARE

Paging and parallel processing represent only two hardware features to which a sort must be adapted. Other problems facing sort development include the following.

1. *Development of a method for writing simultaneously on multiple channels without reducing string size.* A sorting technique, derived from the tournament and from ideas of string extension, that can write multiple strings in parallel is known. Given N output paths, the last element of each string is kept, and a new element is compared with the set of all last elements. The new element joins the first string on which it fits. For example, with four output paths, four strings at a time will be attempted. Initially four elements are ordered. Each successor element is compared with elements on the list. When a high is found, the list element is put out to its string. The difficulty with the method is that, although it produces a very fast dispersion pass with very few comparisons and allows machine overlap of writing, it leads to larger merge phases because of the serious reduction in string size.

2. *Archives and hierarchical storage.* The presence in a large computing system of various levels of storage devices, with different data rates and capacities, makes it necessary to determine which device should be used in the sort. Should a file on a slow tape be sorted on that device, or should it be "percolated" forward to be sorted on faster devices? The issue of mixed devices in very complex environments raises a software and systems issue. Data management and I/O specialists are tending toward concepts of device transparency. A problem program in future will run without any knowledge of the device from which it receives its data. Data interfaces in standard systems shield problem programs from all device and format problems.

This development obviously presents a very difficult problem for a sort, which must know a great deal about device characteristics, number, and arrangement before it can develop meaningful decisions about merge order, buffering, and blocking. In a transparent, one-level storage environment, such decisions are taken away from the sort.

If virtual-memory concepts are expanded and a data-staging strategy is developed, the sort and device transparency may be compatible. But so long as sort technology requires external operations, a sort system cannot effectively optimize a transparent environment.

3. *Microcoding.* It is possible that sorts can be included in a machine at a microcode level, or that an optimum sort architecture can be implemented in microcode and read into microstorage, so that the CPU can be a sorter. The impact of this technique on existing algorithms remains to be determined.

4. *Associative memories.* Sorting and searching algorithms in microcode or hard-wired circuits can be substituted for programmed sorts. Since a sort is performed to fix the location of an object conveniently, devices that allow

transparent reference by name naturally impact the extent of sorting in a system. In addition, small associative memories may be used in support of programmed sorts.

5. *Sorting machines.* A number of attempts have been made in the past to develop sorting machines. The UNIVAC File-Computer system offered an off-line, stand-alone tape device, which was plugboard-programmed for data parameters [23]. IBM's 703 was intended as a sorting device but never made commercially available. The difficulty with hard-wired, off-line "sorting machines" is that, to achieve the flexibility of data management and sort-technique adaptability to data factors, they require very nearly as much logic and power as a general-purpose processor. The implementation of sort algorithms in hardware or some form other than a CPU has by no means been abandoned. Neither has the search for new CPU features that will make sorting more efficient. It is quite possible that N-way comparison logic, which has been suggested frequently, may some day actually be built into a machine. In a machine with eight arithmetic registers, an eight-way comparison instruction may be quite feasible.

The general tendency toward the distribution of capability in a system, whereby control units and channels become specialized programmable processors, can be extended to sorting. A tape or disk control unit can be extended to participate in sorting. The purpose may be not so much to sort faster as to sort more cheaply, permitting a small subsystem to sort while the CPU goes on to other things. The control unit may function as a normal unit when it is not sorting.

An on-line sorting device developed for use in an IBM 360 system and in connection with the OS sort has been offered commercially. The electronic Data Sorter of the Astrodata Corporation undertakes to perform phase one (dispersion phase) of the OS sort merge. The core and CPU cycles of phase one are relieved by attachment of this device, which develops strings as a specialized sort processor. The merge phases of the OS sort operate in the usual manner.

Bibliography

PART 1   INTERNAL SORTING

Sources and Further Readings for Preface and Chapters 1–11

All sources providing for Part I are contained herein except the algorithms from *ACM,* which are presented as Appendix A.

1. Applebaum, F. H., "Variable Word Sorting in RCA 501 System," Association for Computing Machinery 14th National Meeting (1959).

2. Bayes, A., "A Generalized Partial-Pass Block Sort," *Comm. ACM* **11,** 7, 491–493.

3. Bell, D. A., "The Principles of Sorting," *Computer J.* **1** (1959), 71–77. See also *Computer J.* **10** (1968), 123; **1** (1959), 171.

4. Burge, W. H., "Sorting, Trees, and Measures of Order," *Inform. Contr.* **1** (1958), 181–197.

5. Feerst, S., and F. Sherwood, "The Effect of Simultaneity on Sorting Operations," Paper 42, Association for Computing Machinery 14th National Meeting (1959).

6. Flores, I., "Computer Time for Address Calculation Sorting," *Journal ACM* **7,** 389–409.

7. Flores, I., "Analysis of Internal Computer Sorting," *Journal ACM* **8** (1961), 41–80.

8. Frazer, W. D., and C. K. Wong, "Sorting by Natural Selection," *Comm. ACM* **15,** 10, 910–912.

9. Frazer, W. D., and A. C. McKellar, "Samplesort—A Sampling Approach to Minimal Storage in All Sorting," *Journal ACM* **17,** 3, 496–507.

10. Frank, R. M., and R. B. Lazarus, "A High-Speed Sorting Procedure," *Comm. ACM* **3,** 1, 20–23.

11. Gotlieb, C. C., "Sorting on Computers," *Comm. ACM* **6**, 5, 194–201.

12. Hall, M. M., "A Method of Comparing the Time Requirements of Sorting Methods," *Comm. ACM* **6**, 5, 259–263.

13. Hibbard, T., "Some Combinatorial Properties of Certain Trees with Application to Searching and Sorting," *Journal ACM* (1962), 13–28.

14. Hibbard, T. N., "An Empirical Study of Minimal-Storage Sorting," *Comm. ACM* **6**, 5, 206–213.

15. Hildebrant, P., and H. Isbitz, "Radix Exchange—An Internal Sorting Method for Digital Computers," *Journal ACM* **6**, 156–163.

16. Hoare, C. A. R., "Quicksort," *Computer J.* **5** (1962), 10–15.

17. Hosken, J. C., "Evaluation of Sorting Methods," *Proceedings EJCC*, 1955.

18. IBM Corporation, "Sorting Techniques," C20-1639 (1965).

19. Isaac, E. J., and R. C. Singleton, "Sorting by Address Calculation," *Journal ACM* **3** (1956), 169–174.

20. Jones, B., "A Variation on Sorting by Address Calculation," *Comm. ACM* **13**, 2, 105–107.

21. Knuth, D. E., *The Art of Computer Programming;* Volume 3, *Sorting and Searching.* Reading, Mass.: Addison-Wesley, 1973.

22. Kronmal, R., and M. Tarter, "Cumulative Polygon Address-Calculation Sorting," ACM 20th National Conference (1965), 376–385.

23. Lorin, H., "A Guided Bibliography to Sorting," *IBM Systems Journal* **10**, 3, 244–254.

24. Maclaren, M. D., "Radix Exchange Plus Sifting," *Journal ACM* **13**, 404–411.

25. Martin, W. A., "Sorting," *ACM Computing Surveys,* **3**, 4, 147–174.

26. Nagler, H., "Amphisbaenic Sorting," *Journal ACM* **6**, 4, 459–468.

27. Shell, D. L., "A High-Speed Sorting Procedure," *Comm. ACM* **2**, 7, 30–33.

28. Van Emden, M. H., "Increasing the Efficiency of Quicksort," *Comm. ACM* **13**, 9, 563–567.

29. Windley, P. F., "Trees, Forests and Rearranging," *Computer J.* (1960), 84–88.

30. Woodrum, L. J., "Internal Sorting with Minimal Comparing," *IBM Systems Journal* **8**, 3, 189–203.

PART 2   EXTERNAL SORTING

Sources and Further Readings for Chapters 12–18

1. Bennett, B. T., and W. D. Frazer, "Approximating Optimal Direct-Access Merge Performance," *Proceedings of the IFIP Congress 1971,* Vol. 1, 450–453.

2. Betz, D, K., and W. C. Carter, "New Merge Sorting Techniques," *Proceedings ACM,* 14th National Conference (1959).

3. Black, N. A., "Optimum Merging from Mass Storage," *Comm. ACM* **13**, 12, 745–749.

4. Buhagiar, J., and A. Jons, "A Scan Sort Using Magnetic Tape Units," *Computer Bulletin* **12**, 1, 11–13.

5. Burge, W. H., "Analysis of Compromise Merge Sort Techniques," *Proceedings of the IFIP Congress 1971,* Vol. 1, 454–459.

6. Cooke, W. S., "A Tape File Merge Pattern Generator," *Comm. ACM* **6**, 5, 227–230.

7. Dinsmore, R. J., "Longer Strings from Sorting," *Comm. ACM* **8**, 1, 48.

8. Edwards, L. G., "Flexible Replacement Presorting," *ACM Sort Symposium,* November 1962.

9. Falkin, J., and S. Savastano, Jr., "Sorting with Large-Volume, Random-Access, Drum Storage," *Comm. ACM* **6**, 5, 240–244.

10. Ferguson, D. E., "More on Merging" (letter to editor), *Comm. ACM* **7**, 4, 297.

11. Ferguson, D. E., "Buffer Allocation in Merge Sorting," *Comm. ACM* **14**, 7, 476–478.

12. Frazer, W. D., and B. T. Bennett, "Bounds on Optimal Merge Performance and a Strategy for Optimality," *Journal ACM* **19**, 4, 641–648.

13. French, N. C., "Computer Planned Collates," *Comm. ACM* **6**, 5, 225–226.

14. Friend, E. H., "Sorting on Electronic Computer Systems," *Journal ACM* **3** (1956), 134–168.

15. Gassner, B. J., "Sorting by Replacement Selecting," *Comm. ACM* **10**, 2, 89–93.

16. Gilstad, R. L., "Polyphase Merge Sorting—An Advanced Technique," *Proceedings EJCC,* Vol. 18, December 1960, 143–148.

17. Gilstad, R. L., "Read-Backward Polyphase Sorting," *Comm. ACM* **6**, 5, 220–223.

18. Glicksman, S., "Concerning the Merging of Equal-Length Tape Files," *Journal ACM* **12**, 254–258.

19. Goetz, M. A., "Organization and Structure of Data on Disc-File Memory Systems for Efficient Sorting and Other Data Processing Programs," *ACM Sort Symposium,* November 1962.

20. Goetz, M. A., "Internal and Tape Sorting using the Replacement-Selection Technique," *Comm. ACM* **6**, 5, 201–206.

21. Goetz, M. A., and G. S. Toth, "A Comparison Between the Polyphase and Oscillating Sort Techniques," *Comm. ACM* **6**, 5, 223–225.

22. Goetz, M. A., "Design and Characteristics of a Variable-Length Record Sort using New Fixed-Length Record Sort Techniques," *Comm. ACM* **6**, 5, 264–267.

23. Goetz, M. A., "Three Letters on Merging (II)" (letter to editor), *Comm. ACM* **6**, 10, 586.

24. Goetz, M. A., "More on Sorting Techniques" (letter to editor), *Comm. ACM* **7**, 6, 379–380.

25. Goetz, M. A., "Some Improvements in the Technology of String Merging and Internal Sorting," *AFIPS SJCC* **25** (1964), 599–607.

26. Hooker, W. W., "On the Expected Lengths of Sequences Generated in Sorting by Replacement Selection," *Comm. ACM* **12,** 7, 411–413.

27. Hubbard, G. V., "Some Characteristics of Sorting in Computing Systems using Random-Access Storage Devices," *Comm. ACM* **6,** 5, 248–255.

28. Johnsen, T. L., "Efficiency of the Polyphase Merge and a Related Method," *BIT* **6** (1966), 129–143.

29. Knuth, D. E., "Length of Strings for a Merge Sort," *Comm. ACM* **6,** 11, 685–688.

30. Knuth, D. E., "Three Letters on Merging (I, III)" (letter to editor), *Comm. ACM* **6,** 10, 585.

31. Lorin, H., A. Silverstein, and M. Smith, *A Report on Sorting,* Sperry Rand Corporation, 1963.

32. Malcolm, W. D., Jr., "String Distribution for the Polyphase Sort," *Comm. ACM* **6,** 5, 217–220.

33. Manker, H. H., "Multiphase Sorting," *Comm. ACM* **6,** 5, 214–217.

34. McAllester, R. L., "Polyphase Sorting with Overlapped Rewind," *Comm. ACM* **7,** 3, 158–159.

35. Mendoza, A. G., "A Dispersion Pass Algorithm for the Polyphase Merge," *Comm. ACM* **5,** 10, 502–504.

36. Radke, C. E., "Merge Sort Analysis by Matrix Techniques," *IBM Systems Journal* **5,** 4, 226–247.

37. Reynolds, S. W., "A Generalized Polyphase Merge Algorithm," *Comm. ACM* **4,** 8, 347–349; Addendum, 495.

38. Sackman, B. S., and T. Singer, "A Vector Model for Merge Sort Analysis," *ACM Sort Symposium,* November 1962.

39. Schick, T., "Disk-File Sorting," *Comm. ACM* **6,** 6, 330–331.

40. Shell, D. L., "Optimizing the Polyphase Sort," *Comm. ACM* **14,** 11, 713–719.

41. Sobel, S., "Oscillating Sort—A New Sort Merging Technique," *Journal ACM* **9,** 372–374.

42. Windley, P. F., "The Influence of Storage Access Time on Merging Processes in a Computer," *Computer J.* **2,** 2, 49–53.

43. Woodrum, L. J., "A Model of Floating Buffering," *IBM Systems Journal* **9,** 2, 118.

PART 3   SORTING SYSTEMS

Sources and Further Readings for Chapters 19–20

1. Batcher, K. E., "Sorting Networks and Their Applications," *AFIPS SJCC* Vol. **32,** (1968) 307–314.

2. Beus, H. L., "The Use of Information in Sorting," *Journal ACM* **17,** 482–495.

3. Bose, R. C., and R. J. Nelson, "A Sorting Problem," *JACM* **9,** 2, 282–296.

4. Brawn, B. D., F. G. Gustavson, and E. S. Mankin, "Sorting in a Paging Environment," *Comm. ACM* **13,** 8, 483–494.

5. Control Data Corp., *3400 Computer System, Sort/Merge Reference Manual.*

6. DeFiore, C. R., "Fast Sorting," *Datamation* **16,** 8, 47–51.

7. Falkman, H. H., "Sort/Merge—Its Facilities and Method of Operation," Presentation to Guide Utilities Committee, Session P15 and P26, 27 May 1971.

8. Floyd, R. W., and D. E. Knuth, "Improved Constructions for the Bose–Nelson Sorting Problem," *Notices American Mathematics Society* **14** (1967), 283.

9. _____, "The Bose–Nelson Sorting Problem," *Stanford Computer Science Memo,* STAN–CS–70–177, November 1970.

10. Foster, C. C., "Sorting Almost Ordered Arrays," *Computer J.* **11,** 134–137.

11. Glore, J. B., "Sorting Nonredundant Files—Techniques Used in the FACT Compiler," *Comm. ACM* **6,** 5, 231–240.

12. Hibbard, T. N., "A Simple Sorting Algorithm," *JACM* **10,** 2, 142–150.

13. Johnson, L. R., and R. D. Pratt, "An Introduction to the Complete UNIVAC II Sort–Merge System SESAME," *UNIVAC Review,* Fall 1958.

14. IBM, "IBM System/360 OS Sort/Merge," GC28–6543.

15. IBM, "IBM System/360 OS Sort/Merge Timing Estimates," GC33–4008.

16. IBM, "Sorting Methods for IBM Data Processing Systems," F28–8001 (1958).

17. Paterson, J. B., "The COBOL Sort Verb," *Comm. ACM* **6,** 5, 255–258.

18. Pratt, R. D., "The Sort Program for the UNIVAC III Data Processing System," *ACM Sort Symposium,* 1962.

19. Sperry Rand Corp., "Sort III," *UNIVAC III General Reference Manual,* U3518.

20. Sperry Rand Corp., "Sort/Merge I," *UNIVAC 1107 General Reference Manual,* UP–3959.

21. Sperry Rand Corp., "Sort/Merge Program Programmer's Reference," *UNIVAC 1107 Technical Bulletin,* UT 2576.

22. Sperry Rand Corp., "Sort/Merge," *UNIVAC 490 Real-Time System Technical Bulletin,* UP–3809.

23. Sperry Rand Corp., "UNIVAC File Computer Sort–Collate System," U1489B8.

24. Van Voorhis, D. C., "Efficient Sorting Networks." (author's thesis), Stanford University: University Microfilms, 72–16282, 1972.

Appendix A
Sorting Procedures from CACM Algorithms

The following algorithms and related certifications and remarks are reprinted by permission from *Communications of the ACM,* Copyright © 1960–1971, Association for Computing Machinery, Inc.

MASTER LIST

ALGORITHM 23

MATHSORT
Wallace Feurzeig
Laboratories for Applied Science, University of Chicago, Chicago, Ill.

CERTIFICATION OF ALGORITHM 23

MATHSORT (Wallace Feurzeig, *Comm. ACM,* Nov. 1960)
Russell W. Ranshaw
University of Pittsburgh, Pittsburgh, Pa.

ALGORITHM 63

PARTITION
C. A. R. Hoare
Elliott Brothers Ltd., Borehamwood, Hertfordshire, England

ALGORITHM 64

QUICKSORT
C. A. R. Hoare
Elliott Brothers Ltd., Borehamwood, Hertfordshire, England

ALGORITHM 65

FIND
C. A. R. Hoare
Elliott Brothers Ltd., Borehamwood, Hertfordshire, England

CERTIFICATION OF ALGORITHMS 63, 64, 65

PARTITION, QUICKSORT, FIND (C. A. R. Hoare, *Comm. ACM,* July 1961)
J. S. Hillmore
Elliott Bros. Ltd., Borehamwood, Hertfordshire, England

CERTIFICATION OF ALGORITHMS 63, 64 and 65

PARTITION, QUICKSORT, AND FIND (*Comm. ACM,* July 1961)
B. Randell and L. J. Russell
The English Electric Company Ltd., Whetstone, England

ALGORITHM 76

SORTING PROCEDURES
Ivan Flores
Private Consultant, Norwalk, Conn.

REMARK ON ALGORITHM 76

SORTING PROCEDURES (Ivan Flores, *Comm. ACM* **5,** January 1962)
B. Randell
Atomic Power Div., The English Electric Co., Whetstone, England

ALGORITHM 113

TREESORT
Robert W. Floyd
Computer Associates, Inc., Woburn, Mass.

ALGORITHM 143

TREESORT 1
Arthur F. Kaupe, Jr.
Westinghouse Electric Corp., Pittsburgh, Pa.

ALGORITHM 144

TREESORT 2
Arthur F. Kaupe, Jr.
Westinghouse Electric Corp., Pittsburgh, Pa.

ALGORITHM 175

SHUTTLE SORT
C. J. Shaw and T. N. Trimble
System Development Corporation, Santa Monica, Calif.

CERTIFICATION OF ALGORITHM 175

SHUTTLE SORT (C. J. Shaw and T. N. Trimble, *Comm. ACM*, June 1963)
George R. Schubert
University of Dayton, Dayton, Ohio

REMARK ON ALGORITHM 175

SHUTTLE SORT (C. J. Shaw and T. N. Trimble, *Comm. ACM* **6,** June 1963)
O. C. Juelich
North American Aviation, Inc., Columbus, Ohio

REMARK ON ALGORITHM 175

SHUTTLE SORT (C. J. Shaw and T. N. Trimble, *Comm. ACM* **6** (June 1963), 312; G. R. Schubert, *Comm. ACM* **6** (Oct. 1963), 619; O. C. Juelich, *Comm. ACM* **6** (Dec. 1963), 739)
Otto C. Juelich
North American Aviation, Columbus, Ohio

ALGORITHM 201

SHELLSORT
J. Boothroyd
The English Electric-Leo Computers, Kidsgrove, Staffordshire, England

CERTIFICATION OF ALGORITHM 201

SHELLSORT (J. Boothroyd, *Comm. ACM* **6** (Aug. 1963), 445)
M. A. Batty
The English Electric Co., Whetstone, England

REMARK ON ALGORITHM 201

SHELLSORT (J. Boothroyd, *Comm. ACM* **6** (Aug. 1963), 445)
J. P. Chandler and W. C. Harrison
Department of Physics, Florida State University, Tallahassee, Fla. 32306

ALGORITHM 207

STRINGSORT

J. Boothroyd

The English Electric-Leo Computers, Ltd., Kidsgrove, Staffordshire, England

CERTIFICATION OF ALGORITHM 207

STRINGSORT (J. Boothroyd, *Comm. ACM* **6** (Oct. 1963), 615)

Charles P. Blair

Department of Defense, Washington 25, D.C.

ALGORITHM 232

HEAPSORT

J. W. J. Williams

Elliott Bros. Ltd., Borehamwood, Hertfordshire, England

ALGORITHM 245

TREESORT 3

Robert W. Floyd

Computer Associates, Inc., Wakefield, Mass.

CERTIFICATION OF ALGORITHM 245

TREESORT 3 (Robert W. Floyd, *Comm. ACM* **7** (Dec. 1964), 701)

Philip S. Abrams

Computation Center, Stanford University, Stanford, Cal.

CERTIFICATION OF ALGORITHM 245

TREESORT 3 (Robert W. Floyd, *Comm. ACM* **7** (Dec. 1964), 701):
Proof of Algorithms—A New Kind of Certification

Ralph L. London

Computer Sciences Department and Mathematics Research Center, Universtiy of Wisconsin, Madison, Wis. 53706

ALGORITHM 271

QUICKERSORT

R. S. Scowen

National Physical Laboratory, Teddington, England

CERTIFICATION OF ALGORITHM 271

QUICKERSORT (R. S. Scowen, *Comm. ACM* **8** (Nov. 1965), 669)

Charles R. Blair

Department of Defense, Washington 25, D.C.

ALGORITHM 347

AN EFFICIENT ALGORITHM FOR SORTING WITH MINIMAL STORAGE
Richard C. Singleton
Mathematical Statistics and Operations Research Department, Stanford Research Institute, Menlo Park, Calif. 94025

REMARK ON ALGORITHM 347

AN EFFICIENT ALGORITHM FOR SORTING WITH MINIMAL STORAGE (Richard C. Singleton, *Comm. ACM* **12** (March 1969), 185)
Robin Griffin and K. A. Redish (Recd. 14 April 1969 and 11 Aug. 1969)
McMaster University, Hamilton, Ontario, Canada

REMARK ON ALGORITHM 347

AN EFFICIENT ALGORITHM FOR SORTING WITH MINIMAL STORAGE (Richard C. Singleton, *Comm. ACM* **12** (March 1969), 185)
Richard Peto (Recd. 18 Feb. 1970)
Medical Research Council, 115 Gower Street, London W.C. 1

ALGORITHM 402

INCREASING THE EFFICIENCY OF QUICKSORT
M. H. van Emden
Mathematical Centre, Amsterdam, The Netherlands

REMARK ON ALGORITHM 402 [MI]

INCREASING THE EFFICIENCY OF QUICKSORT (M. H. van Emden, *Comm. ACM* **13**, 693–694)
Robert E. Wheeler
E. I. duPont de Nemours and Company, Wilmington, Del. 19899

ALGORITHM 410

PARTIAL SORTING
J. M. Chambers
Bell Telephone Laboratories, Murray Hill, N.J. 07974

ALGORITHM 426

MERGE SORT ALGORITHM
C. Bron
Technological University, Eindhoven, The Netherlands

ALGORITHMS, CERTIFICATIONS, AND TESTS

ALGORITHM 23
MATH SORT
WALLACE FEURZEIG
Laboratories for Applied Science, University of Chicago, Chicago, Ill.

procedure MATHSORT (INVEC, OUTVEC, TOTEVEC, n, k, SETFUNC) ; **value** n, k ;
    array INVEC, OUTVEC ;
      integer array TOTEVEC ;
      integer procedure SETFUNC ;
    integer n, k ;

begin comment MATHSORT is a fast sorting algorithm which produces a monotone rearrangement of an arbitrarily ordered set of n numbers (represented by the vector INVEC) by a surprising though familiar device. The resultant sorted set is represented by the vector OUTVEC. The key field, i.e. the ordered set of bits (or bytes) on which the sort is to be done, is obtained by some extraction-justification function denoted SETFUNC. The key field allows the representation of k possible values denoted $0, 1, \ldots, k-1$.

The procedure determines first of all the exact frequency distribution of the set with respect to the key, i.e. the number of elements of INVEC with key field value precisely equal to j for all j between 0 and $k-1$. The cumulative frequency distribution TOTEVEC $[i] \equiv \sum_{j=0}^{i}$ (Number of elements of INVEC with key value = j) is then computed for $0 \leq i \leq k-1$. This induces the direct assignment (storage mapping function) of each element of INVEC to a unique cell in OUTVEC. This assignment (like the determination of the frequency distribution) requires just one inspection of each element of INVEC. Thus the algorithm requires only 2n "look and do" operations plus $k-1$ additions (to get the cumulative frequency distribution).

The algorithm can be easily and efficiently extended to handle alphabetic sorts or multiple key sorts. To sort on another key the same algorithm is applied to each new key field with the new INVEC designated as the last induced ordering (i.e. the current OUTVEC). The algorithm has been used extensively at LAS on binary as well as decimal machines

both for internal memory sorts and (with trivial modification)
for large tape sorts   ;
for   i := 1 **step** 1 **until** n **do**
    TOTEVEC[SETFUNC(INVEC[i])] := TOTEVEC
    [SETFUNC(INVEC[i])] + 1   ;
for   i := 1 **step** 1 **until** k−1 **do**
    TOTEVEC[i] := TOTEVEC[i] + TOTEVEC[i−1]   ;
for   i := 1 **step** 1 **until** n **do**
    begin   OUTVEC[TOTEVEC[SETFUNC(INVEC[i])]]
        := INVEC[i]   ;
        TOTEVEC[SETFUNC(INVEC[i])] :=
        TOTEVEC[SETFUNC(INVEC[i])] − 1   ;
    end
end   MATHSORT.

CERTIFICATION OF ALGORITHM 23
MATHSORT (Wallace Feurzeig, *Comm. ACM*, Nov., 1960)
RUSSELL W. RANSHAW
University of Pittsburgh, Pittsburgh, Pa.

The MATHSORT procedure as published was coded for the
IBM 7070 in FORTRAN. Two deficiencies were discovered:

1. The TOTVEC array was not zeroed within the procedure.
This led to some difficulties in repeated use of the procedure.

2. Input vectors already in sort on nonsort fields were unsorted.
That is, given the sequence

$$31, 21, 32, 22, 33,$$

Mathsort would produce, for a sort on the 10's digit:

$$22, 21, 33, 32, 31,$$

which is definitely out of sequence.

The following modified form of the procedure corrects these
difficulties. Note the transformation of symbols.

```
procedure   MATHSORT (I, O, T, n, k, S); value n, k;
            array I, O;  integer array T;  integer procedure S;
            integer n, k;
begin       for i := 0 step 1 until k − 1 do T[i] := 0;
            for i := 1 step 1 until n do T[S(I[i])] := T[S(I[i])] + 1;
            for i := k − 2 step −1 until 0 do T[i] := T[i] +
            T[i + 1];
            for i := 1 step 1 until n do
                begin   O[n + 1 − T[S(I[i])]] := I[i];
                        T[S(I[i])] := T[S(I[i])] − 1;
                end
end MATHSORT.
```

Using the MATHSORT procedure ten times and having the procedure S supply each digit in order, 1000 random numbers of 10 digits each were sorted into sequence in 31 seconds. The method of locating the lowest element, interchanging with the first element, and continuing until the entire list has been so examined yielded a complete sort on the same 1000 random numbers in 227 seconds. Using the Table-Lookup-Lowest command in the 7070 yielded 56 seconds for the same set of random numbers.

ALGORITHM 63
PARTITION
C. A. R. HOARE
Elliott Brothers Ltd., Borehamwood, Hertfordshire, Eng.

procedure partition (A,M,N,I,J); **value** M,N;
        array A; **integer** M,N,I,J;

comment I and J are output variables, and A is the array (with subscript bounds M:N) which is operated upon by this procedure. Partition takes the value X of a random element of the array A, and rearranges the values of the elements of the array in such a way that there exist integers I and J with the following properties:

$$M \leq J < I \leq N \text{ provided } M < N$$
$$A[R] \leq X \text{ for } M \leq R \leq J$$
$$A[R] = X \text{ for } J < R < I$$
$$A[R] \geq X \text{ for } I \leq R \leq N$$

The procedure uses an integer procedure random (M,N) which chooses equiprobably a random integer F between M and N, and also a procedure exchange, which exchanges the values of its two parameters;

```
begin    real X;  integer F;
         F := random (M,N);  X := A[F];
         I := M;  J := N;
up:      for I := I step 1 until N do
                 if X < A [I] then go to down;
         I := N;
down:    for J := J  step −1 until M do
                 if A[J]<X then go to change;
         J := M;
change:  if I < J then begin exchange (A[I], A[J]);
                           I := I + 1; J := J − 1;
                           go to up
                     end
```

else   **if** I < F **then begin** exchange (A[I], A[F]);
                              I := I + 1
                  end
else   **if** F < J **then begin** exchange (A[F], A[J]);
                              J := J − 1
                  end;
end   partition

ALGORITHM 64
QUICKSORT
C. A. R. HOARE
Elliott Brothers Ltd., Borehamwood, Hertfordshire, Eng.

procedure quicksort (A,M,N): **value** M,N;
              array A; **integer** M,N;
comment Quicksort is a very fast and convenient method of
sorting an array in the random-access store of a computer. The
entire contents of the store may be sorted, since no extra space is
required. The average number of comparisons made is 2(M − N) in
(N − M), and the average number of exchanges is one sixth this
amount. Suitable refinements of this method will be desirable for
its implementation on any actual computer;
begin   **integer** l,J:
              if M < N **then begin** partition (A,M,N,I,J);
                              quicksort (A,M,J);
                              quicksort (A, I, N)
                  end
end   quicksort

ALGORITHM 65
FIND
C. A. R. HOARE
Elliott Brothers Ltd., Borehamwood, Hertfordshire, Eng.

procedure find (A,M,N,K); **value** M,N,K;
              array A; **integer** M,N,K;
comment Find will assign to A [K] the value which it would
have if the array A [M:N] had been sorted. The array A will be
partly sorted, and subsequent entries will be faster than the first;

```
begin        integer I,J;
             if M < N then begin partition (A, M, N, I, J);
                            if K≤I then find (A,M,I,K)
                            else if J≤K then find (A,J,N,K)
                            end
end          find
```

CERTIFICATION OF ALGORITHMS 63, 64, 65 PARTITION, QUICKSORT, FIND [C. A. R. Hoare, *Comm. ACM*, July 1961]

J. S. HILLMORE

Elliott Bros. (London) Ltd., Borehamwood, Herts., England

The body of the procedure find was corrected to read:

```
begin integer I, J;
if M < N then begin partition (A, M, N, I, J);
                  if K ≤ I then find (A, M, J, K)
                  else if J ≤ K then find (A, I, N, K)
                  end
end find
```

and the trio of procedures was then successfully run using the Elliott ALGOL translator on the National-Elliott 803.

The author's estimate of $\frac{1}{3}(N-M)\ln(N-M)$ for the number of exchanges required to sort a random set was found to be correct. However, the number of comparisons was generally less than $2(N-M)\ln(N-M)$ even without the modification mentioned below.

The efficiency of the procedure quicksort was increased by changing its body to read:

```
begin integer I, J;
if M < N−1 then begin partition (A, M, N, I, J);
                  quicksort (A, M, J);
                  quicksort (A, I, N)
                  end
else if N−M = 1 then begin if A[N] < A[M] then
                            exchange (A[M], A[N])
                  end
end quicksort
```

This alteration reduced the number of comparisons involved in sorting a set of random numbers by 4–5 percent, and the number of entries to the procedure partition by 25–30 percent.

CERTIFICATION OF ALGORITHMS 63, 64 AND 65, PARTITION, QUICKSORT, AND FIND, [*Comm. ACM*, July 1961]

B. RANDELL AND L. J. RUSSELL

The English Electric Company Ltd., Whetstone, England

Algorithms 63, 64, and 65 have been tested using the Pegasus ALGOL 60 Compiler developed at the De Havilland Aircraft Company Ltd., Hatfield, England.

No changes were necessary to Algorithms 63 and 64 (Partition and Quicksort) which worked satisfactorily. However, the comment that Quicksort will sort an array without the need for any extra storage space is incorrect, as space is needed for the organization of the sequence of recursive procedure activations, or, if implemented without using recursive procedures, for storing information which records the progress of the partitioning and sorting.

A misprint ('if' for '**if**' on the line starting '**else if** $J \leq K$ **then** $\cdots$') was corrected in Algorithm 65 (Find), but it was found that in certain cases the sequence of recursive activations of *Find* would not terminate successfully. Since *Partition* produces as output two integers J and I such that elements of the array $A[M:N]$ which lie between $A[J]$ and $A[I]$ are in the positions that they will occupy when the sorting of the array is completed, *Find* should cease to make further recursive activations of itself if K fulfills the condition $J < K < I$.

Therefore the conditional statement in the body of *Find* was changed to read

> **if** $K \leq J$ **then** find (A,M,J,K)
> **else if** $I \leq K$ **then** find (A,I,N,K)

With this change the procedure worked satisfactorily.

ALGORITHM 76
SORTING PROCEDURES

IVAN FLORES

Private Consultant, Norwalk, Connecticut

comment The following ALGOL 60 algorithms are procedures for the sorting of records stored within the memory of the computer. These procedures are described in detail, flow-charted, compared, and contrasted in "Analysis of Internal Computer Sorting" by Ivan Flores [*J. ACM 8* (Jan. 1961)]. Although sorting is usually a business computer application, it can be described completely in ALGOL if we stretch our imagination a little. Sorting is ordering with respect to a key contained within the record. If the key *is* the active record, the sorting is trivial. A means is required to extract the key from the record. This is essentially string manipulation, for which no provision, as yet, has been made in ALGOL. We circumambulate this difficulty by defining an **integer procedure** K(I) which "creates" a key from the record, I. ALGOL *does* provide for machine language code substitutions, which is one way to think of K(I). This could be more accurately represented by using the string notation proposed by Julien Green ["Remarks on ALGOL and Symbol Manipulation," *Comm. ACM 2* (Sept. 1959), 25–27]. The function **sub** ($,i,g) represents the procedure, K(I). $ corresponds to the record I,   i corresponds to the starting position of the key and g corresponds to the length of the key. Both i and g are **values** which must be specified when the sort procedure is called for as a statement instead of a declaration.

Another factor, which might vex some, is that the key might be alphabetic instead of numeric. Then, of course, K(I) would not be integer. It would, however, be string when such is defined eventually. Note, also, that keys are frequently compared. This is done using the ordering relations ">" for "greater than," etc. These are not really defined in the ALGOL statement [NAUR, PETER, ET AL. "Report on the Algorithmic Language ALGOL 60". *Comm. ACM 3* (May 1960), 294–314]. They can simply be defined so that $Z > Y > \cdots > A > 9 > \cdots > 1 > 0$. Also the assignment X[i] := z should be interpreted as "Assign the key 'z' which is larger than any other key." For any sort procedure (I,N,S),   "I" is the set of unsorted records, "N" is their number, and "S" the sorted set of records.

Caution, these algorithms were developed purely for the love of it: No one was available with the combined knowledge of sorting and ALGOL to check this work. Hence each algorithm should be carefully checked before use. I will be glad to answer any questions which may arise;

```
Sort insert (I,N,S);  value N;  array I[1:N], S[1:N];
  integer procedure K(I);  integer N;
  begin integer i, j, k;
    S[1] := I[1];
    for i := 2 step 1 until N do begin
      for j := i − 1,  j − 1 while K(I[i]) > K(S[j]) do
        for k := i step − 1 until j + 1 do
          S[k] := S[K − 1];
      S[j + 1] := I[i] end end
```

```
Sort count (I,N,S);  value N;   array I[1:N], S[1:N];
   integer procedure K(I);  integer N;
   begin integer array C[1:N];  integer i,j;
      for i := 1 step 1 until N do C[i] := 0;
      for i := 2 step 1 until N do
         for j := 1 step 1 until i − 1 do
            if K(I[i]) > K(I[j]) then C[i] := C[i] + 1
            else C[j] := C[j] + 1;
      for i := 1 step 1 until N do
         S[C[i]] := I[i] end
Sort select (I,N,S); value N;   array I[1:N], S[1:N];
   integer procedure K(I);  integer N;
   begin integer i,j,A,h;
      for i := 1 step 1 until N do begin
      h := K(I[1]);
      for j := 2 step 1 until N do
      if h > K(I[j]) then begin h := K(I[j]);  A := j end;
      S[i] := I[A];
      I[A] := z end end
Sort select exchange (I,N);  value N;   array I[1:N];
   integer procedure K(I);  integer N;
   begin integer h,i,j,H;   real T;
      for i := 1 step 1 until N do begin
         H := K(I[i]);  h := i;
         for j := i + 1 step 1 until N do
            if K(I[j]) < H then begin
            H := K(I[j]);  h := j end
         T := I[i];  I[i] := I[h];  I[A] := T end
   end
Sort binary insert (I,N,S);  value N;   array I[1:N], S[1:N];
   integer procedure K(I);  integer N;
   begin integer i,k,j,l;
      if K(I[1]) < K(I[2]) then begin
         S[1] := I[1];  S[2] := I[2] end
         else begin S[1] := I[2];  S[2] := I[1] end;
start:   for    i := 3 step 1 until N do begin
               j := (i + 1) ÷ 2;
find spot:     for k := (i + 1) ÷ 2, (k + 1) ÷ 2 while k > 1 do
                  if K(I[i]) < K(S[j]) then j := j − k
                  else j := j + k;
               if K(I[i]) ≥ K(S[j]) then j := j − 1;
move items:    for l := i step − 1 until j do
               S[l + 1] := S[l];
enter this
one:           S[j] := I[i] end end
```

```
Sort address calculation (I,N,S,F);  value N;
  array S[1:M], I[1:N];  integer procedure F(K), K(I);
  integer N,M;
              begin integer i,j,G,H,F,M;
              M := entier(2.5 × N)
              for i := 1 step 1 until M do S[i] = 0;
Address:      for i := 1 step 1 until N do begin
              F := F(K(I[i]));
              if S[F] = 0 then begin S[F] := I[i];
                go to NEXT end
              else if K(S[F]) > K(I[i]) then go to SMALLER;
LARGER:       for H := F, H + 1 while K(S[H]) < K(I[i]) do
              for G := H, G + 1 while K(S[G]) ≠ 0 do
              for j := G step −1 until H + 1 do
                S[j] := S[j − 1];
              S[H] := I[i];  go to NEXT;
SMALLER:      for H := F, H − 1 while K(S[H]) > K(I[i]) do
              for G := H, G − 1 while K(S[G]) ≠ 0 do
              for j := G step 1 until H − 1 do
                S[j] = S[j + 1];
              S[H] := I[i];
NEXT:     end end

Sort quadratic select (I,N,S);  value N;  array I[1:N], S[1:N];
  integer procedure K(I);  integer N;
                begin integer i,j,k,C,D,J,M;
                integer array C[1:M], D[1:M];
                array I[1:M, 1:M];
Divide inputs:  M := entier (sqrt (N)) + 1;  j := k := 1;
                for i := 1 step 1 until N do begin
                I[j,k] := I[i];  k := k + 1;
                if k > M then begin k := 1;
                  j := j + 1 end end
Fill up inputs: I[j,k] := z;  k := k + 1;
                if k > M then begin k := 1;  j := j + 1 end
                if j ≤ M then go to Fill up inputs;
Set controls:   for j := 1 step 1 until M do begin
                C[j] := K(I[j, 1]);  D[j] := 1;
                for k = 2 step 1 until M do
                  if C[j] > K(I[j,k]) then begin
                  C[j] := K(I[j,k]);  D[j] := k end end;
                i := 1;
Find least:     C := C[1];  D := D[1];  J := 1;
                for j := 2 step 1 until M do
                  if C > C[j] then begin C := C[j];
                  D := D[j];  J := j end;
```

```
Fill file:          S[i] := I[J,D];  i := i + 1;  I[J,D] := z;
                    if i = N + 1 go to STOP;
Reset controls:     for j := J do begin
                    C[j] := K(I[j, 1]);  D[j] := 1;
                    for k := 2 step 1 until M do
                      if C[j] > K(I[j,k]) then begin C[j] :=
                      K(I[j,k];  D[j] := k end end;
                    go to Find least;
STOP:               end


Presort quadratic selection (I,N,S);  value N;
  array I[1:N], S[1:N];  integer procedure K(I);  integer N;
                begin integer i,j,k,C,J,M;
                    integer array C[1:M], D[1:M];
                    array I[1:M,1:M];
Divide inputs:  M := entier (sqrt(N)) + 1;  j := k := 1;
                    for i := 1 step 1 until N do begin
                    I[j,k] := I[i];  k := k + 1;
                    if k > M then begin k := 1;
                    j := j + 1 end end
Fill up inputs: I[j,k] := z;  k := k + 1;
                    if k > M then begin k := 1;  j = j + 1 end
                    if j ≤ M then go to Fill up inputs;
First sort:     for j := 1 step 1 until M do
                    sort select exchange (I[j,k],M);
Set controls:   for j := 1 step 1 until M do begin
                    C[j] := K(I[j,1]);  D[j] := 1 end
                    i := 1;
Find least:     C := C[1];  J := 1;
                    for j := 1 step 1 until M do
                      if C > C[j] then begin C := C[j];
                      J := j end;
Fill file:      S[i] := I[J,D[J]];  i := i + 1;
                    if i = N + 1 go to STOP
Reset control:  for j := J do begin
                    D[j] := D[j] + 1;
                    if D[j] > M then C[j] := z else C[j] :=
                    K(I[j, D[j]]) end
                    go to Find least;
STOP:               end


Sort binary merge (I,N,S);  value N; array I[1:N];
  integer procedure K(I);  integer N;
  begin real array S[1:N];
```

```
integer array A[0:1, 0:J[a]], B[0:1, 0:K[b]], Aloc[0:1, 0:J[a]],
  Bloc[0:1, 0:K[b]], J[0:1], K[0:1], j[0:1], k[0:1];
  integer a,b,i,j,k;
distribute:        a := b := j[0] := j[1] := 1;
                   for i := 1 step 1 until N do begin
                     if K(I[i]) < K(I[i−1] then
                       if a = 1 then a := 0 else a := 1;
                     A[a, j[a]] := K(I[i]);  Aloc[a, j[a]] := i;
                     j[a] := j[a] + 1 end;
                   J[0] := j[0];  J[1] := j[1];
next sort:         begin a := b := j[0] := j[1] := k[0] :=
                   k[1] := 1;
two inputs:        if A[1, j[1]] ≤ A[0, j[0]] then a := 1 else
                   a := 0;
                   B[b, k[b]] := A[a, j[a]];
                     Bloc[b, k[b]] := Aloc[a, j[a]];
                   j[a] := j[a] + 1;  k[b] := k[b] + 1;
                   if A[a, j[a]] ≥ A[a, j[a] − 1] then go to two
                     inputs else
                   if a = 1 then a := 0 else a := 1;
single step:       B[b, k[b]] := A[a, j[a]];
                     Bloc[b, k[b]] := Aloc[a, j[a]];
                   j[a] := j[a] +1;  k[b] := k[b] + 1;
                   if A[a, j[a]] ≥ A[a, j[a] − 1] then go to
                     single step;
switch file:       if b = 1 then b := 0 else b := 1;
check rollout:     for a := 0, 1 do
                     if j[a] = J[a] then go to rollout;
                   go to two inputs;
rollout:           B[b, k[b]] := A[a, j[a]];
                     Bloc[b, k[b]] := Aloc [a, j[a]];
                   k[b] := k[b] + 1;  j[a] := j[a] + 1;
                   if j[a] = J[a] then go to interchange files;
                   if A[a, j[a]] < A[a, j[a] − 1] then
                     if b = 1 then b := 0 else b := 1;
                   go to rollout;
interchange files: K[0] := k[0];  K[1] := k[1];
                   if K[0] = 1 then go to output end
                   for b := 1, 0 do begin
                     for k[b] := 1 step 1 until K[b] do begin
                       A[b, k[b]] := B[b, k[b]];
                         Aloc[b, k[b]] := Bloc[b, k[b]];
                       J[b] := K[b] end end
                   go to next sort;
```

output: **for** i := 1 **step** 1 **until** N **do**
                       S[i] := I[Bloc[0, i]];
                   end

REMARK ON ALGORITHM 76
SORTING PROCEDURES (Ivan Flores, *Comm. ACM* **5**, Jan. 1962)
B. RANDELL
Atomic Power Div., The English Electric Co., Whetstone, England

The following types of errors have been found in the Sorting Procedures:
1. Procedure declarations not starting with **procedure.**
2. Bound pair list given with array specification.
3. = used instead of :=, in assignment statements, and in a **for** clause.
4. A large number of semicolons missing (usually after **end**).
5. Expressions in bound pair lists in array declarations depending on local variables.
6. Right parentheses missing in some procedure statements.
7. Conditional statement following a **then.**
8. No declarations for A, or Z, which is presumably a misprint.
9. In several procedures attempt is made to use the same identifier for two different quantities, and sometimes to declare an identifier twice in the same block head.
10. In the Presort quadratic selection procedure an array, declared as having two dimensions, is used by a subscripted variable with only one subscript.
11. At one point a subscripted variable is given as an actual parameter corresponding to a formal parameter specified as an array.
12. In several of the procedures, identifiers used as formal parameters are redeclared, and still assumed to be available as parameters.
13. In every procedure K is given in the specification part, with a parameter, whilst not given in the formal parameter list.

No attempt has been made to translate, or even to understand the logic of these procedures. Indeed it is felt that such a grossly inaccurate attempt at ALGOL should never have appeared as an algorithm in the *Communications.*

ALGORITHM 113
TREESORT
ROBERT W. FLOYD
Computer Associates, Inc., Woburn, Mass.

procedure *TREESORT* (*UNSORTED*, *n*, *SORTED*, *k*); **value**
n, *k*;
integer *n*, *k*; **array** *UNSORTED*, *SORTED*;
comment *TREESORT* sorts the smallest *k* elements of the *n*-component array *UNSORTED* into the *k*-component array *SORTED* (the two arrays may be the same). The number of operations is on the order of $2 \times n + k \times \log_2(n)$. The number of auxiliary storage cells required is on the order of $2 \times n$. It is assumed that procedures are available for finding the minimum of two quantities, for packing one real number and one integer into a word, and for obtaining the left and right half of a packed word. The value of infinity is assumed to be larger than that of any element of *UNSORTED*;
begin integer *i*, *j*; **array** $m[1:2 \times n - 1]$;
for $i := 1$ **step** 1 **until** *n* **do** $m[n + i - 1] := pack$ (*UNSORTED* $[i], n + i - 1)$;
for $i := n - 1$ **step** $- 1$ **until** 1 **do** $m[i] := minimum$ ($m[2 \times i]$, $m[2 \times i + 1])$;
for $j := 1$ **step** 1 **until** *k* **do**
  begin *SORTED* $[j] := left\ half$ ($m[1]$); $i := right\ half$ ($m[1]$); $m[i] := $ infinity;
  for $i := i \div 2$ **while** $i > 0$ **do** $m[i] := minimum$ ($m[2 \times i], m[2 \times i + 1])$
  end
end TREESORT

ALGORITHM 143
TREESORT 1
ARTHUR F. KAUPE, JR.
Westinghouse Electric Corp., Pittsburgh, Penn.

procedure *TREESORT* 1 (*UNSORTED*, *n*, *SORTED*, *k*);
  value *n*, *k*;
integer *n*, *k*; **array** *UNSORTED*, *SORTED*;

comment *TREESORT* 1 is a revision of *TREESORT* (ALGORITHM 113) which requires neither the "packed" array *m* nor the machine procedures *pack*, *left half*, *right half*, and *minimum*. The identifier *infinity* is used as nonlocal **real** variable with value greater than any element of *UNSORTED*;

begin integer *i*, *j*; **array** $m1\ [1:2 \times n - 1]$;
  integer array $m2\ [1:2 \times n - 1]$;

procedure *minimum*; **if** $m1\ [2\times i] \le m1[2\times i+1]$ **then**
  begin $m1[i] := m1[2\times i];\;\; m2[i] := m2[2\times i]$ **end else**
  begin $m1[i] := m1[2\times i+1];\;\; m2[i] := m2[2\times i+1]$ **end** *mini-*
   mum;
for $i := n$ **step** 1 **until** $2 \times n - 1$ **do begin** $m1[i] := UNSORTED$
  $[i-n+1];\;\; m2[i] := i$ **end**
for $i := n - 1$ **step** -1 **until** 1 **do** *minimum*;
for $j := 1$ **step** 1 **until** k **do**
  begin $SORTED\ [j] := m1[1];\;\; i := m2[1];\;\; m1[i] := $ *infinity*;
   for $i := i \div 2$ **while** $i > 0$ **do** *minimum* **end**
end *TREESORT* 1

ALGORITHM 144
TREESORT 2
Arthur F. Kaupe, Jr.
Westinghouse Electric Corp., Pittsburgh, Penn.

procedure *TREESORT* 2 $(UNSORTED, n, SORTED, k, ordered)$;
  value n, k;
integer n, k; **array** $UNSORTED, SORTED$; **Boolean proce-**
  dure *ordered*;

comment *TREESORT* 2 is a generalized version of *TREESORT*
  1. The **Boolean procedure** *ordered* is to have two **real** arguments. The array *SORTED* will have the property that *ordered*
  $(SORTED[i], SORTED[j])$ is **true** when $j > i$ if *ordered* is a
  linear order relation;

begin integer i, j; **array** $m1\ [1:2\times n-1]$; **integer array** $m2$
  $[1:2\times n-1]$;
procedure *minimum*; **if** *ordered* $(m1[2\times i], m1[2\times i+1])$ **then**
  begin $m1[i] := m1[2\times i];\;\; m2[i] := m2[2\times i]$ **end else**
  begin $m1[i] := m1[2\times i+1];\;\; m2[i] := m2[2\times i+1]$ **end** *mini-*
   mum;
for $i := n$ **step** 1 **until** $2 \times n - 1$ **do begin** $m1[i] := UNSORTED$
  $[i-n+1];\;\; m2[i] := i$ **end**
for $i := n - 1$ **step** -1 **until** 1 **do** *minimum*;
for $j := 1$ **step** 1 **until** k **do**
  begin $SORTED[j] := m1[1];\;\; i := m2[1];\;\; m1[i] := $ *infinity*;
   for $i := i \div 2$ **while** $i > 0$ **do** *minimum* **end**
end *TREESORT* 2

ALGORITHM 175
SHUTTLE SORT
C. J. SHAW AND T. N. TRIMBLE
System Development Corporation, Santa Monica, Calif.

procedure *shuttle sort* (m, *Temporary*, N);
value m; **integer** m; **array** $N[1:m]$;
comment This procedure sorts the list of numbers $N[1]$ through $N[m]$ into numeric order, by exchanging out-of-order number pairs. The procedure is simple, requires only *Temporary* as extra storage, and is quite fast for short lists (say 25 numbers) and fairly fast for slightly longer lists (say 100 numbers). For still longer lists, though, other methods are much swifter. The actual parameters for *Temporary* and N should, of course, be similar in type;
begin integer i, j;
for $i := 1$ **step** 1 **until** $m - 1$ **do**
  begin
  for $j := i$ **step** -1 **until** 1 **do**
    begin
    if $N[j] \leq N[j+1]$ **then go to** *Test*;
Exchange: *Temporary* $:= N[j]$; $N[j] := N[j+1]$;
        $N[j+1] :=$ *Temporary*; **end of** j loop;
Test: **end** of i loop
  end shuttle sort

CERTIFICATION OF ALGORITHM 175
SHUTTLE SORT [C. J. Shaw and T. N. Trimble, *Comm. ACM*, June 1963]
GEORGE R. SCHUBERT*
University of Dayton, Dayton, Ohio

Algorithm 175 was translated into BALGOL and ran successfully on the Burroughs 220. The following actual sorting times were observed:

Number of Items	Average Time (sec)
25	1.6
50	6.2
100	25.8
250	181
500	684

* Undergraduate research project, Computer Science Program, Univ. of Dayton.

The algorithm can be extended so that the sort is made on one array, while retaining a one-to-one correspondence to a second array. This is done by inserting immediately before **end** of the j loop the following:

$Temporary := Y[j]$; $Y[j] := Y[j + 1]$; $Y[j + 1] := Temporary$;   where $Y[k]$ is the element to be associated with $N[k]$. Other variations are obviously possible.

REMARK ON ALGORITHM 175
SHUTTLE SORT [C. J. Shaw and T. N. Trimble, *Comm. ACM 6*, June 1963]

O. C. JUELICH
North American Aviation, Inc., Columbus, Ohio

The authors of this algorithm do well to remind the reader that "Shuttle Sort" is not an efficient procedure, except for lists of items so short that they do not justify the housekeeping apparatus needed by the usual sorting routines.

The algorithm as published is not free from errors. The statement

    for $j := i$ **step** -1 **until** 1 **do**

should be replaced by either:

    for $j := m - 1$ **step** -1 **until** i **do**

or

    for $j := 1$ **step** 1 **until** $m - i$ **do**

In the former case the process can be visualized as placing the ith smallest element in place on the ith pass; in the latter the ith largest element is put in place on the ith pass.

The label *"Test"* should precede the delimiter "**end** of j loop" rather than the "**end** of i loop". The algorithm can be slightly accelerated by rewriting the body of the procedure:

```
begin integer i, j, j max;
    i := m - 1;
loop:  j max := 1;
    for j := 1 step 1 until i do
      begin
      compare:  if N[j] > N[j + 1] then
        begin Exchange:  Temporary := N[j];
          N[j] := N[j + 1];
          N[j + 1] := Temporary;
          j max := j
        end Exchange;
      end of j loop;
    i := j max;
```

if $i > 1$ **then go to** *loop*;
end *shuttle sort*

The revised procedure body will eliminate redundant iterations when some of the data is already ordered.

It was studied in this form by R. L. Boyell and the writer on the ORDVAC at Ballistics Research Laboratories, Aberdeen Proving Ground, in 1955. For randomly ordered data the i-loop may be expected to be executed about $m - \sqrt{m}$ times.

REMARK ON ALGORITHM 175

SHUTTLE SORT [C. J. Shaw and T. N. Trimble, *Comm. ACM 6* (June 1963), 312; G. R. Schubert, *Comm. ACM 6* (Oct. 1963), 619; O. C. Juelich, *Comm. ACM 6* (Dec. 1963), 739]

OTTO C. JUELICH (Recd. 18 Dec. 1963)

North American Aviation, 4300 E. Fifth Ave., Columbus, Ohio

The appearance of Schubert's certification has caused me to restudy the algorithm. What I supposed were errors amount to a rearrangement of the order in which the comparisons are carried out. The efficiency of the algorithm is not much affected by the rearrangement, since the number of executions of the statements labeled *Exchange* remains the same.

ALGORITHM 201
SHELLSORT
J. BOOTHROYD

English Electric-Leo Computers, Kidsgrove, Staffs, England

procedure *Shellsort* (a, n); **value** n; **real array** a; **integer** n;
comment $a[1]$ through $a[n]$ of $a[1:n]$ are rearranged in ascending order. The method is that of D. A. Shell, (A high-speed sorting procedure, *Comm. ACM* 2 (1959), 30–32) with subsequences chosen as suggested by T. N. Hibberd (An empirical study of minimal storage sorting, SDC Report SP-982). Subsequences depend on m_1 the first operative value of m. Here $m_1 = 2^k - 1$ for $2^k \leq n < 2^{k+1}$. To implement Shell's original choice of $m_1 = [n/2]$ change the first statement to $m := n$;

```
begin integer i, j, k, m; real w;
  for i := 1 step i until n do m := 2 × ι − 1;
  for m := m ÷ 2 while m ≠ 0 do
    begin k := n − m;
      for j := 1 step 1 until k do
        begin for i := j step −m until 1 do
          begin if a[i+m] ≧ a[i] then go to 1;
            w := a[i];  a[i] := a[i+m];  a[i+m] := w;
          end i;
        1: end j
      end m
end Shellsort;
```

CERTIFICATION OF ALGORITHM 201
SHELLSORT [J. BOOTHROYD, *Comm. ACM 6* (Aug. 1963), 445]

M. A. Batty   (Recd 27 Jan. 1964)

English Electric Co., Whetstone, Nr. Leicester, England

This algorithm has been tested successfully using the Deuce Algol Compiler. When the first statement of the algorithm was replaced by the statement

$$m := n;$$

to implement Shell's original choice of $m_1 := n/2$, a slight increase in sorting time was observed with most of the cases tested.

REMARK ON ALGORITHM 201 [M1]
SHELLSORT [J. Boothroyd, *Comm. ACM 6* (Aug. 1963), 445]

J. P. Chandler and W. C. Harrison* (Recd. 19 Sept. 1969)

Department of Physics, Florida State University, Tallahassee, FL 32306

KEY WORDS AND PHRASES: sorting, minimal storage sorting, digital computer sorting
CR CATEGORIES: 5.31

* This work was supported in part by AEC Contract No. AT-(40-1)-3509. Computational costs were supported in part by National Science Foundation Grant GJ 367 to the Florida State University Computing Center.

Hibbard [1] has coded this method in a way that increases the speed significantly. In SHELLSORT, each stage of each sift consists of successive pair swaps. The modification replaces each set of n pair swaps by one "save," n − 1 moves, and one insertion.

Table I gives timing information for ALGOL, FORTRAN, and COMPASS (assembly language) versions of SHELLSORT and the modified version (called SHELLSORT2), for the CDC 6400 computer. The savings in time achieved by the modification are 32%, 17%, and 21%, respectively. The savings are greater than this when vectors of more than one word each are being sorted.

The comparative execution times of the ALGOL and FORTRAN versions, for these compilers, are quite interesting.

TABLE I. SORTING TIMES IN SECONDS FOR 10,000 RANDOMLY ORDERED NUMBERS ON THE CDC 6400 COMPUTER

Algorithm	Source Language		
	ALGOL	FORTRAN	COMPASS
SHELLSORT	53.40	7.18	2.38
SHELLSORT2	36.56	5.98	1.87

REFERENCES:

1. HIBBARD, T. N.   An empirical study of minimal storage sorting. *Comm. ACM 6* (May 1963), 206.

ALGORITHM 207
STRINGSORT
J. BOOTHROYD

English Electric-Leo Computers, Ltd.

Staffordshire, England

procedure *stringsort* (a, n); **comment** elements $a[1] \cdots a[n]$ of $a[1:2n]$ are sorted into ascending sequence using $a[n+1] \cdots a[2n]$ as auxiliary storage. Von Neumann extended string logic is employed to merge input strings from both ends of a sending area into output strings which are sent alternately to either end of a receiving area. The procedure takes advantage of naturally occurring ascending or descending order in the original data;

```
value n;   integer n;   array a;
begin integer d, i, j, m, u, v, z;   integer array c[−1:1];
   switch p := jz1, str i;   switch q := merge, jz2;
oddpass:  i := 1;   j := n;   c[−1] := n + 1;   c[1] := 2 × n;
allpass:  d := 1;   go to firststring;
merge:   if a[i] ≧ a[z]
         then begin go to p[v];
            jz1:   if a[j] ≧ a[z]
                   then ij:  begin if a[i] ≧ a[j]
                                   then str j: begin a[m] := a[j];
                                      j := j − 1 end
                                   else str i: begin a[m] := a[i]:
                                      i := i + 1 end
                             end
                   else begin v := 2;   go to str i end
            end
         else begin u := 2;
            jz2:   if a[j] ≧ a[z]
                   then go to str j
                   else begin d := −d;   c[d] := m;
                      firststring:   m := c[−d];
                         v := u := 1;
                      go to ij
                   end
            end;
z := m;   m := m + d;   if j ≧ i then go to q[u];
if m > n + 1 then begin  comment evenpass;   i := n + 1;
               j := 2 × n;   c[−1] := 1;   c[1] := n;   go to
            allpass end
         else if m < n + 1 then go to oddpass
end stringsort;
```

CERTIFICATION OF ALGORITHM 207 [M1]
STRINGSORT [J. Boothroyd, *Comm. ACM* 6 (Oct. 1963), 615]

CHARLES R. BLAIR (Recd. 31 Jul. 1964)
Department of Defense, Washington 25, D. C.

STRINGSORT compiled and ran successfully without correction on the ALDAP translator for the CDC 1604A. The following sorting times were observed.

Number of Items	Time in Seconds
10	0.03
20	0.05
50	0.20
100	0.38
200	1.03
500	3.22
1000	6.43
2000	12.85
5000	38.72
10000	90.72

ALGORITHM 232

HEAPSORT

J. W. J. WILLIAMS   (Recd 1 Oct. 1963 and, revised, 15 Feb. 1964)

Elliott Bros. (London) Ltd., Borehamwood, Herts, England

comment   The following procedures are related to *TREESORT* [R. W. Floyd, Alg. 113, *Comm. ACM 5* (Aug. 1962), 434, and A. F. Kaupe, Jr., Alg. 143 and 144, *Comm. ACM 5* (Dec. 1962), 604] but avoid the use of pointers and so preserve storage space. All the procedures operate on single word items, stored as elements 1 to n of the array A. The elements are normally so arranged that $A[i] \leq A[j]$ for $2 \leq j \leq n$, $i = j \div 2$. Such an arrangement will be called a heap. $A[1]$ is always the least element of the heap.

The procedure *SETHEAP* arranges n elements as a heap, *INHEAP* adds a new element to an existing heap, *OUTHEAP* extracts the least element from a heap, and *SWOPHEAP* is effectively the result of *INHEAP* followed by *OUTHEAP*. In all cases the array A contains elements arranged as a heap on exit.

SWOPHEAP is essentially the same as the tournament sort described by K. E. Iverson—*A Programming Language*, 1962, pp. 223–226—which is a top to bottom method, but it uses an improved storage allocation and initialisation. *INHEAP* resembles *TREESORT* in being a bottom to top method. *HEAPSORT* can thus be considered as a marriage of these two methods.

The procedures may be used for replacement-selection sort-ing, for sorting the elements of an array, or for choosing the current minimum of any set of items to which new items are added from time to time. The procedures are the more useful because the active elements of the array are maintained densely packed, as elements $A[1]$ to $A[n]$;

procedure *SWOPHEAP* (A,n,in,out);
    value in,n;   **integer** n;   **real** in,out;   **real array** A;
    comment *SWOPHEAP* is given an array A, elements $A[1]$ to $A[n]$ forming a heap, $n \geq 0$. *SWOPHEAP* effectively adds the element *in* to the heap, extracts and assigns to *out* the value of the least member of the resulting set, and leaves the remaining elements in a heap of the original size. In this process elements 1 to $(n+1)$ of the array A may be dis-turbed. The maximum number of repetitions of the cycle labeled *scan* is $log_2 n$;
  begin integer i,j;   **real** *temp*, *temp* 1;
    if $in \leq A[1]$ **then** $out := in$ **else**
    begin $i := 1$;
      $A[n+1] := in$;   **comment** this last statement is only necessary in case $j = n$ at some stage, or $n = 0$;
      $out := A[1]$;
    scan: $j := i+i$;
      if $j \leq n$ **then**
      begin *temp* $:= A[j]$;
        temp 1 $:= A[j+1]$;
        if *temp* 1 < *temp* **then**
        begin *temp* $:= temp$ 1;
          $j := j+1$
        end;
        if *temp* < *in* **then**
        begin $A[i] := temp$;
          $i := j$;
          go to *scan*
        end
      end;
      $A[i] := in$
    end
  end *SWOPHEAP*;

procedure *INHEAP* (A, n, in);
    value in;   **integer** n;   **real** in;   **real array** A;
    comment *INHEAP* is given an array A, elements $A[1]$ to $A[n]$ forming a heap and $n \geq 0$. *INHEAP* adds the element *in* to the heap and adjusts n accordingly. The cycle labeled

scan may be repeated log_2n times, but on average is repeated twice only;
begin integer i,j;
  $i := n :=n+1$;
scan: **if** $i>1$ **then**
  begin $j := i\div2$;
    if $in<A[j]$ **then**
    begin $A[i] := A[j]$;
      $i := j$;
      go to *scan*
      end
    end;
    $A[i] := in$
  end *INHEAP*;
procedure *OUTHEAP* (A,n,out);
  integer n; **real** *out*; **real array** A;
  comment given array A, elements 1 to n of which form a heap, $n\geq1$, *OUTHEAP* assigns to *out* the value of $A[1]$, the least member of the heap, and rearranges the remaining members as elements 1 to $n-1$ of A. Also, n is adjusted accordingly;
  begin *SWOPHEAP* $(A,n-1, A[n],out)$;
    $n := n-1$
  end *OUTHEAP*;
procedure *SETHEAP* (A,n);
  value n; **integer** n; **real array** A;
  comment *SETHEAP* rearranges the elements $A[1]$ to $A[n]$ to form a heap;
  begin integer j;
    $j := 1$;
  L: $INHEAP(A,j,A[j+1])$;
    if $j<n$ **then go to** L
  end *SETHEAP*

ALGORITHM 245
TREESORT 3 [M1]
Robert W. Floyd (Recd. 22 June 1964 and 17 Aug. 1964)
Computer Associates, Inc., Wakefield, Mass.

procedure *TREESORT* 3 (M, n);
  value n; **array** M; **integer** n;
  comment *TREESORT* 3 is a major revision of *TREESORT* [R. W. Floyd, Alg. 113, *Comm. ACM 5* (Aug. 1962), 434] sug-

gested by *HEAPSORT* [J. W. J. Williams, Alg. 232, *Comm. ACM 7* (June 1964), 347] from which it differs in being an in-place sort. It is shorter and probably faster, requiring fewer comparisons and only one division. It sorts the array $M[1:n]$, requiring no more than $2 \times (2\uparrow p-2) \times (p-1)$, or approximately $2 \times n \times (\log_2(n)-1)$ comparisons and half as many exchanges in the worst case to sort $n = 2\uparrow p - 1$ items. The algorithm is most easily followed if M is thought of as a tree, with $M[j\div2]$ the father of $M[j]$ for $1 < j \leqq n$;

```
begin
  procedure exchange (x,y); real x,y;
    begin real t; t := x; x := y; y := t
    end exchange;
  procedure siftup (i,n); value i, n; integer i, n;
  comment   M[i] is moved upward in the subtree of M[1:n] of
    which it is the root;
  begin real copy; integer j;
    copy := M[i];
loop: j := 2 × i;
    if j ≦ n then
    begin if j < n then
        begin if M[j+1] > M[j] then j := j + 1 end;
      if M[j] > copy then
        begin M[i] := M[j];  i := j;  go to loop end
    end;
    M[i] := copy
  end siftup;
  integer i;
  for i := n÷2 step −1 until 2 do siftup (i,n);
  for i := n step −1 until 2 do
  begin siftup (1,i);
    comment   M[j÷2] ≧ M[j] for 1 < j ≦ i;
    exchange (M[1], M[i]);
    comment   M[i:n] is fully sorted;
  end
end TREESORT 3
```

CERTIFICATION OF ALGORITHM 245 [M1]

TREESORT 3 [Robert W. Floyd, *Comm. ACM 7* (Dec. 1964), 701]

PHILIP S. ABRAMS (Recd. 14 Jan. 1965)

Computation Center, Stanford University, Stanford, California

The procedure *TREESORT* 3 was translated into B5000 Extended Algol and tested on the Burroughs B5500. Tests were run on arrays of length 50 to 1000 in steps of 50. For each array size, 50 random arrays were generated, sorted, timed and checked for sequencing. No corrections were required and the procedure gave correct results for all cases tested.

exchange is unnecessary as a separate procedure, since it is used at only one place in *TREESORT* 3. Sorts were found to run significantly faster when the body of *exchange* was inserted in the appropriate place, than when run with the algorithm as published.

CERTIFICATION OF ALGORITHM 245 [M1]
TREESORT 3 [Robert W. Floyd, *Comm. ACM 7* (Dec. 1964), 701]: PROOF OF ALGORITHMS—A NEW KIND OF CERTIFICATION

Ralph L. London* (Recd. 27 Feb. 1969 and 8 Jan. 1970)
Computer Sciences Department and Mathematics Research Center, University of Wisconsin, Madison, WI 53706

ABSTRACT: The certification of an algorithm can take the form of a proof that the algorithm is correct. As an illustrative but practical example, Algorithm 245, *TREESORT 3* for sorting an array, is proved correct.

KEY WORDS AND PHRASES: proof of algorithms, debugging, certification, metatheory, sorting, in-place sorting
CR CATEGORIES: 4.42, 4.49, 5.24, 5.31

Certification of algorithms by proof. Since suitable techniques now exist for proving the correctness of many algorithms [for example, 3–7], it is possible and appropriate to certify algorithms with a proof of correctness. This certification would be in addition to, or in many cases instead of, the usual certification. Certification by testing still is useful because it is easier and because it also provides, for example, timing data. Nevertheless the existence of a proof should be welcome additional certification of an algorithm. The proof shows that an algorithm is debugged by showing conclusively that no bugs exist.

* This work was supported by NSF Grant GP-7069 and the Mathematics Research Center, US Army under Contract Number DA-31-124-ARO-D-462.

It does not matter whether all users of an algorithm will wish to, or be able to, verify a sometimes lengthy proof. One is not required to accept a proof before using the algorithm any more than one is expected to rerun the certification tests. In both cases one could depend, in part at least, upon the author and the referee.

As an example of a certification by proof, the algorithm *TREESORT 3* [2] is proved to perform properly its claimed task of sorting an array $M[1:n]$ into ascending order. This algorithm has been previously certified [1], but in that certification, for example, no arrays of odd length were tested. Since *TREESORT 3* is a fast practical algorithm for in-place sorting and one with sufficient complexity so that its correctness is not immediately apparent, its use as the example is more than an abstract exercise. It is an example of considerable practical importance.

Outline of TREESORT 3 and method of proof. The algorithm is most easily followed if the array is viewed as a binary tree. $M[k \div 2]$ is the parent of $M[k]$, $2 \leq k \leq n$. In other words the children of $M[j]$ are $M[2j]$ and $M[2j+1]$ provided one or both of the children exist.

The first part of the algorithm permutes the M array so that for a segment of the array, each parent is larger than both of the children (one child if the second does not exist). Each call of the auxiliary procedure *siftup* enlarges the segment by causing one more parent to dominate its children. The second part of the algorithm uses *siftup* to make the parents larger over the whole array, exchanges $M[1]$ with the last element and repeats on an array one element shorter. The above statements are motivation and not part of the formal proof.

That *TREESORT 3* is correct is proved in three parts. First the procedure *siftup* is shown to perform as it is formally defined below. Then the body of *TREESORT 3*, which uses *siftup* in two ways, is shown to sort the array into ascending order. (The proof of the procedure *exchange* is omitted.) The proofs are by a method described in [3, 4, 7]: assertions concerning the progress of the computation are made between lines of code, and the proof consists of demonstrating that each assertion is true each time control reaches that assertion, under the assumption that the previously encountered assertions are true. Finally termination of the algorithm is shown separately.

The lines of the original algorithm have been numbered and the assertions, in the form of program comments, are numbered correspondingly. The numbers are used only to refer to code and to assertions and have no other significance. One extra begin-end pair has been inserted into the body of *TREESORT 3* in order that the control points of two assertions (3.1 and 4.1) could be distinguished. In *siftup* the assertions 10.1 and 10.2 express the correct result; in the body of *TREESORT 3* the assertions 9.3 and 9.4 do likewise.

Definition of siftup and notation. We now define formally the procedure $siftup(i,n)$, where n is a formal parameter and not the length of the array M. Let $A(s)$ denote the set of inequalities $M[k \div 2] \geq M[k]$ for $2s \leq k \leq n$. (If $s > n \div 2$, then $A(s)$ is a vacuous statement.) If $A(i+1)$ holds before the call of $siftup(i,n)$ and if $1 \leq i \leq n \leq$ *array size*, then after $siftup(i,n)$:

(1) $A(i)$ holds;

(2) the segment of the array $M[i]$ through $M[n]$ is permuted; and

(3) the segment outside $M[i]$ through $M[n]$ is unaltered.

In order to prove these properties of *siftup*, some notation is required. The formal parameter i will be changed inside *siftup*. Since i is called by value, that change will be invisible outside *siftup*. Nevertheless it is necessary to use the initial value of i as well as the current value of i in the proof of *siftup*. Let i_0 denote the value of i upon entry to *siftup*.

Similarly let M_0 denote the M array upon entry to *siftup*. The notation "$M = p(M_0)$ with $M := copy$" means "if $M[i] := copy$ were done, M is some permutation of M_0 as described in (2) and (3) of the definition of *siftup*." "$M = p(M_0)$" means the same without the reference to $M[i] := copy$ being done.

Code and assertions for siftup.

```
0    procedure siftup(i, n);  value i, n;  integer i, n;
1    begin real copy;  integer j;
        comment
            1.1: 1 ≤ i₀ = i ≤ n ≤ array size
            1.2: A(i₀+1)
            1.3: M = p(M₀);
2    copy := M[i];
3    loop: j := 2 × i;
        comment
            3.1: i ≤ n
            3.2: 2i = j
            3.3: i = i₀ or i ≥ 2i₀
            3.4: M = p(M₀) with M[i] := copy
            3.5: A(i₀) or (i = i₀ and A(i₀+1))
            3.6: M[i÷2] > copy or i = i₀
            3.7: M[i÷2] ≥ M[i] or i = i₀;
```

```
4      if j ≤ n then
5      begin if j < n then
6a       begin if M[j+1] > M[j] then
6b          j := j + 1 end;
```
         comment
            6.1: $i = j \div 2$
            6.2: $2i \leq j \leq n$
            6.3: $i = i_0$ or $i \geq 2i_0$
            6.4: $M = p(M_0)$ with $M[i] := copy$
            6.5: $A(i_0)$ or $(i = i_0$ and $A(i_0+1))$
            6.6: $M[i \div 2] > copy$ or $i = i_0$
            6.7: $M[i \div 2] \geq M[i]$ or $i = i_0$
            6.8: $(2i < n$ and $M[j] = \max(M[2i], M[2i+1]))$ or
                 $(2i = n$ and $M[j] = M[n])$
            6.9: $M[i] \geq M[j]$ or $i = i_0$;
```
7      if M[j] > copy then
8a       begin M[i] := M[j];
```
         comment
            8.1: $i = i_0$ or $i \geq 2i_0$
            8.2: $2i \leq j \leq n$
            8.3: $M[j \div 2] = M[i] = M[j] > copy$
            8.4: $M[i \div 2] \geq M[j]$ or $i = i_0$
            8.5: $M = p(M_0)$ with $M[j] := copy$
            8.6: $A(i_0)$;
```
8b         i := j;
```
         comment
            8.7: $i \geq 2i_0$
            8.8: $i = j \leq n$
            8.9: $M[i \div 2] > copy$
            8.10: $M[i \div 2] \geq M[i]$
            8.11: $M = p(M_0)$ with $M[i] := copy$
            8.12: $A(i_0)$;
```
8c         go to loop end
9      end;
```
         comment
            9.1: $M[j] \leq copy$ if reached from 7 or
                 $2i = j > n$ if reached from 4;
```
10     M[i] := copy;
```
         comment
            10.1: $M = p(M_0)$
            10.2: $A(i_0)$;
```
11   end siftup;
```

Verification of the assertions of siftup.   Reasons for the truth of each assertion follow:

1.1–1.2: Assumptions for using *siftup*.

1.3: p is the identity permutation.

3.1–3.7: If reached from 2,

> 3.1: 1.1.
>
> 3.2: 3.
>
> 3.3, 3.5–3.7: $i = i_0$ by 1.1.   3.5 also requires 1.2.
>
> 3.4: 1.3 and 2.

> If reached from 8, respectively, 8.8, 3, 8.7, 8.11, 8.12, 8.9 and 8.10.

6.1: At 3.2 $j = 2i$ and by 6b, j might be $2i + 1$. $i = j \div 2$ in either case.

6.2: After 4, $j \leq n$. j is altered from 3.1 to 6.2 only at 6b. Before 6b, $j < n$ by 5. Hence $j \leq n$ at 6.2. $2i \leq j$ by 6.1.

6.3–6.7: 3.3–3.7, respectively.

6.8: If 4 is **true** and 5 is **false**, $j = 2i = n$ (using 3.2) so the second clause of 6.8 holds. If 4 is **true** and 5 is **true**, then at 6a, $2i = j < n$ (using 3.2) so $M[j+1] = M[2i+1]$ is defined. Now at 6.8, $j = 2i$ or $j = 2i+1$. In either case, by 6a and 6b, the first clause of 6.8 holds.

6.9: By 6.5 $i \neq i_0$ gives $A(i_0)$. $2i_0 \leq 2i \leq j \leq n$ by 6.3 and 6.2. Hence $A(i_0)$ and 6.1 give $M[i] = M[j \div 2] \geq M[j]$.

8.1: 6.3.

8.2: 6.2.

8.3: $i = j \div 2$ by 6.1, $M[i] = M[j]$ by 8a and $M[j] > copy$ by 7.

8.4: 6.7 and 6.9.

8.5: 6.4 requires that $M[i]$ be replaced by *copy*. Since $M[i] = M[j]$ by 8a, $M[j]$ may equally well be replaced with *copy*. 8.1 and 8.2 give $i_0 \leq i \leq n$ so that the change to M at 8a is in the segment $M[i_0]$ through $M[n]$.

8.6: By 8a and if 6.8 (first clause) holds, $M[i] \geq M[2i]$ and $M[i] \geq M[2i+1]$. By 8a and if 6.8 (second clause) holds, $M[i] = M[j] = M[n] = M[2i]$ and $M[2i+1]$ does not exist for this call of *siftup*. $A(i_0+1)$ holds at 6.5 since $A(i_0)$ implies $A(i_0+1)$. If $i = i_0$, $A(i_0+1)$ and the relations above on $M[i]$ give $A(i_0)$. If $i \neq i_0$, then 8a, 8.4, $A(i_0)$ at 6.5 and the relations above on $M[i]$ give $A(i_0)$ at 8.6.

8.7: 8b, 8.1 and 8.2.

8.8: 8b and 8.2.

8.9: 8b and 8.3.

8.10: At 8.6, $2i_0 \leq j \leq n$ by 8.1 and 8.2. Hence by 8.6, $M[j \div 2] \geq M[j]$. Use 8b on $M[j \div 2] \geq M[j]$.

8.11: 8b and 8.5.

8.12: 8.6.

9.1: 9.1 is reached only if 7 is **false** or if 4 is **false**. $2i = j$ by 3.2.

10.1–10.2:   If reached from 7,

10.1: 6.4 and 10. (6.2 and 6.3 give $i_0 \leq i \leq n$ ensuring the change to M at 10 is in the segment $M[i_0]$ through $M[n]$.)

10.2: By 10, 9.1, 6.2 and 6.8, $M[i] = copy \geq M[j] \geq M[2i]$ and, if $M[2i+1]$ exists, $M[j] \geq M[2i+1]$. If $i = i_0$, 10.2 follows as in 8.6. If $i \neq i_\bullet$, 6.6 and 10 give $M[i \div 2] > copy = M[i]$. $A(i_0)$ at 6.5 now gives $A(i_0)$ at 10.2.

If reached from 4,

10.1: 3.4 and 10. (3.1 and 3.3 give $i_0 \leq i \leq n$.)

10.2: $2i > n$ means no relations in $A(i_\bullet)$ of the form $M[i] \geq \cdots$. If $i = i_0$, 3.5 gives 10.2. If $i \neq i_\bullet$, 3.6 and 10 give $M[i \div 2] > copy = M[i]$. $A(i_0)$ at 3.5 now gives 10.2.

Code and assertions for the body of TREESORT 3.

```
0    integer i;
     comment
        0.1: A(n÷2+1);
1    for i := n÷2 step −1 until 2 do
2    begin
        comment
           2.1: A(i+1)
           2.2: Assumptions of siftup satisfied;
3       siftup(i,n);
        comment
           3.1: A(i);
4    end;
     comment
        4.1: M[p] ≤ M[p+1] for n + 1 ≤ p ≤ n − 1
        4.2: A(2), i.e. M[k÷2] ≥ M[k] for 4 ≤ k ≤ n;
5    for i := n step −1 until 2 do
6    begin
        comment
           6.1: M[p] ≤ M[p+1] for i + 1 ≤ p ≤ n − 1
           6.2: M[k÷2] ≥ M[k] for 4 ≤ k ≤ i
           6.3: M[i+1] ≥ M[r] for 1 ≤ r ≤ i
           6.4: Assumptions of siftup satisfied;
7       siftup (1,i);
        comment
           7.1: M[p] ≤ M[p+1] for i + 1 ≤ p ≤ n − 1
           7.2: M[k÷2] ≥ M[k] for 2 ≤ k ≤ i
           7.3: M[1] ≥ M[r] for 2 ≤ r ≤ i
           7.4: M[i+1] ≥ M[1];
```

8   *exchange* $(M[1], M[i])$;
       comment
           8.1: $M[i] \geq M[r]$ for $1 \leq r \leq i - 1$
           8.2: $M[p] \leq M[p+1]$ for $i \leq p \leq n - 1$
           8.3: $M[k \div 2] \geq M[k]$ for $4 \leq k \leq i - 1$;
9   **end**;
       comment
           9.1: $M[p] \leq M[p+1]$ for $2 \leq p \leq n - 1$
           9.2: $M[2] \geq M[1]$
           9.3: $M[p] \leq M[p+1]$ for $1 \leq p \leq n - 1$, i.e. M is fully ordered
           9.4: M is a permutation of M_0;

Verification of the assertions for the body of TREESORT 3.
Reasons for the truth of each assertion follow:

0.1:   Vacuous statement since $2(n \div 2 + 1) > n$.
2.1:   If reached from 0.1, by 1 substitute $i = n \div 2$ in 0.1.
       If reached from 3.1, by 1 substitute $i = i + 1$ in 3.1 to account for the change in i from 3.1 to 2.1.
2.2:   2.1, the bound on i implied by 1 and the array size being n.
3.1:   2.1 and the definition of *siftup* (i, n).
4.1:   Vacuous statement.
4.2:   If $n \geq 4$, 3 is executed; hence 3.1 with $i = 2$. If $n \leq 3$, vacuous statement.
6.1–6.3: If reached from 4.1,
               6.1–6.2: By 5 substitute $i = n$ in 4.1 and 4.2.
               6.3: Vacuous statement for $i = n$.
           If reached from 8.1, by 5 substitute $i = i + 1$ in 8.2, 8.3 and 8.1, respectively.
6.4:   5 and 6.2, i.e. $A\,(2)$ for the subarray $M[1:i]$.
7.1:   6.1 and (3) of *siftup*.
7.2:   6.2 and (1) of *siftup*.
7.3:   7.2 noting that $M[1] = M[k \div 2]$ if $k = 2$ and using the transitivity of $\geq$.
7.4:   Vacuous for $i = n$. Otherwise 6.3 for the appropriate r since by (2) of *siftup*, $M[1]$ at 7.3 is one of the $M[r]$, $1 \leq r \leq i$, at 6.3.
8.1:   7.3 with the changes caused by 8 (only $M[1]$ and $M[i]$ are altered by 8).
8.2:   By 8 substitute $M[i]$ for $M[1]$ in 7.4; then 7.1 also holds for $p = i$.
8.3:   7.2 excluding only the one or two relations $M[1] \geq \cdots$, and the one relation $\cdots \geq M[i]$.

9.1–9.3: If $n \geq 2$, 8 is executed;
  9.1: 8.2 with $i = 2$.
  9.2: 8.1 with $i = 2$.
  9.3: 9.1 and 9.2.
  If $n \leq 1$, 9.1–9.3 are vacuous statements.
9.4: The only operations done to M are *siftup* and *exchange* all of
  which leave M as a permutation of M_0 .

Proof of termination of TREESORT 3. Provided *siftup* and *exchange* terminate, it is clear that *TREESORT 3* terminates. Note that each parameter of *siftup* is called by value so that i is not changed in the body of the for loops.

The procedure *exchange* certainly terminates. In *siftup* the only possibility for an unending loop is from 3 to 8b and back to 3. Note that all changes to i (only at 8b) and to j (only at 3 and 6b) occur in this loop and that on each cycle of this loop both i and j are changed. By the test at 4, it is sufficient to show that j strictly increases in value. $i \geq 1$ means $2i > i$. At 8b, $j = i < 2i$ while at 3, $j = 2i$, i.e. j(at 3) $= 2i > i = j$(at 8b). Hence each setting to j at 3 strictly increases the value of j. The only other setting to j (at 6b), if made, similarly increases the value of j.

REFERENCES:

1. ABRAMS, P. S.   Certification of Algorithm 245. *Comm. ACM 8* (July 1965), 445.
2. FLOYD, R. W.   Algorithm 245, TREESORT 3. *Comm. ACM 7* (Dec. 1964), 701.
3. FLOYD, R. W.   Assigning meanings to programs. Proc. of a Symposium in Applied Mathematics, Vol. 19—Mathematical Aspects of Computer Science, J. T. Schwartz (Ed.), American Math. Society, Providence, R. I., 1967, pp. 19–32.
4. KNUTH, D. E.   *The Art of Computer Programming, Vol. 1—Fundamental Algorithms.* Addison-Wesley, Reading, Mass., 1968, Sec. 1.2.1.
5. McCARTHY, J.   A basis for a mathematical theory of computation. In *Computer Programming and Formal Systems*, P. Braffort and D. Hirschberg (Eds.), North Holland, Amsterdam, 1963, pp. 33–70.
6. McCARTHY, J., AND PAINTER, J. A.   Correctness of a compiler for arithmetic expressions. Proc. of a Symposium in Applied Mathematics, Vol. 19—Mathematical Aspects of Computer Science, J. T. Schwartz (Ed.), American Math. Society, Providence, R. I., 1967, pp. 33–41.
7. NAUR, P.   Proof of algorithms by general snapshots. *BIT 6* (1966), 310–316.

ALGORITHM 271
QUICKERSORT [M1]
R. S. Scowen* (Recd. 22 Mar. 1965 and 30 June 1965)
National Physical Laboratory, Teddington, England

~   * The work described below was started while the author was
at English Electric Co. Ltd, completed as part of the research
programme of the National Physical Laboratory and is published
by permission of the Director of the Laboratory.

procedure *quickersort*(a, j);
   value j;   **integer** j;   **array** a;
begin integer i, k, q, m, p;   **real** t, x;   **integer array** ut,
   $lt[1:ln(abs(j)+2)/ln(2)+0.01]$;

comment   The procedure sorts the elements of the array $a[1:j]$
   into ascending order. It uses a method similar to that of QUICK-
   SORT by C. A. R. Hoare [1], i.e., by continually splitting the
   array into parts such that all elements of one part are less than
   all elements of the other, with a third part in the middle consisting of a single element. I am grateful to the referee for point-
   ing out that QUICKERSORT also bears a marked resemblance
   to sorting algorithms proposed by T. N. Hibbard [2, 3]. In particular, the elimination of explicit recursion by choosing the
   shortest sub-sequence for the secondary sort was introduced by
   Hibbard in [2].
      An element with value t is chosen arbitrarily (in QUICKER-
   SORT the middle element is chosen, in QUICKSORT a random
   element is chosen). i and j give the lower and upper limits of
   the segment being split. After the split has taken place a value
   q will have been found such that $a[q] = t$ and $a[I] \leq t \leq a[J]$
   for all I, J such that $i \leq I < q < J \leq j$. The program then
   performs operations on the two segments $a[i:q-1]$ and $a[q+1:j]$
   as follows. The smaller segment is split and the position of the
   larger segment is stored in the lt and ut arrays (lt and ut are
   mnemonics for lower temporary and upper temporary). If the
   segment to be split has two or fewer elements it is sorted and
   another segment obtained from the lt and ut arrays. When no
   more segments remain, the array is completely sorted.

   $i := m := 1$;
N: **if** $j-i > 1$ **then**
   begin comment   This segment has more than two elements,
      so split it;
      $p := (j+i) \div 2$;

comment p is the position of an arbitrary element in the segment $a[i:j]$. The best possible value of p would be one which splits the segment into two halves of equal size, thus if the array (segment) is roughly sorted, the middle element is an excellent choice. If the array is completely random the middle element is as good as any other.

If however the array $a[1:j]$ is such that the parts $a[1:j \div 2]$ and $a[j \div 2+1:j]$ are both sorted the middle element could be very bad. Accordingly in some circumstances $p := (i+j) \div 2$ should be replaced by $p := (i+3\times j) \div 4$ or $p := RANDOM(i, j)$ as in QUICKSORT;

$t := a[p]$;
$a[p] := a[i]$;
$q := j$;
for $k := i + 1$ **step** 1 **until** q **do**
begin comment Search for an element $a[k] > t$ starting from the beginning of the segment;
 if $a[k] > t$ **then**
 begin comment Such an $a[k]$ has been found;
  for $q := q$ **step** -1 **until** k **do**
  begin comment Now search for $a[q] < t$ starting from the end of the segment;
   if $a[q] < t$ **then**
   begin comment $a[q]$ has been found, so exchange $a[q]$ and $a[k]$;
   $x := a[k]$;
   $a[k] := a[q]$;
   $a[q] := x$;
   $q := q-1$;
   comment Search for another pair to exchange;
   go to L
   end
  end for q;
  $q := k - 1$;
  comment q was undefined according to Para. 4.6.5 of the Revised ALGOL 60 Report [*Comm. ACM 6* (Jan. 1963), 1–17];
  go to M
 end;
L: **end for** k;
comment We reach the label M when the search going upwards meets the search coming down;
M: $a[i] := a[q]$;
 $a[q] := t$;

comment The segment has been split into the three parts (the middle part has only one element), now store the position of the largest segment in the lt and ut arrays and reset i and j to give the position of the next largest segment;

if $2 \times q > i + j$ **then**

begin

$\quad lt[m] := i;$

$\quad ut[m] := q-1;$

$\quad i := q+1$

end

else

begin

$\quad lt[m] := q+1;$

$\quad ut[m] := j;$

$\quad j := q-1$

end;

comment Update m and split this new smaller segment;

$m := m+1;$

go to N

end

else if $i \geq j$ **then**

begin comment This segment has less than two elements;

$\quad$ **go to** P

end

else

begin comment This is the case when the segment has just two elements, so sort $a[i]$ and $a[j]$ where $j = i + 1;$

$\quad$ **if** $a[i] > a[j]$ **then**

$\quad$ **begin**

$\qquad x := a[i];$

$\qquad a[i] := a[j];$

$\qquad a[j] := x$

$\quad$ **end**;

comment If the lt and ut arrays contain more segments to be sorted then repeat the process by splitting the smallest of these. If no more segments remain the array has been completely sorted;

$P:\ m := m-1;$

if $m > 0$ **then**

begin

$\quad i := lt[m];$

$\quad j := ut[m];$

```
      go to N
   end;
  end
end quickersort
```

REFERENCES:

1. HOARE, C. A. R. Algorithms 63 and 64. *Comm. ACM 4* (July 1961), 321.
2. HIBBARD, THOMAS N. Some combinatorial properties of certain trees with applications to searching and sorting. *J. ACM 9* (Jan. 1962), 13.
3. ——. An empirical study of minimal storage sorting. *Comm. ACM 6* (May 1963), 206–213;

CERTIFICATION OF ALGORITHM 271 (M1)
QUICKERSORT [R. S. Scowen, *Comm. ACM 8* (Nov. 1965), 669]

CHARLES R. BLAIR (Recd. 11 Jan. 1966)
Department of Defense, Washington, D.C.

QUICKERSORT compiled and ran without correction through the ALDAP translator for the CDC 1604A. Comparison of average sorting times, shown in Table I, with other recently published algorithms demonstrates *QUICKERSORT*'s superior performance.

TABLE I. AVERAGE SORTING TIMES IN SECONDS

Number of items	Algorithm 201 Shellsort		Algorithm 207 Stringsort		Algorithm 245 Treesort 3		Algorithm 271 Quickersort	
	Integers	Reals	Integers	Reals	Integers	Reals	Integers	Reals
10	0.01	0.01	0.03	0.03	0.02	0.02	0.01	0.01
20	0.02	0.02	0.05	0.05	0.04	0.04	0.02	0.02
50	0.08	0.08	0.20	0.20	0.11	0.12	0.06	0.06
100	0.19	0.22	0.39	0.40	0.26	0.27	0.13	0.13
200	0.48	0.53	1.0	1.1	0.59	0.62	0.28	0.30
500	1.5	1.7	2.8	2.9	1.7	1.8	0.80	0.85
1000	3.7	4.2	6.6	6.9	3.7	4.0	1.8	1.9
2000	9.1	10.	13.	14.	8.2	8.7	3.9	4.1
5000	27.	30.	40.	41.	23.	24.	11.	12.
10000	65.	72.	93.	97.	49.	52.	23.	25.

ALGORITHM 347
AN EFFICIENT ALGORITHM FOR SORTING WITH
MINIMAL STORAGE [M1]

Richard C. Singleton* (Recd. 17 Sept. 1968)
Mathematical Statistics and Operations Research De-
partment, Stanford Research Institute, Menlo Park,
CA 94025

KEY WORDS AND PHRASES: sorting, minimal storage sorting, digital computer sorting
CR CATEGORIES: 5.31

procedure $SORT(A, i, j)$;
  value i, j; **integer** i, j;
  array A;
comment This procedure sorts the elements of array A into
ascending order, so that

$$A[k] \leq A[k+1], \quad k = i, i + 1, \cdots, j - 1.$$

The method used is similar to $QUICKERSORT$ by R. S. Scowen
[5], which in turn is similar to an algorithm given by Hibbard
[2, 3] and to Hoare's $QUICKSORT$ [4]. $QUICKERSORT$ is used
as a standard, as it was shown in a recent comparison to be the
fastest among four ACM algorithms tested [1]. On the Burroughs B5500 computer, the present algorithm is about 25
percent faster than $QUICKERSORT$ when tested on random uniform numbers (see Table I) and about 40 percent
faster on numbers in natural order $(1, 2, \cdots, n)$, in reverse
order $(n, n-1, \cdots, 1)$, and sorted by halves
$(2, 4, \cdots, n, 1, 3, \cdots, n-1)$. $QUICKERSORT$ is slow in sorting
data with numerous "tied" observations, a problem that can be
corrected by changing the code to exchange elements $a[k] \geq t$
in the lower segment with elements $a[q] \leq t$ in the upper segment. This change gives a better split of the original segment,
which more than compensates for the additional interchanges.

  In the earlier algorithms, an element with value t was selected
from the array. Then the array was split into a lower segment
with all values less than or equal to t and an upper segment with
all values greater than or equal to t, separated by a third segment of length one and value t. The method was then applied
recursively to the lower and upper segments, continuing until
all segments were of length one and the data were sorted. The

* This work was supported by Stanford Research Institute with
Research and Development funds.

TABLE I. Sorting Times in Seconds for *SORT* and *QUICKERSORT*, on the Burroughs B5500 Computer—Average of Five Trials

Original order and number of items	Algorithm	
	SORT	QUICKERSORT
Random uniform:		
500	0.48	0.63
1000	1.02	1.40
Natural order:		
500	0.29	0.48
1000	0.62	1.00
Reverse order:		
500	0.30	0.51
1000	0.63	1.08
Sorted by halves:		
500	0.73	1.15
1000	1.72	2.89
Constant value:		
500	0.43	10.60
1000	0.97	41.65

present method differs slightly—the middle segment is usually missing—since the comparison element with value t is not removed from the array while splitting. A more important difference is that the median of the values of $A[i]$, $A[(i+j) \div 2]$, and $A[j]$ is used for t, yielding a better estimate of the median value for the segment than the single element used in the earlier algorithms. Then while searching for a pair of elements to exchange, the previously sorted data (initially, $A[i] \leq t \leq A[j]$) are used to bound the search, and the index values are compared only when an exchange is about to be made. This leads to a small amount of overshoot in the search, adding to the fixed cost of splitting a segment but lowering the variable cost. The longest segment remaining after splitting a segment o n has length less than or equal to $n - 2$, rather than $n - 1$ as in *QUICKERSORT*.

For efficiency, the upper and lower segments after splitting should be of nearly equal length. Thus t should be close to the median of the data in the segment to be split. For good statistical properties, the median estimate should be based on an odd number of observations. Three gives an improvement over one and the extra effort involved in using five or more observations may be worthwhile on long segments, particularly in the early stages of a sort.

Hibbard [3] suggests using an alternative method, such as Shell's [6], to complete the sort on short sequences. An experimental investigation of this idea using the splitting algorithm adopted here showed no improvement in going beyond the final stage of Shell's algorithm, i.e. the familiar "sinking" method of sorting by interchange of adjacent pairs. The minimum time was obtained by sorting sequences of 11 or fewer items by this method. Again the number of comparisons is reduced by using the data themselves to bound the downward search. This requires

$$A[i-1] \leq A[k], \qquad i \leq k \leq j.$$

Thus the initial segment cannot be sorted in this way. The initial segment is treated as a special case and sorted by the splitting algorithm. Because of this feature, the present algorithm lacks the pure recursive structure of the earlier algorithms.

For n elements to be sorted, where $2^k \leq n < 2^{k+1}$, a maximum of k elements each are needed in arrays IL and IU. On the B5500 computer, single-dimensional arrays have a maximum length of 1023. Thus the array bounds [0:8] suffice.

This algorithm was developed as a FORTRAN subroutine, then translated to ALGOL. The original FORTRAN version follows:

```
      SUBROUTINE SORT(A,II,JJ)
C   SORTS ARRAY A INTO INCREASING ORDER, FROM A(II) TO A(JJ)
C   ORDERING IS BY INTEGER SUBTRACTION, THUS FLOATING POINT
C     NUMBERS MUST BE IN NORMALIZED FORM.
C   ARRAYS IU(K) AND IL(K) PERMIT SORTING UP TO 2**(K+1)-1 ELEMENTS
      DIMENSION A(1),IU(16),IL(16)
      INTEGER A,T,TT
      M=1
      I=II
      J=JJ
    5 IF(I .GE. J) GO TO 70
   10 K=I
      IJ=(J+I)/2
      T=A(IJ)
      IF(A(I) .LE. T) GO TO 20
      A(IJ)=A(I)
      A(I)=T
      T=A(IJ)
   20 L=J
      IF(A(J) .GE. T) GO TO 40
      A(IJ)=A(J)
      A(J)=T
      T=A(IJ)
      IF(A(I) .LE. T) GO TO 40
      A(IJ)=A(I)
      A(I)=T
      T=A(IJ)
      GO TO 40
   30 A(L)=A(K)
      A(K)=TT
   40 L=L-1
      IF(A(L) .GT. T) GO TO 40
      TT=A(L)
```

```
 50 K=K+1
    IF(A(K) .LT. T) GO TO 50
    IF(K .LE. L) GO TO 30
    IF(L-I .LE. J-K) GO TO 60
    IL(M)=I
    IU(M)=L
    I=K
    M=M+1
    GO TO 80
 60 IL(M)=K
    IU(M)=J
    J=L
    M=M+1
    GO TO 80
 70 M=M-1
    IF(M .EQ. 0) RETURN
    I=IL(M)
    J=IU(M)
 80 IF(J-I .GE. 11) GO TO 10
    IF(I .EQ. II) GO TO 5
 90 I=I+1
    IF(I .EQ. J) GO TO 70
    T=A(I+1)
    IF(A(I) .LE. T) GO TO 90
    K=I
100 A(K+1)=A(K)
    K=K-1
    IF(T .LT. A(K)) GO TO 100
    A(K+1)=T
    GO TO 90
    END
```

This FORTRAN subroutine was tested on a CDC 6400 computer. For random uniform numbers, sorting times divided by $n \log_2 n$ were nearly constant at 20.2×10^{-6} for $100 \leq n \leq 10{,}000$, with a time of 0.202 seconds for 1000 items. This subroutine was also hand-compiled for the same computer to produce a more efficient machine code. In this version the constant of proportionality was 5.2×10^{-6}, with a time of 0.052 seconds for 1000 items. In both cases, integer comparisons were used to order normalized floating-point numbers.

REFERENCES:
1. BLAIR, CHARLES R. Certification of algorithm 271. *Comm. ACM 9* (May 1966), 354.
2. HIBBARD, THOMAS N. Some combinatorial properties of certain trees with applications to searching and sorting. *J. ACM 9* (Jan. 1962), 13–28.
3. HIBBARD, THOMAS N. An empirical study of minimal storage sorting. *Comm. ACM 6* (May 1963), 206–213.
4. HOARE, C. A. R. Algorithms 63, Partition, and 64, Quicksort. *Comm. ACM 4* (July 1961), 321.
5. SCOWEN, R. S. Algorithm 271, Quickersort. *Comm. ACM 8* (Nov. 1965), 669.
6. SHELL, D. L. A high speed sorting procedure. *Comm. ACM 2* (July 1959), 30–32;

```
begin
  real t, tt;
  integer ii, ij, k, L, m;
  integer array IL, IU[0:8];
  m := 0;  ii := i;  go to L4;

L1:  ij := (i+j) ÷ 2;  t := A[ij];  k := i;  L := j;
  if A[i] > t then
    begin A[ij] := A[i];  A[i] := t;  t := A[ij] end;
  if A[j] < t then
  begin
    A[ij] := A[j];  A[j] := t;  t := A[ij];
    if A[i] > t then
      begin A[ij] := A[i];  A[i] := t;  t := A[ij] end
  end;

L2:  L := L − 1;
  if A[L] > t then go to L2;
    tt := A[L];

L3:  k := k + 1;
  if A[k] < t then go to L3;
  if k ≤ L then
    begin A[L] := A[k];  A[k] := tt;  go to L2 end;
  if L − i > j − k then
    begin IL[m] := i;  IU[m] := L;  i := k end
  else
    begin IL[m] := k;  IU[m] := j;  j := L end;
  m := m + 1;

L4:  if j − i > 10 then go to L1;
    if i = ii then
    begin if i < j then go to L1 end;
    for i := i + 1 step 1 until j do
    begin
    t := A[i];  k := i − 1;
    if A[k] > t then
    begin

L5:  A[k+1] := A[k];  k := k − 1;
    if A[k] > t then go to L5;
    A[k+1] := t
    end
  end;
    m := m − 1;  if m ≥ 0 then
      begin i := IL[m];  j := IU[m];  go to L4 end
  end SORT
```

REMARK ON ALGORITHM 347 [M1]
AN EFFICIENT ALGORITHM FOR SORTING WITH MINIMAL STORAGE

[Richard C. Singleton, *Comm. ACM 12* (Mar. 1969), 185]

ROBIN GRIFFIN AND K. A. REDISH (Recd. 14 Apr. 1969 and 11 Aug. 1969)

McMaster University, Hamilton, Ontario, Canada

KEY WORDS AND PHRASES: sorting, minimal storage sorting, digital computer sorting
CR CATEGORIES: 5.31

The algorithm was tested on the CDC 6400 ALGOL compiler (version 1.1, running under the SCOPE operating system, version 3.1.4). One trial was made using an array of 5000 pseudorandom numbers; the results were correct.

The central processor time was about 6.9 seconds corresponding to a value for K (defined below) of about 110 microseconds.

It would be more in the spirit of ALGOL to follow QUICKER-SORT [1] and give arrays *IL* and *IU* dynamic bounds. This involves changing line 4 on page 187 from

integer array *IL*, *IU*[0:8];

to

integer array *IL*, *IU*[0:$ln(j-i+1)/ln(2)-0.9$];

The FORTRAN subroutine given in the comments to the algorithm was tested on a CDC FORTRAN compiler (the RUN compiler version 2.3, running under the SCOPE operating system, version 3.1.4). Tests were made with each of the five initial orderings described with the algorithm for a variety of array lengths from 500 to 40,000. For integer arrays, the results were correct; but when the actual argument corresponding to the dummy argument A was a real array containing large positive and negative numbers, errors occurred. This does not invalidate the subroutine, but the comments should be changed to

```
C  SORTS INTEGER ARRAY A INTO INCREASING OR-
      DER, FROM A(II) TO A(IJ)
C  ARRAYS IU(K) AND IL(K) PERMIT SORTING UP TO
      2**(K+1) − 1 ELEMENTS
C  THE USER SHOULD CONSIDER THE POSSIBILITY OF
      INTEGER OVERFLOW
C  THE ONLY ARITHMETIC OPERATION ON THE ARRAY
      ELEMENTS IS SUBTRACTION
```

This gives enough information (and a hint) but leaves the responsibility for any abuse of American National Standards Institute (formerly USASI) FORTRAN where it belongs—with the user.

The subroutine was also tested on the IBM 7040 FORTRAN compiler (the IBFTC compiler running under the IBSYS operating system, version 9 level 10). The results were correct. The statement

INTEGER A, T, TT

was removed and the amended subroutine tested using similar, but real, arrays. The results were again correct; running times increased by up to 5 percent on the CDC 6400 and were unchanged on the IBM 7040.

Tables I and II summarize the information on running times in terms of K, where

$$\text{time} = Kn \log_2 n$$

(runs of other lengths are omitted for brevity).

For use as a library routine one slight change is recommended: $JJ - II$ should be tested on entry and a suitable error message produced if negative. It would be possible to transfer "work" arrays to replace IU and IL thus allowing the user more control of storage allocation, but the additional instructions needed to handle the extra arguments reduce the saving and this is hardly worthwhile.

The authors would like to thank the referee for his helpful comments.

REFERENCE:
1. SCOWEN, R. S. Algorithm 271, Quickersort. *Comm. ACM 8* (Nov. 1965), 669–670.

REMARK ON ALGORITHM 347 [M1]
AN EFFICIENT ALGORITHM FOR SORTING WITH MINIMAL STORAGE [Richard C. Singleton, *Comm. ACM 12* (Mar. 1969), 185]

RICHARD PETO (Recd. 18 Feb. 1970)
Medical Research Council, 115 Gower Street, London W. C. 1
KEY WORDS AND PHRASES: sorting, ranking, minimal storage sorting, digital computer sorting
CR CATEGORIES: 5.31

If the values of ij, instead of always being $(i+j) \div 2$, are at varying positions between i and j, then there is less likelihood of peculiar initial structure causing failure of the algorithm to perform rapidly. The position of ij can be made to vary by replacing

TABLE I. SORTING TIMES
K in microseconds where time $= Kn \log_2 n$

Test	Method			
Original order and number of items	Burroughs 5500 ALGOL*	CDC 6400 FORTRAN (REAL)	CDC 6400 FORTRAN (INTEGER ARRAY)	IBM 7040 FORTRAN
Random uniform				
500	107	21.2	20.5	
1000	102	21.7	20.5	
5000		21.1	20.2	269
10000		21.1	20.1	263
40000		21.2	20.1	
Natural order				
500	65	12.9	12.5	
1000	62	13.1	12.4	146
5000		12.6	11.9	148
10000		12.7	12.0	
40000		12.9	12.1	
Reverse order				
500	67	14.3	13.4	
1000	63	13.9	13.4	
5000		13.4	12.7	158
10000		13.4	12.7	158
40000		13.5	12.8	
Sorted by halves				
500	163	34.8	32.6	
1000	173	37.1	35.1	
5000		39.5	37.2	465
10000		41.8	39.3	491
40000		46.6	44.1	
Constant value				
500	96	19.2	18.5	
1000	97	19.4	18.7	
5000		19.4	18.7	237
10000		19.9	19.0	241
40000		20.2	19.5	

TABLE II. VALUES OF $n \log_2 n$

n	500	1000	5000	10000	40000
$10^{-6} n \log_2 n$	0.00448	0.00996	0.0614	0.1329	0.6115

* Calculated from Singleton's results

the statements

$m := 0;\quad ii := i;\quad$ **go to** $L4;\quad L1: ij := (i+j) \div 2;$

by

  real $r;\quad r := 0.375;\quad m := 0;\quad ii := i;\quad$ **go to** $L4;$

  $L1: r :=$ **if** $r > 0.58984375$ **then** $r - 0.21875$ **else** $r + 0.0390625;$

  $ij := i + (j-i) \times r;$

comment These four decimal constants, which are respectively 48/128, 75.5/128, 28/128, and 5/128, are rather arbitrary. On most compilers their binary representations will be exact, and the use of them in the statement $L1$ causes r to vary cyclically over the 33 values 48/128 $\cdots$ 80/128. Therefore ij takes a variable position somewhere within the middle quarter of the segment to be sorted. Wider variation of ij would be undesirable in the special case of a partially presorted array;

In sorting an array of N elements which are initially in random order this will waste (on ICL Atlas) less than $N/10^5$ seconds, but if the array is, for example, composed initially of two equal presorted halves, then the use of the original rather than the modified version would more than double the sorting time required if $N > 10^4$.

As the author points out, the published version could fail if used to sort arrays of 1024 or more elements because the upper bounds of IU and IL might be inadequate. For a standard procedure the declaration IL, IU [0:8] should be replaced by the declaration IL, IU [0:20]. This permits the sorting of arrays of up to 4 million elements, which is, with present core store sizes, sufficient.

The statement $tt := a[L]$ which precedes $L3$: will be executed less frequently if it is transferred into the next conditional statement, which then reads

  if $k \le L$ **then begin** $tt := a[L];\quad a[L] := a[k];\quad a[k] := tt;$

  go to $L2$ **end**

ALGORITHM 402

INCREASING THE EFFICIENCY OF

  QUICKSORT* [M1]

M. H. van Emden (Recd. 15 Dec. 1969 and 7 July 1970)

Mathematical Centre, Amsterdam, The Netherlands

---

* The algorithm is related to a paper with the same title and by the same author, which was published in *Comm. ACM 13* (Sept. 1970), 563–567.

KEY WORDS AND PHRASES: sorting, quicksort
CR CATEGORIES: 5.31, 3.73, 5.6, 4.49

procedure *qsort*(*a*, *l*1, *u*1);
  value *l*1, *u*1; **integer** *l*1, *u*1; **array** *a*;
comment This procedure sorts the elements $a[l1]$, $a[l1+1]$, $\cdots$, $a[u1]$ into nondescending order. It is based on the idea described in [1]. A comparison of this procedure with another procedure, called *sortvec*, obtained by combining C. A. R. Hoare's *quicksort* [2] and R. S. Scowen's *quickersort* [3], in such a way as to be optimal for the Algol 60 system in use on the Electrologica X-8 computer at the Mathematical Centre is shown below. Here "repetitions" denotes the number of times the sorting of a sequence of that "length" is repeated; "average time" is the time in seconds averaged over the repetitions; "gain" is the difference in time relative to time taken by *sortvec*.

procedure	length	repetitions	average time	gain
sortvec	30	23	.09	
qsort	30	23	.06	+.37
sortvec	300	16	1.25	
qsort	300	16	1.03	+.17
sortvec	3000	9	17.43	
qsort	3000	9	15.25	+.13
sortvec	30000	2	232.46	
qsort	30000	2	197.96	+.15

REFERENCES:
1. VAN EMDEN, M. H. Increasing the efficiency of quicksort. *Comm. ACM 13* (Sept. 1970), 563–567.
2. HOARE, C. A. R. Algorithm 64, quicksort. *Comm. ACM 4* (July 1961), 321–322.
3. SCOWEN, R. S. Algorithm 271, quickersort. *Comm. ACM 8* (Nov. 1965), 669;

```
begin
  integer p, q, ix, iz;
  real x, xx, y, zz, z;
  procedure sort;
  begin
    integer l, u;
    l := l1;  u := u1;
part:
    p := l;  q := u;  x := a[p];  z := a[q];
    if x > z then
    begin y := x;  a[p] := x := z;  a[q] := z := y end;
    if u − l > 1 then
    begin
      xx := x;  ix := p;  zz := z;  iz := q;
```

left:

```
        for p := p + 1 while p < q do
        begin
          x := a[p];
          if x ≥ xx then go to right
        end;
        p := q - 1; go to out;
```

right:

```
        for q := q - 1 while q > p do
        begin
          z := a[q];
          if z ≤ zz then go to dist
        end;
        q := p;  p := p - 1;  z := x;  x := a[p];
```

dist:

```
        if x > z then
        begin
          y := x;  a[p] := x := z;
          a[q] := z := y
        end;
        if x > xx then
        begin xx := x;  ix := p end;
        if z < zz then
        begin zz := z;  iz := q end;
        go to left;
```

out:

```
        if p ≠ ix ∧ x ≠ xx then
        begin a[p] := xx;  a[ix] := x end;
        if q ≠ iz ∧ z ≠ zz then
        begin a[q] := zz;  a[iz] := z end;
        if u - q > p - l then
        begin l1 := l;  u1 := p - 1;  l := q + 1 end
        else
        begin u1 := u;  l1 := q + 1;  u := p - 1 end;
        if u1 > l1 then sort;
        if u > l then go to part
      end
    end of sort;
    if u1 > l1 then sort
end of qsort
```

REMARK ON ALGORITHM 402

INCREASING THE EFFICIENCY OF QUICKSORT [M. H. van Emden, *Comm. ACM* **13** (Nov. 1970), 693–694]

ROBERT E. WHEELER (Recd. 6 July 1971)
E. I. du Pont de Nemours and Company, Wilmington, Del. 19899
KEY WORDS AND PHRASES: sorting, quicksort
CR CATEGORIES: 3.73, 4.49, 5.31, 5.6

It will happen during execution of this algorithm that sequences will be encountered which are already in nondescending order and which should not be further sorted. Changes to the algorithm which accomplish this are indicated below. For a FORTRAN version of this algorithm running on a UNIVAC 1108, these changes decreased running time by 1.25 percent when sorting random arrays of length 500 and by 2.7 percent when sorting random arrays of length 50.

Line	Change to:
2	**integer** p, q, ix, iz, i, j;
9	$p := 1; q := u; x := a[p]; z := a[q]; i := 0; j := q - p - 1;$
36	**begin** $xx := x; i := i + 1; ix := p$ **end;**
38	**begin** $zz := z; i := i + 1; iz := q$ **end;**
48.5	**if** $i \neq j$ **begin**
50.5	**end;**

ALGORITHM 72
PARTIAL SORTING [M1]
J. M. Chambers [Recd. 15 July 1970]
Bell Telephone Laboratories, Murray Hill, NJ 07974
Key Words and Phrases: sorting, partial sorting, order statistics
CR Categories: 5.31

Description
We introduce the notion of partial sorting as follows. Given an array A of N elements the result of sorting the array (in place) is to arrange the elements of A so that
$$A(1) \leq A(2) \leq \cdots \leq A(N).$$
An equivalent statement is that, for $J = 1, 2, \cdots, N$, $A(J)$ is a value such that for $1 \leq I < J < K \leq N$
$$A(I) \leq A(J) \leq A(K) \tag{1}$$
This property is also equivalent to the statement that $A(J)$ is the Jth order statistic [4] of A, for all J.

Partial sorting is a procedure which rearranges A so that (1) holds for some selected values of J, but not necessarily for all J. The advantage of using partial sorting, where possible, is that the cost is substantially less than for sorting, when the number of order statistics required is small compared to N.

Such will frequently be the case, for example, in statistical applications, when the sample is to be summarized using some of the order statistics. For large N only a portion of the sample would be needed, even for displays such as the empirical distribution function.

Specifically, in the algorithm *PSORT* below, the user supplies the array A of size N and a set of indices *IND* of size *NI*. On return, A will have been rearranged so that relation (1) holds, i.e. $A(J)$ has the value it would have if A were sorted, for $J = IND(1), IND(2), \dots, IND(NI)$.

For example, suppose A is the vector (10., 8., 3., 5., 7., 2.) and *IND* is the vector (2, 5). Then after a partial sort of A with *IND*, $A(2) = 3.$ and $A(5) = 8.$.

The method used is based on Hoare's method [1, 2] as developed by Singleton [3]. Hoare's method consists of choosing an element $A(m)$ and splitting the array into three portions which are respectively smaller than, equal to, and larger than this element. The method is then applied recursively to the first and third portions, until the data is completely sorted. Successive versions leading to [2] alter the method in four important respects. (i) instead of choosing $A(m)$ arbitrarily, the median of the first, last and middle element are chosen; (ii) the recursion is simulated, rather than explicit; (iii) short sequences (less than 10 in [3]) are sorted by a "sinking" sort; (iv) a different treatment of "tied" observations is introduced.

Hoare's method is very well suited to handle the partial sorting problem. The algorithm is modified simply by passing over the portion of A in which none of the indices in *IND* are found. Once we have established a segment of A which is known not to contain any of the desired order statistics, there is no need to sort it further. The special case of $NI = 1$ was treated in procedure *FIND* of [1].

For a fixed number of indices, the cost of applying *PSORT* is very nearly proportional to N, as opposed to the full sort, with cost of the order of $N\log(N)$. Because of the simplicity of the modified algorithm, the cost of *PSORT* will almost always be significantly less than the cost of the full sort, providing *NI* is substantially less than N. Notice, however, that a full sort will be carried out unless some adjacent elements of *IND* differ by more than 10.

The following restrictions are to be noted: it is assumed that *IND* is initially sorted into ascending order; *A* is of type *REAL*; if *N* is the dimension of the *A* array then the arrays *INDU*, *INDL*, *IU*, *IL* must have dimension *K* where $N < 2^{K+1}$, (see [3]);

Examples. Table I gives some examples of the performance of *PSORT* on various size arrays with various initial orderings. The examples were constructed as follows. Samples of *N* were simulated with a standard normal marginal distribution, and a correlation ρ with an ordered normal sample. (Specifically we generated a_i, b_i for $i = 1, \cdots, N$ as independent standard normal variates, then formed $y_i = \rho a_i + (1 - \rho^2)^{\frac{1}{2}} b_i$ and sorted the y_i, carrying along the a_i. The resulting a_i are the desired input to *PSORT*.)

Computations were carried out in two ways. By replacing the comparisons of elements in *A* by special functions, the number of comparisons required was counted, and is shown in the columns of Table I headed *C*. This gives a machine independent result, but does not include the costs of transposition, logic, etc. Therefore, we also give timings for the original algorithm, on a GE 635 computer, in the columns headed *T*. The unit of time is one millisecond.

The results of Table I suggest, as one would expect, that the most expensive case, for given value of *NI*, is for the desired order statistics to be evenly spaced; i.e. $jN/(NI+1)$ for $j = 1, \cdots, NI$. For this worst case, the cost does grow proportionately to *N* (a little less than that, in the table).

A comparison with the full sort, using Singleton's algorithm, is included for sample size 500.

References

1. Hoare, C. A. R. Algorithms 63, Partition; 64, Quicksort; and 65, Find. *Comm ACM 4* (July 1961), 321–322.
2. Hoare, C. A. R. Quicksort. *Comput. J. 5* (1962), 10–15.
3. Singleton, R. S. Algorithm 347 Sort. *Comm. ACM 12* (1969), 185–186.
4. Wilks, S. S. *Mathematical Statistics*. Wiley, New York, 1962, p. 234.

Algorithm

```
      SUBROUTINE PSORT(A,N,IND,NI)
C PARAMETERS TO PSORT HAVE THE FOLLOWING MEANING
C A     ARRAY TO BE SORTED
C N     NUMBER OF ELEMENTS IN A
C IND   ARRAY OF INDICES IN ASCENDING ORDER
C NI    NUMBER OF ELEMENTS IN IND
      DIMENSION A(N),IND(NI)
      DIMENSION INDU(16),INDL(16)
      DIMENSION IU(16),IL(16)
      INTEGER P
      JL=1
      JU=NI
      INDL(1)=1
      INDU(1)=NI
```

Table I. Examples of *PSORT*. C = number of comparisons, T = time in 10^{-3} sec.

			Correlation with ordered data									
			−1.0		−0.5		0.0		+0.5		−1.0	
N	*NI*	*IND*	*C*	*T*	*C*	*T*	*C*	*T*	*C*	*T*	*C*	*T*
100	2	33 67	303	11.0	384	14.2	392	14.2	386	13.6	291	9.0
100	3	25 50 75	323	11.8	468	17.5	429	15.9	470	16.7	329	10.2
500	2	33 67	1122	36.0	1356	43.7	1169	36.7	1362	41.3	1121	27.8
500	3	25 50 75	1182	37.9	1414	46.3	1307	41.5	1406	43.1	1181	29.9
500	3	125 250 375	1628	49.3	2213	70.1	2184	71.3	2205	67.2	1748	43.9
500	Call to *SORT*			151.3		151.2		150.2		150.4		150.9
1000	3	250 500 750	3258	96.9	4870	145.1	4438	137.27	4725	140.4	3503	85.6

```
C ARRAYS INDL, INDU KEEP ACCOUNT OF THE PORTION OF IND RELATED TO THE
C CURRENT SEGMENT OF DATA BEING ORDERED.
      I=1
      J=N
      M=1
5     IF(I.GE.J) GO TO 70
C FIRST ORDER A(I),A(J),A((I+J)/2), AND USE MEDIAN TO SPLIT THE DATA
10    K=I
      IJ=(I+J)/2
      T=A(IJ)
      IF(A(I).LE.T) GO TO 20
      A(IJ)=A(I)
      A(I)=T
      T=A(IJ)
20    L=J
      IF(A(J).GE.T) GO TO 40
      A(IJ)=A(J)
      A(J)=T
      T=A(IJ)
      IF(A(I).LE.T) GO TO 40
      A(IJ)=A(I)
      A(I)=T
      T=A(IJ)
      GO TO 40
30    A(L)=A(K)
      A(K)=TT
40    L=L-1
      IF(A(L).GT.T) GO TO 40
      TT=A(L)
C SPLIT THE DATA INTO A(I TO L).LT.T, A(K TO J).GT.T
50    K=K+1
      IF(A(K).LT.T) GO TO 50
      IF(K.LE.L) GO TO 30
      INDL(M)=JL
      INDU(M)=JU
      P=M
      M=M+1
C SPLIT THE LARGER OF THE SEGMENTS
      IF(L-I.LE.J-K) GO TO 60
      IL(P)=I
      IU(P)=L
      I=K
C SKIP ALL SEGMENTS NOT CORRESPONDING TO AN ENTRY IN IND
55    IF(JL.GT.JU) GO TO 70
      IF(IND(JL).GE.I) GO TO 58
      JL=JL+1
      GO TO 55
58    INDU(P)=JL-1
      GO TO 80
60    IL(P)=K
      IU(P)=J
      J=L
65    IF(JL.GT.JU) GO TO 70
      IF(IND(JU).LE.J) GO TO 68
      JU=JU-1
      GO TO 65
68    INDL(P)=JU+1
      GO TO 80
70    M=M-1
      IF(M.EQ.0) RETURN
      I=IL(M)
      J=IU(M)
      JL=INDL(M)
      JU=INDU(M)
      IF(JL.GT.JU) GO TO 70
80    IF(J-I.GT.10) GO TO 10
      IF(I.EQ.1) GO TO 5
      I=I-1
90    I=I+1
      IF(I.EQ.J) GO TO 70
      T=A(I+1)
      IF(A(I).LE.T) GO TO 90
      K=I
100   A(K+1)=A(K)
      K=K-1
      IF(T.LT.A(K)) GO TO 100
      A(K+1)=T
      GO TO 90
      END
```

ALGORITHM 426

MERGE SORT ALGORITHM [M1]
C. Bron (Recd. 4 Feb. 1970 and 10 May 1971)
Technological University, Eindhoven, The Netherlands

KEY WORDS AND PHRASES: sort, merge
CR CATEGORIES: 5.31

Description Sorting by means of a two-way merge has a reputation of requiring a clerically complicated and cumbersome program. This ALGOL 60 procedure demonstrates that, using recursion, an elegant and efficient algorithm can be designed, the correctness of which is easily proved [2]. Sorting n objects gives rise to a maximum recursion depth of $[\log_2(n - 1) + 2]$. This procedure is particularly suitable for sorting when it is not desirable to move the n objects physically in store and the sorting criterion is not simple. In that case it is reasonable to take the number of compare operations as a measure for the speed of the algorithm. When n is an integral power of 2, this number will be comprised between $(n \times \log_2 n)/2$ when the objects are sorted to begin with and $(n \times \log_2 n - n + 1)$ as an upper limit. When n is not an integral power of 2, the above formulas are approximate.

It is assumed that each object can in some way be uniquely identified by one of the integers from 1 to n. This correspondence has to be supplied in the call by replacing hi and lo by two integer variables and the Jensen parameter loafterhi by a Boolean expression that yields the value **true** if the object identified by lo has to follow the object identified by hi in the ordered sequences, and **false** otherwise. Let e_i be the identifying integer of the ith object in the ordered sequence. Upon return from the procedure sort delivers the value of e_1 and the pointer array put will be filled in such a way that put$[e_i] = e_{i+1}$, $1 \le i < n$, and put$[e_n] = 0$. Therefore the bounds of the actual array supplied for put will have to include the range $[1 : n]$. Sorted subsequences that arise during the sorting process will have a similar chain structure.

The essence of the algorithm is to be found in the procedure head. It has the duty to form an ordered chain of desired length (deslen) from the objects identified by count + 1 through count + deslen. It does so by introducing a chain of length l, consisting of object count + 1, and then repeatedly doubling the length of that chain by merging it with a chain of equal length the creation of which is left to a recursive call on head. If deslen is not an integral power of 2, a chain of length deslen cannot be built by repeatedly doubling. In that case, before the last merge operation, a chain of length (desired length − present length) is created and merged with the present chain to produce the required result.

As an example of a call on the sorting procedure we supply sort(10 000, chain, i, j, $A[i] > A[j]$) Although it should be stressed that the present version of the algorithm is not efficient when the sorting criterion is as simple as a comparison of two array elements. In such a case one does not only gain by replacing the calls on the formal parameter loafterhi by $A[lo] > A[hi]$ and declaring lo and hi as local variables of the procedure sort, but also one might resort to *in situ* sorting techniques like [1] that do not need the auxiliary array put. A comparison of this algorithm with QUICKERSORT [1] conducted under equivalent circumstances on the ALGOL system for the EL X8 showed no significant difference in speed when sorting arrays containing random numbers.

Acknowledgment. The author is grateful to Prof. E. W. Dijkstra for his contributions to this version of the algorithm, and to the referee for his careful analysis and valuable suggestions.

References
1. Scowen, R. S. Quickersort, *Comm. ACM* **8** (Oct. 1965), 669–670.
2. Bron, C., Proof of a merge sort algorithm, May 1971 (unpublished).

Algorithm
Integer procedure sort(n, put, lo, hi, loafterhi);
    value n; **integer** n, lo,. hi; **integer array** put;
    Boolean loafterhi;
begin
    integer count, link;
    comment link is a working location for merging;
    integer procedure head(deslen);
        value deslen; **integer** deslen;
     comment The value of head will be the identifying integer of the object
        leading the sorted chain;
    begin
        integer beg, len, nextlen;
INTRODUCE NEW CHAIN OF LENGTH L:
SUPPLY WITH END MARKER:
MAKE beg POINT TO ITS HEAD:
        beg : = count : = count + 1; put[beg] : = 0; len : = 1;
TEST: TO SEE WHETHER DESIRED LENGTH HAS BEEN REACHED:
        if len < deslen **then**
        begin
           nextlen : = **if** len < deslen − len **then** len
            else deslen − len;
INTRODUCE NEW CHAIN:
        hi : = head(nextlen);
AND START MERGING:

```
FIND LEADING OBJECT OF MERGED CHAIN:
      lo : = beg;
      if loafterhi then
      begin beg : = hi; hi : = lo; lo : = beg end;
INITIALIZE CHAIN ON MECHANISM:
      link : = lo;
CHAIN ON:
      lo : = put[link];
TEST FOR END OF lo CHAIN:
      if lo ≠ 0 then
      begin
ADD LINK TO CHAIN:
         if loafterhi then
         begin
SWITCH LINK TO hi CHAIN:
            put[link] : = link : = hi; hi : = lo
         end
         else
STEP DOWN IN lo CHAIN:
         link : = lo;
         go to CHAIN ON
      end;
APPEND REMAINING TAIL:
      put[link] : = hi;
      len : = len + nextlen;
      go to TEST
   end;
   head : = beg
 end head;
 count : = 0; sort : = head(n)
end sort;
```

Appendix B
Sorting Procedures in PL/I

The algorithms presented here are intended only to illustrate the Sort technique.

LINEAR SELECTION

```
LINSEL: PROCEDURE(TOSORT,SORTED,NUMBER);
DECLARE
TOSORT(*) FIXED BINARY(15,0),
SORTED(*) FIXED BINARY(15,0),
NUMBER FIXED BINARY(15,0),
SOURCE FIXED BINARY(15,0),
I FIXED BINARY(15,0),
J FIXED BINARY(15,0),
LOW FIXED BINARY(15,0);
GROW: DO J=1 TO NUMBER BY 1;
LOW=TOSORT(1);
SOURCE=1;
SELECT: DO I=2 TO NUMBER BY 1;
IF LOW>TOSORT(I) THEN DO
LOW=TOSORT(I);
SOURCE=I;
```

```
END;
ELSE; END SELECT;
PLACE: SORTED(J)=LOW;
TOSORT(SOURCE)=99;
END GROW;
END LINSEL;
```

BASIC EXCHANGE

```
BASICEXC:PROCEDURE(TOSORT,NUMBER);
DECLARE
TOSORT(*) FIXED BINARY(15,0),
NUMBER FIXED BINARY(15,0),
EXCOUNT FIXED BINARY(15,0),
ADJUST FIXED BINARY(15,0),
ELIMIT FIXED BINARY(15,0),
OLIMIT FIXED BINARY(15,0),
ODDEVE FIXED BINARY(15,0),
LIMIT FIXED BINARY(15,0),
TEMP FIXED BINARY(15,0),
PASSW FIXED BINARY(1,0),
I FIXED BINARY(15,0),
ADJUST=2*TRUNC(NUMBER/2);
IF ADJUST<NUMBER THEN DO;
ELIMIT=ADJUST;
OLIMIT=ADJUST-1;
END;
ELSE DO;
ELIMIT=NUMBER-2;
OLIMIT=NUMBER-1;
END;
ODD: PASSW=1;
LIMIT=OLIMIT;
ODDEVE=1;
PASS: EXCOUNT=0;
DO I=ODDEVE TO LIMIT BY 2;
IF TOSORT(I)>TOSORT(I+1) THEN DO;
TEMP=TOSORT(I);
TOSORT(I)=TOSORT(I+1);
TOSORT(I+1)=TEMP;
EXCOUNT=1;
```

```
END;
END;
IF EXCOUNT=0 THEN GO TO EXIT:
IF PASSW=1 THEN DO;
PASSW=0;
ODDEVE=2;
LIMIT=ELIMIT;
GO TO PASS;
END;
ELSE GO TO ODD;
EXIT: END BASICEXC:
```

STANDARD EXCHANGE

```
STEXCH:PROCEDURE(TOSORT,NUMBER);
DECLARE
TOSORT(*) FIXED BINARY(15,0),
NUMBER FIXED BINARY(15,0),
EXCOUNT FIXED BINARY(15,0),
TEMP FIXED BINARY(15,0),
I FIXED BINARY(15,0),
PASS: EXCOUNT=0;
PASSLOOP: DO I=1 TO NUMBER-1 BY 1;
IF TOSORT(I)>TOSORT(I+1) THEN DO;
 TEMP=TOSORT(I);
TOSORT(I)=TOSORT(I+1);
TOSORT(I+1)=TEMP;
EXCOUNT=1;
END;
END PASSLOOP;
 IF EXCOUNT=0 THEN GO TO EXIT;
GO TO PASS;
EXIT: END STEXCH;
```

SIMPLE INSERTION

```
SIMSERT: PROCEDURE (TOSORT,NEW);
DECLARE
TOSORT(*) FIXED BINARY(15,0),
NEW FIXED BINARY(15,0),
```

```
I FIXED BINARY(15,0),
J FIXED BINARY(15,0),
DO I=1 TO LENGTH BY 1;
IF NEW<=TOSORT(I) THEN DO;
DO J=LENGTH TO I BY -1;
TOSORT(J+1)=TOSORT(J);
END;
GO TO SETIN;
END;
END:
SETIN: TOSORT(I)=NEW;
END SIMSERT;
END FEED;
```

SIMPLE SIFTING (SHUTTLESORT)

```
SHUTTLESORT: PROCEDURE (TOSORT,NUMBER);
/* PUBLISHED IN ALGOL AS ALGORITHM 175, COMMUNICATIONS OF ACM,
  VOL 6, NO 6, P 312 */
DECLARE
TOSORT(*) FIXED BINARY(15,0),
NUMBER FIXED BINARY(15,0),
TEMP FIXED BINARY(15,0),
I FIXED BINARY(15,0),
J FIXED BINARY(15,0);
PASS: DO I=1 TO NUMBER-1 by 1;
EXCH: DO J=I TO 1 BY -1;
IF TOSORT(J)<=TOSORT(J+1) THEN GO TO EXIT;
TEMP=TOSORT(J);
TOSORT(J)=TOSORT(J+1)
TOSORT(J+1)=TEMP;
END EXCH;
EXIT: END PASS;
END SHUTTLESORT;
```

SHELLSORT

```
BSHELLSORT:PROCEDURE(TOSORT,NUMBER);
/* ALGORITHM 201, SHELLSORT, PUBLISHED IN ALGOL PUBLICATION
  LANGUAGE, COMMUNICATIONS OF ACM, VOL 6, NO. 8, AUGUST 1963 */
DECLARE
TOSORT(*) FIXED BINARY (31,0),
```

```
DISTANCE FIXED BINARY (31,0),
LIMIT FIXED BINARY (31,0),
TEMP FIXED BINARY (31,0),
I FIXED BINARY (31,0),
J FIXED BINARY (31,0),
LOGNMBR FIXED BINARY (31,0),
NUMBER FIXED BINARY (31,0),
LOGNMBR=LOG2(NUMBER);
DISTANCE=2**LOGNMBR-1;
DIST: DO WHILE (DISTANCE>0);
LIMIT=NUMBER-DISTANCE;
SETS: DO J=1 TO LIMIT BY 1;
ELTS: DO I=J TO 1 by -DISTANCE;
IF TOSORT(I+DISTANCE)>=TOSORT(I) THEN GO TO OUT;
TEMP=TOSORT(I);
TOSORT(I)=TOSORT(I+DISTANCE);
TOSORT(I+DISTANCE)=TEMP;
END ELTS;
OUT: END SETS;
DISTANCE=DISTANCE/2;
END DIST;
END BSHELLSORT
```

TREESORT

```
TREESORT3: PROCEDURE(TOSORT,NUMBER);
/* ALGORITHM 245, PUBLISHED IN COMMUNICATIONS ACM, VOL 7, NO 12,
  P 701 */
DECLARE
TOSORT(*) FIXED BINARY (31,0),
ANCESTOR FIXED BINARY (31,0),
TEMP FIXED BINARY (31,0),
I FIXED BINARY (31,0),
LIMNODE FIXED BINARY (31,0),
FATHER FIXED BINARY (31,0),
NUMBER FIXED BINARY (31,0),

DO I=TRUNC(NUMBER/2) TO 2 BY -1;
CALL UPTREE;
END;
I=1;
DO LIMNODE=NUMBER TO 2 BY -1;
CALL UPTREE;
```

```
TEMP=TOSORT(1);
TOSORT(1)=TOSORT(LIMNODE);
TOSORT(LIMNODE)=TEMP;
END:
UPTREE: PROCEDURE;
ANCESTOR=I;
TEMP=TOSORT (ANCESTOR);
INLOOP: FATHER=2*ANCESTOR;
IF FATHER>LIMNODE THEN GO TO SETIN;
IF FATHER=LIMNODE THEN GO TO TEMPTEST;
IF TOSORT(FATHER+1)>TOSORT(FATHER) THEN FATHER=FATHER+1;
TEMPTEST: IF TOSORT(FATHER)>TEMP THEN DO;
TOSORT(ANCESTOR)=TOSORT(FATHER);
ANCESTOR=FATHER;
GO TO INLOOP;
END;
SETIN: TOSORT(ANCESTOR)=TEMP;
END UPTREE;
END TREESORT3
```

QUICKERSORT

```
QUICKERSORT: PROCEDURE(TOSORT,NUMBER);
/* ALGORITHM 271, COMMUNICATIONS ACM, VOL 8, NO 11, p 669 */
DECLARE
TOSORT(*) FIXED BINARY (31),
NUMBER FIXED BINARY (31,0),
ORIGIN FIXED BINARY (31,0),
LOWLIM FIXED BINARY (31,0),
HIGHLIM FIXED BINARY (31,0),
PARTIND FIXED BINARY (31,0),
PIVOT FIXED BINARY (31,0),
TEMP FIXED BINARY (31,0),
EXCH FIXED BINARY (31,0);
LIMIT=20;
SORT: BEGIN;
DECLARE
PARTTABLOW(LIMIT) FIXED BINARY (31,0),
PARTTABHIGH(LIMIT) FIXED BINARY (31,0);
ORIGIN=1;
PARTIND=1;
TESTSIZE: IF NUMBER—ORIGIN>1 THEN DO;
```

```
PIVOT=TRUNC ((NUMBER+ORIGIN)/2);
TEMP=TOSORT(PIVOT);
TOSORT(PIVOT)=TOSORT(ORIGIN);
HIGHLIM=NUMBER;
FINDHIGH: DO LOWLIM=ORIGIN+1 TO HIGHLIM BY 1;
IF TOSORT(LOWLIM)>TEMP THEN DO;
FINDLOW: DO HIGHLIM=HIGHLIM TO LOWLIM BY —1;
IF TOSORT(HIGHLIM)<TEMP THEN DO;
EXCH=TOSORT(LOWLIM);
TOSORT(LOWLIM)=TOSORT(HIGHLIM);
TOSORT(HIGHLIM)=EXCH;
HIGHLIM=HIGHLIM—1;
GO TO ENDHIGH:
END;
END FINDLOW;
HIGHLIM=LOWLIM—1;
GO TO LIMSMET;
END;
ENDHIGH: END FINDHIGH;
LIMSMET: TOSORT(ORIGIN)=TOSORT(HIGHLIM);
TOSORT(HIGHLIM)=TEMP;
IF 2*HIGHLIM>ORIGIN+NUMBER THEN DO;
PARTTABLOW(PARTIND)=ORIGIN;
PARTTABHIGH(PARTIND)=HIGHLIM—1;
ORIGIN=HIGHLIM+1;
END;
ELSE DO;
PARTTABLOW(PARTIND)=HIGHLIM+1;
PARTTABHIGH(PARTIND)=NUMBER;
NUMBER=HIGHLIM—1;
END;
PARTIND=PARTIND+1;
GO TO TESTSIZE;
END:
IF ORIGIN=NUMBER THEN GO TO SETPART;
IF TOSORT(ORIGIN)>TOSORT(NUMBER) THEN DO;
EXCH=TOSORT(ORIGIN)
TOSORT(ORIGIN)=TOSORT(NUMBER);
TOSORT(NUMBER)=EXCH;
END;
SETPART: PARTIND=PARTIND—1;
IF PARTIND>0 THEN DO;
```

```
ORIGIN=PARTTABLOW(PARTIND);
NUMBER=PARTTABHIGH(PARTIND);
GO TO TESTSIZE;
END;
END SORT;
END QUICKERSORT;
```

SINGSORT

```
SINGSORT: PROCEDURE(TOSORT,NUMBER);
/* ALGORITHM 347, COMMUNICATIONS OF ACM, VOL 12, NO 3, P 185 */
DECLARE
TOSORT(*) FIXED BINARY (31,0),
PIVOT FIXED BINARY (31,0),
TEMP2 FIXED BINARY (31,0),
LIMDEX FIXED BINARY (31,0),
INITIAL FIXED BINARY (31,0),
MEDIAN FIXED BINARY (31,0),
BOTIND FIXED BINARY (31,0),
TOPIND FIXED BINARY (31,0),
LIMITS FIXED BINARY (31,0),
I FIXED BINARY (31,0),
NUMBER FIXED BINARY (31,0),
PARTOP FIXED BINARY (31,0) INITIAL (1);

LIMITS=20;
SORT: BEGIN;
DECLARE
TOPS(LIMITS) FIXED BINARY (31,0),
BOTTOMS(LIMITS) FIXED BINARY (31,0),

LIMDEX=1;
INITIAL=PARTOP;
GO TO SINKTEST;
SPLIT: MEDIAN=TRUNC(PARTOP+NUMBER)/2);
PIVOT=TOSORT(MEDIAN);
TOPIND=PARTOP;
BOTIND=NUMBER;
IF TOSORT(PARTOP)>PIVOT THEN DO;
TOSORT(MEDIAN)=TOSORT(PARTOP);
TOSORT(PARTOP)=PIVOT;
PIVOT=TOSORT(MEDIAN);
```

```
END;
IF TOSORT(NUMBER)<PIVOT THEN DO;
TOSORT(MEDIAN)=TOSORT(NUMBER);
TOSORT(NUMBER)=PIVOT;
PIVOT=TOSORT(MEDIAN);
IF TOSORT(PARTOP)>PIVOT THEN DO;
TOSORT(MEDIAN)=TOSORT(PARTOP);
TOSORT(PARTOP)=PIVOT;
PIVOT=TOSORT(MEDIAN);
END:
END;
FINDSMALL: BOTIND=BOTIND-1;
IF TOSORT(BOTIND)>PIVOT THEN GO TO FINDSMALL;
TEMP2=TOSORT(BOTIND);
FINDLARGE: TOPIND=TOPIND+1;
IF TOSORT(TOPIND)<PIVOT THEN GO TO FINDLARGE;
IF TOPIND<=BOTIND THEN DO;
TOSORT(BOTIND)=TOSORT(TOPIND);
TOSORT(TOPIND)=TEMP2;
GO TO FINDSMALL;
END;
IF BOTIND-PARTOP<NUMBER-TOPIND THEN DO;
TOPS(LIMDEX)=PARTOP;
BOTTOMS(LIMDEX)=BOTIND;
PARTOP=TOPIND;
END;
ELSE DO;
TOPS(LIMDEX)=TOPIND;
BOTTOMS(LIMDEX)=NUMBER;
NUMBER=BOTIND;
END;
LIMDEX=LIMDEX+1;
SINKTEST: IF NUMBER-PARTOP>10 THEN GO TO SPLIT;
IF INITIAL=PARTOP THEN DO;
IF PARTOP<NUMBER THEN GO TO SPLIT;
END;
DO I=PARTOP+1 TO NUMBER BY 1;
PIVOT=TOSORT(I);
TOPIND=I-1;
IF TOSORT (TOPIND)>PIVOT THEN DO;
SINK: TOSORT(TOPIND+1)=TOSORT(TOPIND);
TOPIND=TOPIND-1;
```

```
IF TOSORT(TOPIND)>PIVOT THEN GO TO SINK;
TOSORT(TOPIND+1)=PIVOT;
END;
END;
LIMDEX=LIMDEX-1;
IF LIMDEX>=1 THEN DO;
PARTOP=TOPS(LIMDEX);
NUMBER=BOTTOMS(LIMDEX);
GO TO SINKTEST;
END;
END SORT;

END SINGSORT;
END FEED;
```

STRINGSORT

```
STRINGSORT: PROCEDURE(TOSORT,NUMBER);
/* ALGORITHM 207, COMMUNICATIONS ACM, VOL 5, NO 10, P 215 */
DECLARE
TOSORT(*) FIXED BINARY (31,0),
NUMBER FIXED BINARY (31,0),

SORT: BEGIN;
DECLARE
WORK(2*NUMBER) FIXED BINARY (31,0),
TOPST FIXED BINARY (31,0),
BOTST FIXED BINARY (31,0),
LIMITS (2) FIXED BINARY (31,0),
ADVANCE FIXED BINARY (31,0),
NEXT FIXED BINARY (31,0),
LAST FIXED BINARY (31,0),
K FIXED BINARY (31,0),
PASSW FIXED BINARY (1,0),
EXTEND LABEL;
INITIAL: DO I=1 TO NUMBER BY 1;
WORK(I)=TOSORT(I);
END INITIAL;
ODDPASS: TOPST=1;
BOTST=NUMBER;
LIMITS(1)=NUMBER+1;
LIMITS(2)=2*NUMBER;
```

```
K=1;
ADVANCE=1;
PASSW=1;
FIRSTST: EXTEND=NONDOWN;
NEXT=LIMITS(K);
IF WORK(TOPST)>=WORK(BOTST) THEN GO TO BOTTOM;
ELSE GO TO TOP;
TOP: WORK(NEXT)=WORK(TOPST);
TOPST=TOPST+1;
GO TO NEWNEXT;
BOTTOM: WORK(NEXT)=WORK(BOTST);
BOTST=BOTST-1;
NEWNEXT: LAST=NEXT;
NEXT=NEXT+ADVANCE;
IF BOTST>=TOPST THEN GO TO EXTEND;
IF PASSW=0 THEN IF NEXT=NUMBER+1 THEN GO TO EXIT;
ELSE GO TO ODDPASS;
ELSE IF NEXT=2*NUMBER+1 THEN GO TO EXIT;
ELSE GO TO EVENPASS;
JDOWN: IF WORK(TOPST)>=WORK(LAST) THEN GO TO TOP;
ELSE GO TO BOTHDOWN;
IDOWN: IF WORK(BOTST)>=WORK(LAST) THEN GO TO BOTTOM;
ELSE GO TO BOTHDOWN;
NONDOWN: IF WORK(TOPST)>=WORK(LAST) THEN
IF WORK(BOTST)>=WORK(LAST) THEN
IF WORK(BOTST)>=WORK(TOPST) THEN GO TO TOP;
ELSE GO TO BOTTOM;
ELSE DO;
EXTEND=JDOWN;
GO TO TOP;
END;
ELSE DO;
EXTEND=IDOWN;
GO TO IDOWN;
END;
BOTHDOWN:
LIMITS(K)=NEXT;
IF K=1 THEN K=2;
ELSE K=1;
ADVANCE= -ADVANCE;
GO TO FIRSTST;
EVENPASS: TOPST=NUMBER+1;
```

```
BOTST=2*NUMBER;
LIMITS(1)=1;
LIMITS(2)=NUMBER;
ADVANCE=1;
K=1;
PASSW=0;
GO TO FIRSTST;
EXIT: DO I=1 TO NUMBER BY 1;
TOSORT(I)=WORK(LIMITS (1) —1+I);
END;
END SORT;
END STRINGSORT;
```

DIGIT COUNTING (MATHSORT)

```
RMATHSORT: PROCEDURE(TOSORT,SORTED,NUMBER,RANGE);
/* ALGORITHM 23 PUBLISHED IN CACM, VOL 3, NO 11, NOV, 1960.
THIS VERSION DERIVED FROM MODIFICATION OF RANSHAW. CACM VOL 4 NO
  5, MAY 1961. CODING HERE ASSUMES KEY DEVELOPMENT FUNCTION TO BE
  NULL. THAT IS, KEYS PRESENTED IN TOSORT CAN BE USED DIRECTLY AS
  THEY FALL WITHIN RANGE. */
DECLARE
TOSORT(*) FIXED BINARY (15,0),
SORTED(*) FIXED BINARY (15,0),
NUMBER FIXED BINARY (15,0),
RANGE FIXED BINARY (15,0),
I FIXED BINARY (15,0),
J FIXED BINARY (15,0),
K FIXED BINARY (15,0),
SORT: BEGIN;
DECLARE
TOTALS (RANGE) FIXED BINARY INITIAL (RANGE)0);
COUNT: DO I=1 to NUMBER BY 1;
TOTALS(TOSORT(I))=TOTALS(TOSORT(I))+1;
END COUNT;
/* COUNT DEVELOPS IN TOSORT THE NUMBER OF OCCURRENCES OF ALL KEYS
  IN TOSORT */
CUMULATE: DO J=RANGE—1 TO 1 BY —1;
TOTALS(J)=TOTALS(J)+TOTALS(J+1);
END CUMULATE;
```

```
/* CUMULATE DEVELOPS THE DESCENDING CUMULATIVE COUNT
OF KEYS IN TOSORT GREATER THAN J */
PLACE: DO K=1 TO NUMBER BY 1;
SORTED(NUMBER+1-TOTALS(TOSORT(K)))=TOSORT(K);
TOTALS(TOSORT(K))=TOTALS(TOSORT(K))-1;
END PLACE;
END SORT;
END RMATHSORT;
```

PL/I ALGORITHMS DISCUSSED IN TEXT

		Text Chapters	Pages
1.	Linear Selection	1	6, 29
2.	Basic Exchange	2	23, 29
3.	Standard Exchange	2	27, 29
4.	Simple Sifting (SHUTTLESORT, Alg. 175, *CACM*)	2, 11, 12	30, 172, 192, and A-20
5.	Simple Insertion	2, 6	32, 87
6.	Boothroyd's SHELLSORT (*CACM* 201)	3, 11	37, 168, 173 and A-22
7.	Hibbard's SHELLSORT ("Empirical Study of Minimal Storage Sorting," *CACM* **6,** 5, May 1953, pp. 206–213	3, 11, 20	41, 168, 173, 359
8.	TREESORT 3 (Alg. 245, *CACM* **7,** 12, p. 701)	4, 11	173 and A-28
9.	QUICKERSORT (Alg. 271, *CACM* **8,** 11, p. 669)	168 and A-38	
10.	Singleton's Sort (Alg. 347, *CACM* **12,** 3, p. 185)	11	168, 172, and A-42
11.	STRINGSORT (Alg. 207, *CACM* **6,** 10, p. 215)	11	168, A-24
12.	MATHSORT (Alg. 23, *CACM* **3,** 11, p. 601	10	153 and A-6

Appendix C
A Routine for Generating
$K - X$ Way
Merge Tables

The distribution of strings for the compromise merges is not well known. This algorithm, written in PL/I for readability, produces up to *n* levels of distribution for up to *n* tapes for Polyphase, Cascade, and merges in between. Asterisked lines impose level and tape limits and output field lengths.

```
GENERATE STRING DISTRIBUTIONS
DISFEED: PROCEDURE OPTIONS (MAIN);
DECLARE
*LVL FIXED BINARY (15,0) INITIAL (10),
PHS FIXED BINARY (15,0),
MRG FIXED BINARY (15,0),
**DO MRG=2 TO 10 BY 1;
DO PHS=0 TO MRG-1 BY 1;
CALL DISPERSE (LVL,PHS,MRG);
END; END;
DISPERSE: PROCEDURE(LEVELS,PHASES,MERGE);
/* THIS PROCEDURE GENERATES IDEAL LEVELS FOR K-X WAY
MERGES WHICH DEPLETE ALL INPUT BEFORE MERGING BEGINS
WHEN PHASES=0 DISPERSION WILL BE FOR POLYPHASE,
WHEN PHASES=MERGE-1 OR MERGE-2 DISPERSION WILL BE
FOR CASCADE. OTHER VALUES OF PHASES WILL GENERATE
IDEAL DISTRIBUTIONS FOR THE VARIOUS 'COMPROMISE' MERGES. */
```

```
DECLARE
 LEVELS FIXED BINARY (15,0),
 PHASES FIXED BINARY (15,0),
 MERGE FIXED BINARY (15,0),
SUM FIXED BINARY (31,0),
PLACE FIXED BINARY (15,0),
 K FIXED BINARY (15,0),
 I FIXED BINARY (15,0),
WIDTH FIXED BINARY (15,0),
 CURLEV FIXED BINARY (15,0);
WIDTH=MERGE+1;
STRINGS: BEGIN;
DECLARE
 HEADER(WIDTH) FIXED BINARY (15,0),
TAPES(WIDTH,LEVELS) FIXED BINARY (31,0);
HEADER(1)=1;
DO K=2 to MERGE+1;
HEADER (K)=HEADER (K—1)+1;
END;
DO K=1 to WIDTH;
***PUT EDIT (HEADER(K)) (X(6), F(2));
END;
PUT LIST (MERGE,PHASES);
PUT SKIP;
CURLEV=1;
 TAPES=0;
DO K=1 TO MERGE;
TAPES (K,CURLEV)=1; END; /* INITIAL STRINGS */
CYCLE: TAPES(MERGE,CURLEV+1)=TAPES(1,CURLEV);
SUM=TAPES (1,CURLEV);
PLACE=MERGE—1;
CASCADE: DO K=2 TO PHASES+1;
SUM=SUM+TAPES(K,CURLEV);
TAPES(PLACE,CURLEV+1)=SUM;
PLACE=PLACE—1;
END CASCADE;
I=0;
POLYPHASE: DO K=PLACE TO 1 BY —1;
TAPES(K,CURLEV+1)=SUM+TAPES(MERGE—I,CURLEV);
I=I+1;
END POLYPHASE;
CURLEV=CURLEV+1;
```

```
IF CURLEV<LEVELS THEN GO TO CYCLE;
DO J=1 TO LEVELS;
DO K=1 TO LEVELS;
DO K=1 TO WIDTH;
****PUT EDIT TAPES(K,J)) (X(2), F(6));
END: PUT SKIP;
END:
END STRINGS;
END DISPERSE;
END DISFEED:
```

Appendix D
Test
Driving
Routine

The comparison of sorts described in Chapter 11 is based upon sample runnings of various sorts under various data conditions. The algorithm which generated the data and "drove" all sorts is shown here with one of the sorts tested.

```
FEED: PROCEDURE OPTIONS(MAIN);
DECLARE
INT CHARACTER(9),
INTA CHARACTER(9),
INTB CHARACTER(9),
INTC CHARACTER(9),
INPUT(5000) FIXED BINARY (31),
SEED FIXED BINARY (31,0) INITIAL (13579),
FNUM DECIMAL FLOAT (16),
NUM FIXED BINARY (31,0),
TOPNUM DECIMAL FLOAT (16),
RANGE DECIMAL FLOAT (16),
LIMIT FIXED BINARY (21,0) INITIAL (0100),
K FIXED BINARY (15,0) INITIAL (100),
N FIXED BINARY (31,0) INITIAL (100),
WORK (2500) FIXED BINARY (31,0),
EXPON DECIMAL FLOAT (16),
```

```
CYCLES FIXED BINARY (15,0) INITIAL (0),
Y DECIMAL FLOAT (16);
EXPON=10**6;
CYCLEIN: CALL GENERATE;
DO I=1 TO K; PUT EDIT (INPUT (I))(X(1),F(6)); END;
 INT=TIME;
CALL SHUTTLESORT(INPUT,N);
INTA=TIME;
PUT LIST (INT,INTA,INTB,INTC);
DO I=1 TO K; PUT LIST (INPUT(I)); END;
 INT=TIME;
CALL SHUTTLESORT (INPUT,N);
INTA=TIME;
PUT LIST (INT,INTA,INTB,INTC);
DO I=1 TO K/2; WORK (I)=INPUT(I); END;
DO I=1 TO K/2; INPUT(I)=INPUT(K/2+I); END;
DO I=1 TO K/2; INPUT(I+K/2)=WORK(I); END;
DO I=1 TO K; PUT EDIT (INPUT(I))(X(1),F(6)); END;
INT=TIME;
CALL SHUTTLESORT(INPUT,N);
INTA=TIME;
PUT LIST (INT,INTA,INTB,INTC);
DO I=1 TO K; PUT LIST(INPUT(I)); END;
J=1; DO I=K TO 2 BY -2; WORK(J)=INPUT(I); J=J+1; END;
J=1; DO I=2 TO K BY 2; INPUT(I)=WORK(J); J=J+1; END;
DO I=1 TO K; PUT EDIT (INPUT(I))(X(1),F(6)); END;
INT=TIME;
CALL SHUTTLESORT (INPUT,N);
INTA=TIME;
PUT LIST (INT,INTA,INTB,INTC);
DO I=1 TO K; PUT LIST (INPUT(I));END;
EXPON=N/2;
CALL GENERATE;
DO I=1 TO K; PUT LIST (INPUT(I)); END;
INT=TIME;
CALL SHUTTLESORT (INPUT,N);
INTA=TIME;
PUT LIST (INT,INTA,INTB,INTC);
DO I=1 TO K; PUT LIST (INPUT (I)); END;
EXPON=10**6;
CYCLES=CYCLES+1;
```

```
IF CYCLES=1 THEN DO;
LIMIT,K,N=1000; GO TO CYCLEIN; END;
IF CYCLES=2 THEN DO;
LIMIT,K,N=5000; GO TO CYCLEIN; END;
IF CYCLES=2 THEN DO;
LIMIT,K,N=5000; GO TO CYCLEIN: END;
(NOFIXEDOVERFLOW): GNERATE;PROCEDURE:
RANGE=EXPON;
DO I=1 TO 5000;
INPUT(I)=0;
END;
DO I=1 TO LIMIT;
NUM=SEED * 65539;
SEED=NUM;
FNUM=NUM;
TOPNUM=FNUM * (2**-31);
Y=TOPNUM * RANGE;
INPUT(I)=Y;
END;
END GENERATE;
SHUTTLESORT: PROCEDURE (TOSORT,NUMBER);
/* PUBLISHED IN ALGOL AS ALGORITHM 175, CACM JUNE 1963 */
DECLARE
TOSORT(*) FIXED BINARY (31,0),
NUMBER FIXED BINARY (31,0),
TEMP FIXED BINARY (31,0),
I FIXED BINARY (15,0),
J FIXED BINARY (15,0);
INTB=TIME;
PASS: DO I=1 to NUMBER-1 by 1;
EXCH: DO J=I TO 1 BY -1;
IF TOSORT(J)=TOSORT(J=L) THEN GO TO EXIT;
TEMP=TOSORT(J);
TOSORT(J)=TOSORT(J+1);
TOSORT(J+1)=TEMP;
END EXCH;
EXIT: END PASS;
INTC=TIME;
END SHUTTLESORT;
END FEED;
```

ABOUT THE AUTHOR

Harold Lorin has been on the faculty of IBM's Systems Research Institute since 1968. He currently teaches in the areas of Multiprocessor Design, Operating Systems Design, and Minicomputers. He maintains an active relationship with IBM product-development divisions, and is active in SIGARCH.

During his military service (1955–1957), Mr. Lorin was at Omaha with SAC Headquarters, assigned to duty as Comptroller, and served as an analyst-programmer for the IBM 650. He first entered the field with UNIVAC in 1955. In 1959 Mr. Lorin was involved with Strategic Air Command Control System (SACCS–465L) at Lodi, New Jersey, while working for the Systems Development Corporation.

While at the UNIVAC Division of Sperry Rand, he held a variety of professional and management positions in the area of software research. He was associated with the development of a management game, executive systems, report writers, compilers, and I/O support packages. His interest in sorting and sorting systems came about in connection with systems evaluation activities which undertook to predict the behavior of competitive equipment and to study the behavior of proposed machines on various sorting algorithms. The work led to a Sort Evaluation System and a co-authored review of sort techniques copyrighted by Sperry Rand in 1963.

On joining Service Bureau Corporation (then an IBM subsidiary) in 1965, he served as a member of the senior staff responsible for internal software systems development and for customer contracts in the advanced software area. While at SBC he participated in design efforts for a network extension to ASP, an interactive compiler, a design automation system, a computer compiler, and a conversational remote-entry system.

Mr. Lorin's other publications include "Parallelism in Hardware and Software" (Prentice-Hall, 1972); "A Guided Bibliography to Sorting" (IBM Systems Journal, Volume 10, Number 3, 1971); and an article on Parallelism in the "Encyclopedia of Computer Science" (Petrocelli Publishing, 1975).

He has made presentations to I.E.E.E. groups, to UNIVAC Users Association, to SIGARCH and other A.C.M. organizations, and to IBM professional symposia. He is an adjunct faculty member at the Polytechnic Institute of New York.

INDEX

address calculation, 162–166
allocation, 215, 316
amphisbaenic sort, 155
archival storage, 366
assignment phase, 298, 308
associative memory, 331
augmentation tape, 220–221

backward balanced merge, 198
balanced dispersion, 122
balanced disk merge, 273
balanced merge, 197, 287
Basic Exchange, A–62
Batcher's Merge, 358
bin determination, 159
Binary Insertion, 92–96
Binary Radix Exchange, 144–147
Binary Tree (*see also* Tree), 45–50, 60
block sorting, 143
block size, 275–276
blocking, 182, 186, 203–205, 213, 229, 291, 311, 346
bubble sort (*see* Standard Exchange)
buffering, 182–191, 205–213, 291, 311, 345

cache memory, 346
CACM Algorithms, Appendix A

(*see also* MATHSORT, QUICK-
SORT, SHELLSORT, QUICK-
ERSORT, STRINGSORT,
Tournament Sort)
Cascade Merge, 217, 221, 233
centered insertion, 87–90
checkpoint, 215, 305
CKPT parameter, 305
COBOL, 322
collection phase, 149
comparative sorts, definition, 5
comparison of sorts, 36, 166–173, 261–262
computer features, 2, 17, 21, 35, 181, 312, 314
comparison of merges, 261
Compromise Merge, 217, 245
configuration, 314
control phase, 298
counters, 222, 225
counting, with Linear Selection, 13–14
in digit sorting, 152, A–72
Crisscross Merge, 256, 289
cylinder, 270, 275, 290

data passes, 262, 279
(*see also* Merge, external)
demand paging, 332
detached key, 19